RESCUING HUMANITY

Transcending the Limits of Mathematics, Science, and Technology

In *Rescuing Humanity*, Willem H. Vanderburg reminds us that we have relied on discipline-based approaches for human knowing, doing, and organizing for less than a century. During this brief period, these approaches have become responsible for both our spectacular successes and most of our social and environmental crises. At their roots is a cultural mutation that includes secular religious attitudes that veil the limits of these approaches, leading to their overvaluation. Because their use, especially in science and technology, is primarily built up with mathematics, living entities and systems can be dealt with only as if their "architecture" or "design" is based on the principle of non-contradiction, which is true only for non-living entities. This distortion explains our many crises.

Vanderburg begins to explore the limits of discipline-based approaches, which guides the way toward developing complementary ones capable of transcending these limits. It is no different from a carpenter going beyond the limits of his hammer by reaching for other tools. As we grapple with everything from the impacts of social media, the ongoing climate crisis, and divisive political ideologies, *Rescuing Humanity* reveals that our civilization must learn to do the equivalent if humans and other living things are to continue making earth a home.

WILLEM H. VANDERBURG has taught preventive engineering, sociology, and environmental studies at the Centre for Technology and Social Development at the University of Toronto.

T0398415

WILLEM H. VANDERBURG

Rescuing Humanity

Transcending the Limits of Mathematics, Science, and Technology

UNIVERSITY OF TORONTO PRESS
Toronto Buffalo London

© University of Toronto Press 2023
Toronto Buffalo London
utorontopress.com

ISBN 978-1-4875-5110-0 (cloth)
ISBN 978-1-4875-5247-3 (paper)
ISBN 978-1-4875-5370-8 (EPUB)
ISBN 978-1-4875-5317-3 (PDF)

Library and Archives Canada Cataloguing in Publication

Title: Rescuing humanity : transcending the limits of mathematics, science, and technology / Willem H. Vanderburg.
Names: Vanderburg, Willem H., author.
Description: Includes bibliographical references and index.
Identifiers: Canadiana (print) 20220494800 | Canadiana (ebook) 20220494843 | ISBN 9781487551100 (cloth) | ISBN 9781487552473 (softcover) | ISBN 9781487553173 (PDF) | ISBN 9781487553708 (EPUB)
Subjects: LCSH: Humanity. | LCSH: Civilization, Modern – 21st century.
Classification: LCC BJ1533.H9 V36 2023 | DDC 303.48–dc23

Cover design: John Beadle
Cover image: Benjamin Davies/Unsplash

We wish to acknowledge the land on which the University of Toronto Press operates. This land is the traditional territory of the Wendat, the Anishnaabeg, the Haudenosaunee, the Métis, and the Mississaugas of the Credit First Nation.

This book has been published with the help of a grant from the Federation for the Humanities and Social Sciences, through the Awards to Scholarly Publications Program, using funds provided by the Social Sciences and Humanities Research Council of Canada.

University of Toronto Press acknowledges the financial support of the Government of Canada, the Canada Council for the Arts, and the Ontario Arts Council, an agency of the Government of Ontario, for its publishing activities.

Canada Council for the Arts
Conseil des Arts du Canada

Funded by the Government of Canada
Financé par le gouvernement du Canada

Contents

Preface

It was not very long ago that I could not possibly have imagined writing a book on secular religious attitudes in this day and age. I grew up intellectually in engineering and began a dialogue with the social sciences on the postdoctoral level, so this is not the kind of topic usually on my horizon. Nevertheless, here I am tempting you, my reader, to take the topic very seriously. I must confess that I have always felt there was something about engineering and technology that simply did not add up. Of this I will attempt a brief and thus oversimplified explanation.

Prior to the birth of our present civilization, which was launched by the interdependent processes of industrialization, urbanization, rationalization, and secularization, traditional societies, without any exceptions, made gods for themselves, which they served through moralities and religions. The universality of these creations may be interpreted as the essential response of humanity as a symbolic species not anchored in a natural niche in an ecosystem. It thus had to ground human life in an ultimately unknowable universe by means of absolute reference points capable of orienting and unifying the life of each society. Such reference points were established by overvaluing what was most self-evident, obvious, and good in the experiences of the members of a society, to a degree that nothing more certain and good could be imagined or lived. This overvaluation was achieved by symbolizing the unknown as extensions of what was already known and lived in order to create an absolute vantage point and historical orientation. When this kind of grounding of the first generation of industrializing societies was steadily weakened, some observers jumped to the conclusion that human life was becoming rational and secular. What had been happening instead was that human life was becoming grounded in secular equivalents that took on forms we did not recognize. Traditional societies tended to overvalue a social or natural entity, resulting in a divine pharaoh or emperor, for example, or a sacred hippopotamus or serpent embodying a natural power (often of fertility) that sustained the society.

What if our own civilization overvalued our most powerful creations? Such an overvaluation obscures their limits and thus encourages an indiscriminate use beyond them. If a carpenter overvalues his hammer by attributing to it almost limitless powers that include the driving of nails, the cutting and planing of wood, securing screws and more, he would quickly discover such limitations because of the terrible mess he would make. However, what if we overvalued the powers of our screen-based devices, mathematics, science, and technology? Is it possible that our crises today can be directly related to their overvaluation through secular religious attitudes? What if all this overvaluation was rooted in a deeper overvaluation of discipline-based approaches to all human knowing, doing, and organizing, accompanied by an undervaluation of all symbolic approaches based on language and culture? We will explore this possibility by examining how the members of each new generation growing up in our contemporary societies acquire symbolic approaches through a language and culture along with discipline-based approaches that begin with learning to make sense of what appears on many screen-based devices, soon reinforced by some basic mathematics in elementary schools, and still later by studying discipline-based approaches in science or in all technology-like professional activities in high schools and universities. This analysis will reveal the transmission of secular religious attitudes that point to our lives now being grounded in secular myths rather than traditional ones. It will once again confirm the conclusion of earlier analyses: that almost all our current crises stem from our uncritical overvaluation of the capabilities of discipline-based approaches.

We urgently need to draw the practical implications of our failure to find independent foundations for mathematics, science, and technology. Since all human knowing and doing is relative rather than absolute in character, the threats of relativism, nihilism, and anomie are ever-present. Consequently, these discipline-based approaches, like all other human activities, need to be grounded via our lives, in turn grounded in the myths of our culture, as the failure of philosophical foundationalism implies. Growing up in so-called secular societies thus continues to depend on the acquisition of secular religious attitudes, of which those toward discipline-based approaches have vast negative consequences for humanity.

In the first three chapters, we will examine how children are exposed to two streams of experience, one of which extends their ability to make sense of their lives and the world by means of symbolization, language, and culture, and another that builds toward their ability to do so by means of visual approaches and their extensions. In the next three chapters, we will examine what happens when, some time later and especially during their teenage years, they begin to learn discipline-based approaches, including mathematics for learning to make sense of and to map "reality," scientific approaches for all human knowing, and technical approaches for all human doing.

These later chapters confront us with something that is possibly the most decisive issue for our emerging global civilization. We live with our discipline-based approaches to science as if they have no limits. There is no university doing intensive research on where these limits lie and how to transcend them, when they are found, by means of alternate approaches. We are wagering that our discipline-based approaches to knowing have no such limits, or that such limits will turn out to be trivial. In other words, we live with our discipline-based approaches to knowing the way earlier societies lived with their gods – as entities that were autonomous and omnipotent in their spheres. Consequently, the acquisition of discipline-based approaches to knowing is necessarily accompanied by the transmission of secular religious attitudes that will turn out to be as dangerous to human life as were the traditional religious attitudes of past cultures.

In the same vein, we have handed over all human doing to discipline-based approaches, as is evident from the calendars of our universities describing professional programs, applied social sciences, or any other knowledge required for a recognized practice of any kind. Do we really believe that discipline-based approaches are omnipotent in the sphere of human doing? Doesn't this imply that our civilization has increasingly put all of us in a position that may be compared to science being our "collective hammer" for knowing, and all forms of technology acting as our "hammer" for doing? Our civilization appears to have lost the ability to respect the limits when it comes to the use of discipline-based approaches to human knowing and doing. As a result, the acquisition of discipline-based approaches cannot be understood without the implicit but simultaneous transmission of secular religious attitudes toward new entities that we treat the way the gods were treated in the past. No human activity can ever be autonomous, omnipotent, and without limits. If we ignore this, we will be caught out when we take such activities beyond their limits. In earlier volumes, we have shown that most of our current crises can be attributed to our going beyond the limits of our discipline-based approaches. There is no longer any guesswork in stating that our global civilization is as much endangered by the secular gods of its own making as all its predecessors were endangered by their traditional gods. It is therefore essential that we examine the secular religious attitudes that are being passed on from generation to generation and assess the risk they pose to individual and collective human life.

Our dependence on secular religious attitudes is rather surprising, not because we have declared ourselves as having become of age, as rational and secular human beings, but because Western civilization, which gave birth to our present global one, was deeply influenced by two "religious" traditions, both of which taught that the making of gods was possibly the greatest danger to a society. Since the Jewish and Christian traditions have had a vast influence on what we refer to as industrialization, urbanization, rationalization, and secularization, they have been widely critiqued from the "outside," as it were, in ways that were pioneered by Karl Marx and Max Weber. There have been relatively few critiques from "within," as it were, which argued that these traditions had been mostly assimilated by the transformations of individual and collective human life of the last two hundred years. The critiques failed to recognize that the threat of the creation of false gods and the necessity to serve them through a morality and religion had hardly disappeared just because we declared ourselves to be secular and having come of age. Organized Judaism and Christianity have essentially become a personal and private affair, thus leaving public life in the grip of our secular equivalents to the gods, idols, moralities, and religions of the past. Such a critique was advanced in a previous volume entitled *Secular Nations under New Gods*, and has been taken up again in several subsequent volumes.

Although this is the seventh volume of my analysis of human life in our contemporary global civilization, this book can be read on its own. However, the concepts of symbolization and desymbolization may confuse some readers. In my work, symbolization is indissociably linked to our unique abilities as a symbolic species. Growing up to become a member of our society depends on the acquisition of a language as the gateway into a culture. All our languages and cultures are grafted into a unique metalanguage built up in the organizations of the brain-minds of babies and toddlers. It is done through the symbolization, differentiation, and integration of all their experiences by means of neural and synaptic modifications that remain well beyond our intellectual grasp. Desymbolization occurs when these processes are separated from individual and collective human life, to be represented in separate and distinct systems or machine functions, which involves a conceptualization of life as non-life. The Introduction may be helpful in this regard by providing an overview of the relevant previous findings; but readers somewhat familiar with my work may wish to skip it and begin with chapter 1, consulting this Introduction on specific topics when they find they need further background or wish to deepen their understanding of a particular issue. The references have been designed for the same purpose. In many cases, they refer to the relevant parts of previous works, which in their turn provide additional entry points into the literature.

My warm thanks go to my anonymous reviewers for their suggestions to further improve the readability of the manuscript for as many readers with as diverse backgrounds as possible. Given the topic of this book, we all share the importance of this. One of my readers also suggested that I briefly develop my arguments a little further on several strategically important implications. To my delight, this corresponded closely to three volumes already completed on these subjects. I thus restricted myself to briefly describing the most important consequences along with the suggestion that a separate work had been devoted to it. I am reasonably optimistic that my readers will not have to wait very long for their appearance.

I wish to acknowledge the people who were instrumental in helping me write this book. My thanks go to my partner, Rita Vanderburg, who again edited the manuscript, as she did for all my other volumes. I also wish to thank Hannah Wong, who, as a former student, encouraged me to keep teaching so that subsequent generations of students would still hear what I had to say. She later began to word process my manuscripts; and her help has been invaluable, given my limitations in that domain. When she was unable to continue this work, Debbie Lo carried on, and also prepared the index. Thanks, Debbie, for your support as well.

As a former editor of an interdisciplinary journal, I appreciate the difficulties of having a work like this refereed, especially during a time of COVID. I am deeply grateful to Jodi Lewchuk for her dedication and persistence. I wish to thank Barb Porter for accommodating my needs in the production of this manuscript. Last but not least, I again acknowledge the enduring influence of my French mentor, Jacques Ellul, on my life and work. With the completion of this seventh and final volume of my study of technique and culture, my acknowledgments in no way detract from my ultimate responsibility for this work.

Willem H. Vanderburg
Peterborough, Ontario

RESCUING HUMANITY

Introduction

Suspended in Language and Culture

As unbelievable as it may initially appear, I will seek to convince you that the best approach for dealing with the way economic, social, and environmental crises affect our persons and our lives is for us all to revisit our widespread beliefs that we live in a secular society. There may be a small window of opportunity opening up for such a discussion as a result of our growing awareness of what it is to be a symbolic species.[1] Ironically, this rediscovery parallels our awareness that the biosphere is becoming less able to sustain our lives and our societies, to the point that we have not been able to take its support for granted for some time. Similarly, we are now becoming aware of being a symbolic species now that we can no longer take for granted the way individual and collective human lives were sustained by symbolic cultures, because of our impairing them through desymbolization. Our systematic weakening of our material and symbolic supports for the past two hundred years has not yet become regarded as the double root of most (if not all) of our deeply troubling and recurring crises. With an appropriate diagnosis, it will become apparent that all our efforts to address these crises thus far are, almost without exception, wrongheaded and doomed to failure. We will attempt to show that we are actually weakening the material and symbolic ground from underneath our feet, and that the latter can largely explain the former.

Our being a symbolic species permeates everything in individual and collective human life, with the result that it sets us apart from all the other animal species. We do not live in natural niches within the biosphere, even though we depend on it for all matter and energy for living. Instead, we live in "cultural niches" because of our suspension in a language and a culture, made essential by what is referred to as our brain plasticity. It endows us with the ability to symbolize each and every experience derived from our bodies in the world by means of neural and synaptic modifications to the organizations of our brains, which turns them into our brain-minds grown from all prior lived experiences. Our symbolized lives thus work in the background in a manner analogous to a frame of reference relative to which each next experience can be interpreted, to take on a meaning and value relative to those of all prior experiences and thus to our lives as a whole. Because of the exposure of babies and children to the language and culture of those who love and care for them, this symbolization of all their experiences becomes gradually permeated by this language and culture, enabling them to grow up as individually unique expressions of the human community of that time, place, and culture.

Our suspension in a language and a culture makes human communities vulnerable to situations where this suspension cannot develop, and those where it fails to adequately mediate our relationships with others and the world. The former has been documented in children who, for a variety of reasons, did not have access to language. They grew up looking like human beings, but their behaviour was incomparably different from those who lives were suspended in a language and a culture. In other situations where our suspension in a language and culture is threatened by relativism, nihilism, and anomie, the members of a community begin to feel that more and more in their lives no longer makes any sense, and that these lives have been cast adrift without a firm grounding and orientation. If such developments remain unchecked, a community can undergo a "cultural death," as it were, by disintegrating and eventually collapsing, even though its members continue to struggle to stay alive.

Because of our suspension in a language and a culture, the members of our symbolic species cannot understand or deal with anything absolutely – that is, in and of itself and on its own terms, separated from everything else. All our knowing and doing is relative in character, and thus can be defended against relativism, nihilism, and anomie only by creating something absolute through a religion and a morality that can ground and orient human life. It will become evident that a symbolic species is necessarily a religious one as well, even in our so-called secular societies. This relative character of individual and collective human life places severe limits on our discipline-based approaches to knowing and doing, which our emerging global civilization has failed to recognize despite the many crises associated with this failure.

All throughout human history, crises have frequently been dealt with by doing more of the kinds of things that produced them to begin with. For example, a few decades ago, we found that contemporary ways of life place too many demands on the biosphere, thereby making them unsustainable by that biosphere. The response of the universities in general, and that of their engineering, business, and economic departments in particular, has been to make additions to the curriculum, such as a division of the environment or courses on green business practices, but the core curriculum remained largely unaffected.[2] Instead of confronting the situation, we created an escape valve. For example: why continue to experiment with the reorganization of human work by means of autonomous working groups, or contemplate abandoning the assembly line altogether, in order to deal with human and mental health issues and costs, when production facilities could be moved to a part of the world with an abundant source of labour not protected by the kinds of standards found necessary in the course of industrialization?[3] The same holds true for the rethinking of production processes by acknowledging that pollutants are products we produce just like all the others, except that there is no market for them, with the result that the only cost-effective solution is their prevention.[4] Preventive approaches remained on the fringes and never penetrated the core of engineering and business curricula.[5] We convinced ourselves that the best way out of these difficulties was to do more of what had produced them to begin with, namely, a striving for efficiency, productivity, profitability, and increases in GDP. We would "grow" ourselves out of our problems.

There are deep and sound reasons why, throughout human history, societies have tended to deal with crises by doing more of the kinds of things that produced them to begin with. The answer will be shocking, but it can nevertheless be defended. The root cause of much of this kind of ineffective behaviour stems from traditional or secular religious attitudes that in turn are rooted in the way we, as a symbolic species, relate to one another and our surrounding world.[6] Because of our symbolic endowment, we have symbolized all our experiences of one another and the world and have thus interposed a symbolic universe between each group and society and whatever exists "underneath" and "beyond" it.[7] These symbolic universes reflect how everything is related to, and evolves in relation to, everything else. That this is the case is evident from the inventory of the symbolic universes represented by the vocabulary of a language. A dictionary, as a linguistic map of it, defines each word in terms of several (sometimes contradictory) meanings that are each spelled out in terms of other words. These words are in turn spelled out in yet other words, and so on. Simply put: one of the meanings of each word is what all the other words do not mean, and vice versa. The meaning of each word is thus dialectically enfolded into those of all other words as they are in it. The same argument can be made for all the meanings of the phrases and sentences based on this vocabulary, with the result that the language as a whole appears to have a dialectically enfolded architecture.

Such a dialectically enfolded architecture of a language is essential if it is to sustain human communication, relationships, groups, and societies. All human communication may be located on a spectrum of possibilities between two extremes where no communication can take place. On the one end of the spectrum, people are too similar: what one person knows and has experienced is so similar to what the other knows and has experienced that there is nothing to communicate. On the other end, there is nothing that two people know or have experienced that they can share, with the result that if one person says something to the other, there is nothing similar in his or her world. As a result, it cannot take on a meaning relative to it and be dialectically enfolded into the experiences of his or her life. Once again, no communication can take place, but in this case the reason is that the people are totally different. In the middle of this spectrum, we encounter what we experience in our daily lives. For two people to communicate something, they must be different enough that each one can enrich the other's life by bringing something new into it, but not so different that this communication can make little or no sense.

For example, when two people who were initially deeply committed to one another in friendship or love increasingly move into radically different professional and social circles, the sharing of their lives may become a great deal more difficult, and eventually they may find that their commitment is unable to overcome a growing sense that they can no longer enrich one another's lives, so that the relationship becomes a burden rather than a joy. The opposite situation can occur when two people have shared their lives for a very long time, and it becomes more and more difficult to bring something new and enriching to the relationship since there is little they can say that the other does not already know. Human communication thus depends on a dialectical tension between individual differences and similarities. Weakening it will move the relationship toward one or the other end of the spectrum and thus toward non-communication.

The same kind of argument can be made for a social group and society. A social group is viable if each of its members helps to sustain the lives of the others by enriching them with their life experiences, while the others do the same for that person. When the dialectical tension between individual differences and group unity runs down as the members are more and more familiar with one another, there is less and less reason to participate in the group, and only a strong commitment can motivate the members to do something about it. Similarly, a viable society depends on a dialectical tension between individual differences and a cultural unity. If the former imposes itself too strongly on the latter to the point of overwhelming it, a society may weaken and decline, and without some kind of revitalization will eventually disintegrate and collapse. The contrary can also occur. If the cultural unity imposes itself too strongly on individual diversity, it will affect all dialectical tensions that sustain social relationships and groups, and again the society could be threatened by disintegration and collapse.[8]

The dialectical tensions within the social fabric of a society are internalized by each member. That this is the case becomes evident when we recognize that it is impossible that there could be a separate society "out there." Each member helps to constitute the society for all others, as they do for that member. Consequently, each member is thus both individual and society, which is possible since each person is an individually unique manifestation of their society. By growing up in it and acquiring its culture for making sense of, and living in, the world, each member internalizes what constitutes that society. As such, each member becomes aware of what is unique to their person and life, and what has become objectivized by being shared by all members as a part of the working culture of the society. There is thus an internal tension between each person's individual uniqueness and what has become objectivized as culturally typical, and between their person and the society.

The dialectical enfolding of our communications, social relationships, groups, and societies is falling far short of our need to be symbolically sustained by our cultures. Because of the dialectically enfolded architecture of the fabric of all social relationships and thus of all groups, communities, and societies, everything is relative to everything else and nothing can be absolute in order to "anchor" all this relativity. This relativity is always one step away from opening the floodgates to what may begin as a certain arbitrariness. Why deal with a situation in this way rather than in another way if it all amounts to much the same thing? If the differences do not matter, why do anything at all? In other words, a complete relativity is existentially unbearable and socially unliveable. It leads to a condition characterized by relativism, nihilism, and anomie, which will be referred to as cultural chaos. As best as I have been able to determine, every group and society has managed to shut the door to this cultural chaos by symbolizing the unknown as more of what the members of a group or society individually and collectively know and live. In other words, they behave as if what they know and live is identical to what lies "underneath" and "beyond" – minus some "details" that will be discovered and lived in the future. The unknown can only complete and not upset their lives.[9] The members of a society can thus proceed with their lives with complete confidence. They are fundamentally in touch with others and the world because this can no longer be threatened by anything unknown. Everything to be discovered and lived will simply be cumulatively added to their individual and collective lives.[10]

Symbolizing the unknown in this way does not rule out the possibility that some future discoveries or experiences will turn out to be contrary to this symbolization, by not quite fitting into people's lives and their shared way of life. A time will come when things make less sense in the face of what is symbolically anomalous. People first respond by continuing to interpret what is symbolically anomalous as normally as possible. However, when the influx of these kinds of discoveries and experiences becomes too great, their collective lack of fit may make it possible for them to form their own new patterns, and when these grow, people may intuit that the unknown is better symbolized in this new way. These kinds of scenarios are typical when a group or society approaches what may turn out to be the end of a historical epoch. Based on a great many experiences that did not quite fit, a new way of symbolizing the unknown may be intuited. When doing so turns out to be increasingly viable, an alternate way for people to get a better grip on their lives and their world may metaconsciously emerge as the dawn of a new epoch in their historical journey.[11] When this does not occur, growing levels of relativism, nihilism, and anomie will bring the society ever closer to cultural chaos.[12] People then sense that their lives increasingly make little sense, that they are adrift, that they have lost their roots, and thus a society may be headed for a collapse. If it was the creative centre of a civilization, its collapse may also occur in the way our earlier civilizations have disappeared.[13]

It would appear that the symbolization of the unknown as more of what a community knows and lives corresponds exactly to what (in such disciplines as cultural anthropology, the sociology of religion, and depth psychology) is referred to as myths.[14] These turn out to be the central control structure of a culture, somewhat analogous to the DNA on the biological level of an organism and of its cells. This control structure keeps the condition of cultural chaos at bay. Something of this control structure is dialectically enfolded into all aspects of individual collective human life, just as the DNA is enfolded into each cell of an organism. The myths of a culture represent what is most self-evident, reliable, and trustworthy in the life of a community during a historical epoch. These myths absolutize what is known and lived as the only possible and viable vantage point from which to make sense of, and live in, the world. Any radical alternative thus becomes unknowable and unliveable. Even if we read a description of a way of life and culture far removed from our own, we can imagine living that way but we can never fully do so. The intellectual and practical grip a culture has on its life in the world is thus temporarily protected from cultural chaos. Moreover, the greatest good attributed to that element in their experience that is so all-important that life without it is unthinkable becomes the absolute good in all the universe for all times and places. It is usually referred to as the sacred or central myth of a community.[15]

Consequently, its symbolization represents the process by which we have come to understand the way each group and society has always created traditional gods for itself. The group then seeks to serve the gods by means of a religion that assures them that these gods will look favourably on them in return for living good, moral, and religious lives in terms of these cultural creations.[16] All this makes a great deal of sense, since these culturally created gods hold the power of life and death over a community, and it had better bring these powers over to its side.

The above interpretation has become widely accepted as an explanation of why, throughout all of human history, traditional cultures have made gods for themselves. However, this interpretation cannot prove or disprove the existence of a living God. We will avoid these metaphysical debates by simply pointing out that if there are people who *live* their lives out of a revelation they believe comes from a living God beyond their culture, then they must distinguish such a revelation from the religious elements of their culture.[17] There is the well-known distinction between a revelation as recognized in faith and a religion that serves as a means to satisfy the need of a symbolic species to be rooted or "symbolically grounded" in its relative knowing and doing by means of a culture.[18] For those who live as if there is no living God, a revelation is merely another cultural element to satisfy the religious needs of a community. Hence, for the purpose of this work the possibility of a living God is a social and historical question related to the way people live their lives and not a metaphysical issue.

The practical implications of all this are extensive. The symbolization of the unknown as more of what a community knows and lives, and thus the absolutization of the greatest good it knows, ensures that whatever entity in the life of a community is thus absolutized cannot possibly deliver anything that is not good. It cannot possibly have any limits to its power to deliver good, and cannot possibly be affected in its autonomy to do so. What it can do is to withhold that good, which would bring life as it is lived by a community to an end, since anything radically other is unthinkable and unliveable. As a result, when a community faces a crisis, it will bring to bear the greatest good it knows, which amounts to doing more of the kind of thing that probably helped to produce the crisis. In other words, this way of symbolically sustaining our lives in the world tends to lead to end-of-pipe solutions because it is impossible to question the greatest good, which would amount to undoing the symbolization of the unknown.

All of this can shed further light on our history as a symbolic species. When food-gathering and hunting groups were immersed in what we refer to as nature, everything that sustained their lives derived from this nature, as well as every threat. Nature as they conceived it was what had created their lives, what sustained them, and without it these lives would be unliveable. It led to the creation of a natural sacred; and to relate to its divine powers, a natural morality and religion needed to be developed. When the community violated their precepts, these powers could be angered and could threaten the community in a variety of ways. Hence, nature was the first life-milieu in relation to which human life was lived. It was symbolized as a natural sacred served by natural moralities and religions.[19]

When societies began to emerge, it became possible to intervene in nature in very different ways, as was the case for the "irrigation civilizations." The relationship between the group and nature was now mediated by a society; it became the primary life-milieu of the group and nature the secondary life-milieu. The sacred and the system of myths, including the moralities and religions based on them, were transformed accordingly.[20]

The above interpretation of myths as the symbolization of the unknown by the equivalent of symbolically interpolating and extrapolating all the symbolized experiences into an intuitive awareness of living a life is but the culmination of what happens on many levels with a lesser "depth." Myths represent the "deepest" level of metaconscious knowledge that the organization of the brain-mind of a person implies, as a consequence of such an organization symbolically interpolating and extrapolating all the experiences of that person's life.[21] On the lowest level of metaconscious depth, the symbolized experiences of each and every kind of daily-life activity are also interpolated and extrapolated into a "cluster" by the organization of the brain-mind. On a slightly deeper level, metaconscious knowledge is implied in several clusters of symbolized experiences corresponding to several kinds of daily-life activities. As the metaconscious depth increases, so does the scope of a person's life to which it corresponds. For example, the metaconscious knowledge implied in the experiences corresponding to face-to-face conversations with others includes that of an eye etiquette, conversation distance, and protocol for body rhythms, which jointly help to frame what is explicitly said in a manner that is unique to each culture.[22] Whenever a person stands too close to or too far from another person, an emotional reaction ensues, creating an impression of being pushy or aloof. Eventually, the cluster of symbolically differentiated experiences implies an objective value: for others to behave normally, a certain distance must be respected. We all do this without being consciously aware of it as our lives work in the background, helping us deal with each situation. Similarly, all eating experiences will soon imply the objective value that food is not to be spilled. Many clusters of daily-life experiences begin to imply how the members of a culture have learned to situate their lives in time and space and how to relate to their material surroundings.[23]

Continuing our examples: The experiences a person has shared with her friend may one day lead to an outburst of surprise to the effect that she did not expect her friend to behave this way, even to the point of greatly upsetting her. None of us needs to recall all our experiences with a friend, examine the personality type they imply, deduce from this way this particular experience was out of character, and then formulate a response. Our surprise is immediate and spontaneous because we have a metaconscious knowledge of the social self of that friend, as we do of all other people with whom we have shared a great deal of our life.[24] All our experiences with those we know well begin to imply a metaconscious knowledge of our own social self. It is often said that others are the "social mirror" in which we learn to see ourselves. Consequently, our life constantly working in the background as we make sense of a situation and live through it may be interpreted as the symbolic support of a kind of mental map that has a dialectically enfolded map reader in the form of our social self.[25] This is not the "output" of the organization of our brain-mind to a kind of "inner soul." We are entirely physically, socially, and culturally embodied in everything we do.

Similarly, all our experiences of the persons we do not know whom we encounter in terms of the social roles they play in our way of life – such as a teacher, taxi driver, policeman, and others – develop a metaconscious knowledge of what we can and cannot expect from such people.[26] All the experiences of our lives also imply the way we participate in our culture's way of making sense of and living in the world, as the culture-based connectedness of our way of life and community.[27] In sum, "below" the symbolized experiences of our lives there exists a great deal of metaconscious knowledge on various levels of depth related to the way our culture "organizes" every aspect of the lives of the members of the community in which we grew up.

On the conscious level, this formation of metaconscious knowledge may be compared to doing a scientific experiment and plotting the data on a graph. Our confidence in the experimental design is greatly strengthened when we can fit a curve through this data because it confirms all our experiences that show that our world behaves in a regular and orderly fashion. It is this expectation that legitimates our going beyond the experimental data by interpolating and extrapolating it by means of a curve. No one would judge this "going beyond the evidence" as an escape into metaphysics. In the same way, when babies and children grow up in a community, the shared non-subjective order of its cultural ways of making sense of and living in the world is constantly reinforced by the development of metaconscious knowledge in the organization of the brain-mind; and this strengthens their confidence in the symbolic universe they are entering as one that makes total sense and of which they have a sound grasp.

We must avoid conceptualizing a person's life as being lived on parallel levels of metaconscious depth that evolve more or less slowly or quickly. An appropriate metaphor is that of a body of water where the wind comes and goes, producing waves and turbulence of all kinds. When we go a little below the surface, things are already a great deal calmer; and when we get to the bottom, they appear to be almost completely still. In the same vein, just below the turbulence of our daily-life experiences that we symbolize lies the metaconscious knowledge that helps to stabilize them somewhat. As we go deeper, the metaconscious knowledge evolves ever more slowly, while the deepest metaconscious knowledge, that of myths, is hardly affected at all during a historical epoch of our society.

This brief outline of previous works suggests that the implications of brain plasticity are still very far from being adequately developed.[28] As we grow in our mother's womb, our DNA organizes stem cells into becoming the cells of particular tissues and organs. The human brain is one such organ, but with a decisive difference. Shortly before birth, the organization of the brain begins to undergo a constant modification by symbolizing each experience by means of neural and synaptic changes to this organization. The brain develops into a brain-mind. As noted, this organization symbolically interpolates and extrapolates all our experiences into a life lived with others according to the way of life and culture of the community in which we grew up. We are thus relative to a time, place, and culture and dialectically enfolded into our society, as it is into us.

It is not the plasticity of our brain-mind that makes us members of a symbolic species. Initially, the organization of the brain-mind begins to form a "meta-world": the context in which words spoken by others can become vocal signs. However, it is not until these vocal signs become directly differentiated from one another that a secondary frame of reference develops as a "metalanguage" for learning the language of a culture.[29] Vocal signs derive their meaning entirely from the non-linguistic meta-world with which they become associated; and they cannot become symbols until they are directly differentiated from one another as a dialectically enfolded vocabulary, jointly differentiated from all non-verbal experience. This is associated with a new development of the organization of the brain-mind, which gives human life its "lightness of being."[30] It is further amplified by the differentiation of the language Gestalts corresponding to phrases and sentences. This development of a symbolic language is the entryway into a symbolic universe that includes, but cannot be reduced to, our immediate experience. A symbolic universe is not a superstructure erected on our immediate experience, since it entirely transcends and transforms it, thus constituting a meta-world. That this is the case is confirmed by children whose access to language was barred by being brought up by animals, by living in isolation, or by having this access made very difficult as a consequence of deaf-blindness.[31] The case of the so-called wild boy of Aveyron was particularly well documented by Doctor Itard.[32] He developed ingenious methods for attempting to socialize this boy after his living in the woods. Initially, he was barely recognizable as a human being other than by physical appearances; Doctor Itard described him as a "man-plant" because even his dog had a greater social and emotional development than this child.[33] However, all attempts at teaching the boy to understand and use French had so little success that in the end the project of introducing him into the symbolic universe of his society was abandoned. He looked like a boy but could not be helped to live like one. In other words, the symbolic universe is not accessible by sight but only by language. As we will begin to argue in the next section, this impossibility is rooted in the architecture of a symbolic universe having diametrically opposite characteristics from those of the architecture of the visually apprehended world.

The history of the great civilizations that preceded ours provide abundant evidence that symbolization is indissociably linked to creativity and imagination. We have become so accustomed to conceptualizing many aspects of human life in terms of machine-like functions that we commonly use the same words to designate either one. Examples include communication, information, knowledge, expertise, cognition, and memory. For the same reason, it will become apparent that symbolization has nothing in common with information processing. If it did, symbolization would take on the same form in every group, society, and civilization. Instead, symbolization is as unique to a time and place as is the culture of a community. It is a creative, and often imaginative, way of relating to others and the world, which gives it a relative character that can never be entirely eliminated. It is our doing and not that of our gods, or of our computers. The symbolization of human experience is thus inseparable from the culture within which it becomes embedded as people grow up in a community. Initially, babies and toddlers imagine and create their relationships in terms of the meaning and value they have for their lives, but with the onset of language they learn to imagine and create relationships according to the meanings and values they have for their lives lived in relation to those of others by means of the way of life of their community. Without this imagination and creativity, no significant individual diversity could emerge within a cultural unity.

For example, before a fallen branch on the floor of a forest could have become a canoe paddle, a club, or a structural member of a shelter, it had to be imagined as such and symbolized according to its actual and potential meaning and value for the life of a community. Before a tree trunk floating down a river with some birds sitting on it could become a canoe, it had to be imagined as such and symbolized in terms of its actual and potential significance. Before ritual burial could take on any meaning, death had to be imagined and symbolized as much more than what people's immediate experience disclosed of it. It is likely that totemic societies were initially organized by totems as a kind of intellectual division of labour for a "scientific" megaproject of sorts to uncover the actual and potential meaning and value for human life of everything natural in their surroundings. Without such a megaproject it is doubtful that the great civilizations could have emerged.[34]

In daily life, much symbolization shows a high level of routine. However, it is possible to live any daily-life activity in more or less the same way for a very long time and one day be stopped in our tracks by the recognition that, with a little imagination and creativity, it suddenly looks very different; from then on we change our behaviour accordingly. In the extreme, a particular event may trigger a political or religious conversion. We begin to see things in a new light and are struck by a powerful conviction that this may change everything else in our lives. Consequently, according to this insight, we attempt to live everything differently. When we persist, it will result in a comprehensive change in the intelligence via which we apprehend and live in the world, which is the true meaning of conversion. There can be nothing algorithmic about the relative way the organizations of our brain-minds function through our social selves, which are dialectically enfolded in them. That this is the case is confirmed by our response to sensory deprivation and monotony, from which we attempt to escape by daydreaming or falling asleep. Our nervous systems make it impossible for us to rigorously repeat anything for very long. We cannot suppress ourselves in our lives and become an algorithm.

Attempts at Escaping Our Suspension in Language and Culture

Humanity has paid dearly for its symbolic grounding. Every society has become enslaved by the morality and religion it has created for itself. In addition, bestowing the attributes of gods on some elements in their experience through myths warped everything in a society's existence. By absolutizing something that is relative, all societies have distorted the fabrics of their relationships associated with individual and collective human life – relationships through which individual members lived and evolved their lives. Nothing was in its right place, since nothing had its meaning and value established relative to everything else, creating significant distortions. It plunged every society into a symbolic universe of distorted meanings and values, thereby forcing everything out of its "true" relative place. People thus lived by distorted meanings and values that in turn led to misplaced expectations, hopes, fears, anxieties, and guilt inseparable from the living of their lives and plunging them into endless contradictions that could not be faced.

It may be expected, therefore, that each and every civilization would have attempted to deal with at least some of the difficulties that stemmed from the way its constituent societies symbolically grounded themselves. However, this does not appear to be the case until the emergence of Western civilization, whose entire history is marked by diverse attempts to modify its symbolic grounding. It began with the three "perfections" taken to be the "pillars" on which it was founded: the first was Greek philosophy and its invention of universal knowledge based on the principle of non-contradiction; the second was Roman law, an attempt to establish a legal framework for human life through a political authority grounded in a state.[35] The third was Christian revelation, which was initially deeply rooted in the Jewish tradition but soon diverged from it, especially when it became the official religion of the Roman Empire and sought to "ground" its theology. With a great deal of hindsight, it became evident that these three perfections were entirely incompatible with one another; and this incompatibility is frequently associated with the dynamic character of the history of Western civilization.[36] Each of these three pillars departed from human life's suspension in language and culture, a process that we will briefly examine in this Introduction.

All three perfections essentially failed until elements of them became incorporated into a very different development of human cultures that is commonly referred to as industrialization, urbanization, rationalization, and secularization. The introduction of the technical division of labour that paved the way for mechanization and industrialization eventually translated the daily-life activity of work into an architecture of non-life. The economy became the domain of this new architecture, while the remainder of society continued to have a dialectically enfolded cultural architecture. The resulting split in the role of culture led to a first phase of desymbolization.

In addition, the reliance on symbolization and experience of the industrializing societies of Western Europe began to encounter significant limitations, which led to rational, technical, and economic approaches increasingly displacing cultural ones. Combined with an ongoing desymbolization of the cultures of these societies, human knowing, doing, and organization became completely reorganized by means of universal discipline-based approaches, which gradually spread to all aspects of the ways of life of the so-called industrially advanced societies. These eventually gave birth to our global civilization, with its universal science, technology, economy, and discipline-based approaches for organizing and reorganizing everything. Such a global civilization primarily relies on these discipline-based approaches and everything built up with them to replace the role of culture in evolving our ways of life – while continuing to rely on what remains of our highly desymbolized cultures to launch babies, toddlers, and young children towards becoming members of a symbolic species and for personal intimate relationships. These developments have been extensively examined in five volumes previously referred to in this Introduction.

Following brief reviews of the above developments, we will be ready to tackle what will occupy us in the present work. In our contemporary so-called mass societies (more adequately referred to as anti-societies) with their minimal dependence on highly desymbolized cultures, there is a compensating dependence on the mass media and especially the new media. These media are equipping the members of contemporary societies with a digital "umbilical cord" connected to screen-based devices for the exchanges of images and image-words. They present toddlers and children with an entirely new stream of disembodied experiences focused on these images and image-words. The latter take on a meaning not only in the context of a metalanguage, but also in the context of the bath of images displayed on the media, with far-reaching consequences. We will be inquiring into the effects these developments have on us, as human beings suspended in language and culture and symbolically grounded in myths. The possibility of undermining all this at a very early age is historically unprecedented. We will thus have to reflect on what happens in the days of our youth.

This new stream of disembodied experiences distinguishes itself from the realm of symbolic experiences, in that the children's constituent experiences are being divided into two different architectures. The foreground created by focusing our attention on the images and the accompanying image-words has the architecture of non-life; this is also true for whatever of these images and image-words remains in the background. There is a secondary background associated with our embodiment in time, space, and culture which continues to have a dialectically enfolded architecture. We may expect, therefore, that when these disembodied experiences are internalized, they give rise to entirely new developments in the organizations of our brain-minds, including a very different social self: a kind of matching human nature having the characteristics of a *homo informaticus*. Do we now become suspended in images and image-words, and is this suspension still grounded in myths? Will these myths still ground us in an ultimately unknowable universe or have we become truly secular members of a secular society belonging to a global civilization? There appears to be little question that our current situation is very different from anything that preceded us in humanity's history. We have desymbolized our cultures to an unprecedented level,[37] with a corresponding reduction in our dependence on language in favour of images and image-words.[38] Are we undergoing a mutation in what it is to be a symbolic species? There is much about individual and collective human life in our emerging global civilization that must be understood in the context of the above kinds of questions.

Foundationalism and the Architecture of Non-Life

The emergence of Greek philosophy and the invention of universal knowledge based on the principle of non-contradiction mark the first occasion of a Western society responding to difficulties or limitations encountered as a result of its suspension in language and culture. We have already noted that such difficulties may occur whenever a culture is relativized, which can lead to nihilism and anomie. The Greeks faced the threat of relativization by taking recourse to an approach that was ultimately rooted in the way we deal with the world we see.

The significance of this course of action becomes evident when we recognize that the architecture of the world we see may be expressed in terms of five principles that are the diametrical opposite of those characterizing the architecture of the symbolic universe of any culture. These five principles may be summed up as follows. At a particular moment in time, everything we see exists in its own space. Consequently, the architecture of this world exhibits the principle of non-contradiction: something is either a tree or something else, and so on, but never two things at the same time. Since every entity exists in its own space, it can be separated from all the others; it can be defined on its own terms by means of closed definitions; and it is open to measurement, quantification, and mathematical representation. Finally, this architecture of the seen world has a simple complexity, to which any entity can be added as if in a vacant space at any time without affecting all the others. The first four principles can also be interpreted as the attributes of a simple complexity. Although this architecture goes no deeper than the surfaces of the seen entities, it must not be regarded as superficial. For example, in all traditional pre-industrial societies, it included all visual representations of the deities and powers that a community held to be "behind" this world and over which they ruled.

Facts, as we generally conceive them, can exist only relative to a world with this kind of architecture. The reasons flow directly from this world's five characteristics. Once something in this world has been singled out for study, it can be abstracted from it without any loss of information. It can then be defined in a manner that does not depend on anything else in that world. Next, it can be measured, quantified, and mathematically represented. What has thus been established is universally valid. It becomes a fact that, in our popular scientism, has no limits, since it is complete. It is autonomous, since it does not depend on anything else. It is also self-sufficient and value-free. In secular religious language, a fact has many attributes traditionally assigned only to the gods.

From a true scientific perspective, all this is false. The condition for a fact to exist is for the world to have the above architecture, which it rarely exhibits. If the world has a different architecture, Albert Einstein's intuition is valid; namely, it is our theories that determine the facts we can observe.[39] The historical development of scientific disciplines bears this out, as was shown by Thomas Kuhn.[40] During a period of normal science, the status of a fact depends on, and is relative to, the reigning disciplinary matrix; when this disciplinary matrix is replaced by another during a scientific revolution, the status of any fact can be lost, while other observations can gain the status of facts.[41] Consequently, if it turns out that the world we entered as babies and children by listening to others has a very different architecture from the world we see, our scientism regarding facts can plunge us into a great many contradictions.

This current scientism also holds that it is impossible to argue with "the facts." For example, once certain economic facts are established, the rational course of action is to act on them. The possibility that these facts may have been gathered with a flawed theoretical framework does not enter into consideration. It is in this manner that we have become enslaved by our so-called facts. However, this is a scientism that, regrettably, is implicitly and explicitly embraced by a great many economists, scientists, and technical experts. We do not need to look any deeper than our facts, which, from a secular religious perspective, act as our grounding "mini-gods."

It also follows from the above architecture that it, and it alone, is expressible in terms of mathematics, as the only human creation entirely built up with the principle of non-contradiction. If it turns out that the architecture of whatever is being mathematically modelled or simulated does not conform to the architecture of mathematics based on non-contradiction, all manner of distortions will inevitably result and be transferred into the world by applying any results.

We have noted that the symbolic universe of individual and collective human life necessarily has an architecture characterized by dialectical enfolding, because many of its entities are constituted of contradictory elements in a manner that makes these contradictions integral to them. Any attempt to separate these contradictions inevitably leads to the destruction of the entity. This architecture based on dialectical enfolding involves at least three kinds of entities. There can be "zero," or "one," but mostly "zero *and* one."[42] In contrast to the world apprehended by seeing, the world of a time, place, and culture can be entered into only by growing up listening to others talking about it by means of the language of their community. Through this language, they express something of the way they make sense of their world and live in it. For example: behind the visual appearance of a tree is the wood a village in a traditional society may use to cook its food and make the table and chairs to eat from, the log from which to carve a canoe for fishing and travelling, and the branch to fashion the sacred bowl for worshipping the deities and powers the community had intuited as being behind it all. In this way, such a village acquired a great deal of knowledge about trees relative to their actual and potential activities constituent of their way of life and vice versa. It was in this manner that trees and everything else encountered by the village entered into their symbolic universe – which included, but far exceeded, what the members of the village saw. For this and other reasons, such a symbolic universe could not be apprehended by seeing, as noted above.

In sum, the symbolic universe was the world in relation to which life in traditional societies was lived and evolved, which had architectures that could be expressed in terms of principles that were diametrically opposite to those of the seen world. The elements of a symbolic universe were frequently contradictory, non-separable, impossible to define in themselves, unmeasurable, and joined in a dialectically enfolded complexity.

From this perspective, it is possible to shed some new light on the rise and fall of Greek culture. Greek culture was widely admired as a great accomplishment of liveability in the world by the people of their time. However, it began gradually to be relativized by contacts with other cultures whose practices implicitly and explicitly questioned it. For example, the Greeks did not educate their women. When they observed that women of other cultures successfully took charge of trade once their men sailed their merchant ships into Greek ports, questions naturally arose why the Greeks did not educate their women also. Socrates, aided by his students Plato and Aristotle, attempted to arrest this relativization of their culture by questioning people who had a recognized expertise in a particular area in order to elicit the kinds of rules they followed. The conviction that such rules existed, at least in some areas of human life, offered the possibility of putting a part or all of Greek culture on a foundation of universal knowledge that was "eternal," since such rules are valid for all times and places. As such, they would be capable of defending Greek culture against any threat posed by relativization.[43] Of course, at that time no concept of culture existed.

With a great deal of hindsight, it may be clear to us that there was an inherent conflict between the absolute character of universal knowledge and the relative character of a culture. A culture is always relative to a time, place, and itself, although this latter characteristic must be transformed to prevent its plunging into cultural chaos. As noted, doing so requires symbolically grounding itself. There is thus an inherent conflict between universal knowledge and whatever is symbolically absolutized by a culture. In the case of the Greeks, the former was detrimental to the latter, and may well have contributed to a further decline of Greek culture.

It was during a period when the symbolic grounding of Greek culture became progressively weakened that philosophers such as Socrates, Plato, and Aristotle looked for a philosophical grounding, thus marking the beginning of an intellectual project referred to as foundationalism. Philosophical grounding sought to deal with the increasingly apparent limitations to symbolization, language, and culture by essentially extending the way the world was visually apprehended to the intellectual domain of philosophy, thereby gradually transforming the architecture of philosophical thought. In contrast, when societies move from one historical epoch to another by desymbolizing one set of myths and by resymbolizing their experiences to yield another set, any reasoning remains entirely within the architecture of the world implied in symbolization, language, and culture.

Any attempt to express human activities in terms of rules involves their reorganization into an entirely different architecture. When Greek philosophy sought to "ground" its thinking in universal knowledge, it was a denial of the symbolic grounding that every culture relies on. Such a denial can be made only in terms of a higher religious authority. In this case it was the authority of universal and thus eternal knowledge based on the principle of non-contradiction. This religious authority, however, also distorted everything.

All cultures symbolically appropriate the entire universe, the past and the future on behalf of a community by its myths. Philosophy now sought to establish a new religious authority based on universal knowledge that would not be limited by any time, place, or culture, including the latter's divinities. It gradually made philosophy more asocial and ahistorical, as if its practitioners could intellectually "jump" through the myths (and thus the symbolic grounding of their culture) to directly access whatever existed beyond or underneath. The practitioners became the judges as to whether or not any or all elements of a culture required grounding in this "deeper underneath." As such, philosophy had the potential to become the new secular religion displacing all its traditional forms, and its practitioners would become its secular priests of universal knowledge free from all its bonds to a time, place, and culture. We will see that this can only be done in an architecture of non-life.

By way of an example: The Greeks were convinced that without laws a society would be plunged into anomie. This was regarded as a condition of lawlessness void of any meaning, purpose, and direction. It would render impossible any social relationships and thus any groups or communities constituted by them. However, for the Greeks, unlike the Romans, the rule of law in a democracy was essentially philosophically conceived, and the social forms that they invented by this approach had never been practised by any group or society. It raises the question: if these forms were so excellent, why had no group or society ever practically discovered them? Again, a part of the answer would appear to be that political life cannot be translated into another architecture without losing some or all of that life itself.

In the twentieth century, several philosophers broke with foundationalism on the grounds that there were foundations neither of knowledge nor of philosophy. In other words, philosophers were not gods and they could not escape being members of a symbolic species. Their intellectual activities were relative to and dependent on a time, place, and culture, including the latter's symbolic grounding in myths. Consequently, philosophy could at best push what is thinkable and imaginable within these myths to its limits, but it could not transcend them. However, philosophy, like everything else during the twentieth century, was swept up in a massive restructuring of all human knowing and doing by means of discipline-based approaches. We will later see that such approaches intellectually grounded themselves in other ways. The new philosophy was entirely swept up by these new discipline-based developments with all the limitations these embody.

According to Hubert Dreyfus, Martin Heidegger broke with Western philosophy by eliminating five basic assumptions. Dreyfus summarized them in the introduction to his commentary on Heidegger as follows.[44] Explicitness refers to the assumption that all human knowing and doing depend on the application of principles that function as presuppositions. To be a responsible human being requires that we are as clear as we can be about such principles. Mental representation refers to the assumption that we require an internal representation of anything in human life before we can direct our attention toward it in our external world. Theoretical holism refers to a third assumption, which presupposes that our common-sense world can be represented by an implicit theory present in our minds as a network of intentional states connected by rules. This mental background is open to analysis. Detachment and objectivity refer to the assumption that, by withdrawing from our daily-life involvements, we are able to reach a vantage point that is completely detached and thus objective. From this vantage point we are able to discover the reality of everything. Methodological individualism refers to the fifth assumption, which claims that a human being's daily-life practices are independent from his or her social, cultural, and historical context. This assumption rejects the idea that this context can constitute the very basis for the intelligibility of such practices.

With two thousand years of hindsight, Hubert Dreyfus has decisively summed up where Western philosophy was headed all along. He recognized it all around him at the Massachusetts Institute of Technology, where researchers were creating the first generation of artificial intelligence. He correctly predicted that it was doomed to eventual failure because it was based on the above five assumptions.[45]

As someone who has taught engineering in a different way, I would add that the most complete and perfect way of embodying the above five assumptions that I am aware of is in the programming of a robot. I could have cast the above five assumptions into a theoretical framework for doing so. There is nothing more detached than a robot. It "knows" nothing about itself and the world other than whatever logical structure has been coded into its computer memory. It is pure object-ivity. Nothing whatsoever can be assumed or be taken for granted. Everything must be explicitly spelled out by means of rules and algorithms.

This logical approach is in sharp contrast with our daily-life conversations, for example, where a great deal can remain unsaid because we already share it with the other person as people of a time, place, and culture. From a logical perspective, there appear to be gaps in our conversations, but, from a cultural perspective that builds on dialectical enfolding derived from growing up and living in a society, we know very well that filling in these gaps would make no sense at all.

A robot must be coded with a logical set of instructions for the execution of any of the functions it can perform. Its response is a match between what is detected by its sensors and its "mental representation" of its micro-world and the "actions" to be performed in it. It must be the most logical set of instructions possible, in order to constitute objectivized "intentions." All the instruction sets must form a logical whole representing the robot's domain or micro-world, constituted by what it can detect and react to during its operation. Nothing must be able to interfere with any of this, other than a shutdown required for safety reasons, damage to the robot, need for maintenance, or its functioning no longer being required.

It is important to recall the fundamental difference between the computer memory of a robot and the dialectically enfolded organization of a human brain-mind. In the case of the former, only some elements are loaded into the working memory while most are excluded in a particular operation. In contrast, our entire lives work in the background, without which we could not live an experience as a moment of our lives. We would then become like people suffering from short-term memory loss. Each of their experiences becomes like a distinct and separate micro-world incapable of becoming a moment of a life. This person's life ceased to evolve with the onset of the disease, but it continues to work in the background to some extent. Such people are entirely disconnected in space, time, and the social. For example, they cannot follow a conversation because they cannot remember what was said prior to this moment. For the same reasons, they can no longer understand a book or follow a movie. When taken to a particular place, they have no recollection of how they got there or why they are there. These people have no idea when they ate last or took care of any bodily functions. It is a profoundly tragic condition, producing endless anxieties.

With a great deal of hindsight, it is thus becoming apparent that some of the greatest achievements of Western civilization were the result of extending the architecture of a world first encountered through seeing it. These include mathematics, logic, discipline-based science, discipline-based technology and technique, as well as philosophy and, via it, much of Christian theology. There were, of course, notable exceptions. For example, in the case of Christian theology, there are the works of Søren Kierkegaard[46] and Jacques Ellul.[47]

According to Hubert Dreyfus, Martin Heidegger correctly saw that the role of theory in human life could only be very limited.[48] Surely it is impossible to represent anything living with non-living theories. Hence, a shift is required away from a theoretical approach to an interpretive one capable of taking into account meaning, culture, and context. Heidegger introduced a phenomenological hermeneutical approach into philosophy. He showed that our way of being socialized into our culture's way of being and living in the world yields "our background practices in which we dwell." Our lives are thus constantly working in the background in a manner that cannot be made explicit and can be understood only from within, as it were. Since our being in the world can be neither theoretically represented nor formalized, any cognitive approach based on mental representations and intentions, or on computer models, is wrong-headed. Our memories of past events symbolize that we were there and lived them as moments of our lives. Such memories become inseparable from the background, against which subsequent moments of our lives are also lived. Since there can be no theory of our daily-life world, Martin Heidegger proposed a phenomenology of our daily-life activities. However, our lives working in the background must not be mistaken for a tacit belief system by which we live in the world. We cannot say that we have being or that we are, since we do not possess our lives and will eventually die. We dwell in our lives in the world and thus cannot achieve a detached and objective vantage point. Our meaning and cultural context are inseparable from our lives. I believe this captures the thrust of Hubert Dreyfus's interpretation of Martin Heidegger's phenomenology.

As noted, it is important to interpret the latter development of philosophy in the context of what happened all around it within the university and in society. Human knowing and doing were increasingly being reorganized by means of disciplines. Philosophy followed suit and became a discipline itself. Consequently, it had to carve out for itself an autonomy with regard to all other disciplines. A shift from a theoretical to an interpretive approach could not, of course, be extended to include whatever insight all the other disciplines contributed to our understanding of human life and the world. Doing so would involve the resymbolization of all disciplines as an intellectual translation from the architectures of the domains of these disciplines into a lived architecture of the dialectically enfolded and symbolically grounded human life of that time.

The new philosophy raises the question as to what kind of daily life is being interpreted. Is this the daily life characteristic of traditional pre-industrial societies symbolically grounded in a traditional sacred and myths? Is it the daily life of the industrializing societies symbolically grounded in the first generation of secular myths (to be examined in the next section)? Or, is it the daily life that emerged between the two World Wars, which became dominant in the second half of the twentieth century and was symbolically grounded in a second generation of secular myths (to be examined in a later section)? As will soon become apparent, these three forms of daily life are substantially different.

The first generation of secular myths created a hierarchy that gave priority to the new emerging economic approach that governed an increasingly separate and distinct economy, while a desymbolized cultural approach continued to largely govern the remainder of society. Since almost everyone's daily life encompassed the economy as well as the remainder of their society, it was qualitatively different from any daily life that preceded it. Similarly, the second generation of secular myths created a hierarchy that gave priority to the new discipline-based approaches and everything built up with them, which took over from the role of culture throughout the so-called industrially advanced societies. The roles of the even more highly desymbolized cultures were now essentially restricted to the early phase of the bringing up of each new generation and to intimate relationships and family groups. The daily life associated with this development was qualitatively different again. It would appear, therefore, that the new philosophy, like its precursors, was pushed to the limit of what is thinkable and imaginable within the symbolic grounding of a society. In the case of Martin Heidegger, his unique vocabulary and the lack of sociocultural examples that would permit it to be put into a context make it impossible to know which of the above three forms of daily life was being interpreted. This would confirm an apparent interpretive failure or practical unwillingness to discern and take a stand on what was happening all around him in the Germany of his day. In contrast to philosophy, the approach underlying the present work may well be foolish in another way by attempting to resymbolize the findings of all the relevant disciplines for understanding what happened to us in our youth.

It may be argued that all this is hardly relevant, since the new approach to artificial intelligence and so-called deep learning based on neural nets and advanced statistical procedures is much more Heideggerian in character.[49] The more important question is whether or not this approach is also an extension of what we see. Whether trained by a human being or by "big data," neural nets are based on advanced mathematics and run on computers, and thus ultimately have the same architecture as the first generation of artificial intelligence.[50] What these developments do much more successfully is to translate lived human activities into an architecture of non-life. That is their true significance, as we will see in the next section.

The Technical Division of Labour and the Architecture of Non-Life

What foundationalism was able to accomplish for Western philosophical thought the technical division of labour achieved for physical and intellectual human work: the translation of a lived daily-life activity from a dialectically enfolded architecture into one of non-life. The most important difference between this translation and the one underlying philosophical foundationalism is in the consequences for human life, society, and the biosphere. In the case of philosophy, it led to the enslavement of human thought to the principle of non-contradiction and everything built up with it. The consequences were tragic enough but relatively marginal to much of human life and society. The same translation applied to human work soon affected everyone and everything.

It should be noted that, in the most influential religious traditions of Western civilization, the hiring of someone for a wage was fraught with risk. The transaction involved buying a portion of the life of a labourer in return for a wage, which implied the possibility that the effects of doing so could spill over into the remainder of that life and thus possess it, just as a master possesses a slave. Hence, all kinds of precautions were imposed on these kinds of arrangements. The translation of human work into an architecture of non-life can have the same effect, as has been recognized in the social sciences by the concepts of alienated work and the commoditization of labour.

Much of this can once again be interpreted in terms of an extension of the way we deal with one another and the world by seeing. Generally speaking, all societies have constrained the influence of visual apprehension to the contribution it made to making sense of and living in a symbolic universe. Doing so imposed an expectation of reciprocity on everything seen, especially with regard to living entities, thereby ensuring that seeing them would not be degraded into staring at them. When we stare at another human being, for example, what we are truly doing is looking him or her over, much as we would look at an object, neither expecting nor seeking any response. It is surely the reason why we feel profoundly uncomfortable when someone stares at us. It involves a denial of our humanity. The symbolic framing of the visual experience is thus largely suspended in the act of staring. Similarly, when Western societies ceased to symbolize humanity and nature as integral to the Jewish and Christian creation, the emergence of a separate nature "out there" opened up the possibility of new activities such as tourism.[51] The beginning of differentiating people's inner world from nature as a kind of background was symbolically portrayed in the painting of the Mona Lisa.[52] Our staring at another human being or at nature involves a (conscious or unconscious) diminishing of our sustaining reciprocal interdependence with a dialectically enfolded symbolic universe and with an enfolded biosphere. It also involves a measure of objectivization that paves the way for commoditization and reification. It comes close to living as if all the life has been "squeezed out" from a living entity. The way everything is related to, and evolves in relation to, everything else, creating the niches of the biosphere and the occupations of a society, becomes distorted. This results in desymbolization and thus in the diminishing of experience and culture, with a corresponding negative effect on people's grip on their life and their world.[53]

Keeping these potential broader implications in mind, let us begin by examining the technical division of labour in some detail. It reengineers human work in the way a machine carries it out.[54] Work is unfolded into distinct and separate production steps that are designed as much as possible as autonomous domains, whose internal functioning must not be disturbed to any significant extent by the functioning of all the other domains. The domains are externally connected by their desired outputs, transferred as intermediary inputs into the next domain until the technically divided work process working as a sequence of these domains achieves its desired final output. The internal functioning of each domain is limited to a production step that can be endlessly repeated through the application of a single category of phenomena (metallurgical, mechanical, chemical, electrical, and so on), which transforms the intermediary inputs received by the domain into an intermediary desired output.[55]

This unfolding of human work is necessarily accompanied by an unfolding of the involvement of human beings in this technically divided work process. Three separate and distinct technical roles result. The knowers carefully plan the unfolding of the work process, ensure that each production step is optimal, decide on whether to assign it to a human being or machine, optimize the integration of these production steps into the production process, and so on. This knowledge was initially based on practical experience, but was eventually replaced by engineering and business disciplines learned in classrooms. Next, there are the doers, who carry out the work assigned to them in accordance with the production plan devised by the knowers. There are also the supervisors and managers, who mediate between the knowers and the doers to ensure that the latter carry out the work according to plan and who bridge the inevitable gaps between this plan and how it works on the shop floor, deal with all other problems that may arise, and integrate the operations into the organization of the enterprise and (for the higher levels of management) into the economy and society.

No human beings can participate as such in this technically divided work process. They participate only as a "brain," "hand," or intermediary "control." This unfolding makes it impossible for any participant to intervene in a problem at its source (sometimes referred to as negative feedback), thereby imposing end-of-pipe compensating remedies. Unfolding a person's participation into that of his or her brain-mind, body, or coordinating function has significant consequences. In daily-life activities, all three are dialectically enfolded into each other; and their unfolding causes many psychological and health problems, of which nervous fatigue is the best known.[56]

The technically divided work process has also become unfolded from the way of life of a society. In order to shed light on this unfolding, careful attention needs to be paid as to how each and every human activity integral to a way of life can neither create nor destroy the matter and energy on which it depends. All these activities must receive the required inputs directly from the biosphere, or indirectly via a chain of human activities. The use of matter and energy results in their transformations, with the outputs having to be returned to the biosphere, either directly or via other chains of activities. Consequently, all activities contributing to a way of life are joined by a network of flows of matter and a network of flows of energy, which overlap wherever flows of composites of matter and energy occur.[57] There is also a biology-based connectedness that maps the interdependence between the members of a society and their activities according to bonds of kinship. All constituent activities of a way of life take their place in the culture-based connectedness of a symbolic universe according to their relative meanings and values, which are established and evolved by symbolization guided by a culture.[58] In pre-industrial traditional societies, the material- and energy-based connectedness and the biology-based connectedness were enfolded into the culture-based connectedness, with the exceptions of activities mediated by markets and commerce.[59]

With the introduction of the technical division of labour and the installation of machinery of all kinds, the material- and energy-based connectedness of the activities of a way of life of any industrializing society increasingly turned into its technology-based connectedness.[60] For example, consider the transformation of local technology-based connectedness representing the flows of matter, energy, and their composites within a factory. The technology-based connectedness of all factories was soon linked by an encompassing technology-based connectedness representing the exchanges of inputs and outputs between them, as mapped in the input-output tables of a modern economy.

This growing technology-based connectedness had to be unfolded from a society's culture-based connectedness, for practical reasons.[61] Reorganizing a local portion of the material- and energy-based connectedness of industrial activities by introducing the technical division of labour and machinery disturbed the local dynamic equilibrium. It resulted in inadequate levels of inputs in some places and excessive levels of outputs in other places, which required further adjustments in the local networks. All such adjustments could not be managed by symbolizing them according to their relative meaning and value for human life and society, but entrepreneurs were faced by immediate problems such as the output of one production step exceeding the capacity of the next one to absorb it following the introduction of a new machine, or when an upstream production step did not produce adequate output to sustain the next downstream production step. In other words, the constant attempts to rebalance the local dynamic equilibrium required the consideration of the technical significance of inputs and outputs for the local technology-based connectedness, as well as the production costs of doing all this and ensuring that these costs remained below market prices for the final output achieved by the technically divided work process. Moreover, the throughput of matter and energy in the local technology-based connectedness quickly exceeded any correspondence to local needs and social bonds, including patterns of reciprocal obligations. It now corresponded to the needs of an ever-larger geographical area. Cultural mediation became ineffective and had to be replaced by markets for goods. These jointly constituted a Market economy that regulated goods and services one at a time according to the supply and demand in corresponding markets. Market prices thus had a weak correspondence to cultural values. All these developments resulted in the local technology-based connectedness overwhelming the local culture-based connectedness.[62] This divided any industrializing society into a relatively separate and distinct economy, which was regulated by the Market as its organizing principle, and the remainder of society, which continued to be ordered by a now substantially desymbolized culture.[63] In the economy, the technology-based connectedness dominated the culture-based connectedness, while in the remainder of society, the reverse relationship was the case. In the economy, technical reasoning and economic reasoning increasingly distanced themselves from culture-based reasoning. This development eventually imposed a technical and economic rationality.[64]

This technical and economic rationality cannot be conceptualized as a mere narrowing of the context taken into account by reducing it from the culture-based connectedness to the local technology-based connectedness and the markets that govern it. It also involves a change in the architecture that is presupposed: from a dialectically enfolded one to an architecture we encounter when we visually apprehend the world. Every domain in the industrial and economic orders is internally autonomous from all the others, containing a closed cycle of production operations that jointly repeat a production step. It has its own (technical) "goal" independent of those of all the other domains. All such goals belong to these domains since they all express how much intermediary output can be obtained from the inputs supplied to a domain. They have no common denominator with cultural goals. They all measure a domain on its own terms, that is, by output-input ratios. Alternative forms of the production step in question can only be compared with each other, and thus by means of the same output-input ratios. This is the basis for technical and economic "ratio-nality." Technical and economic rationality thus represent an approach that is similar to visually apprehending the world without feeling any need to frame it by symbolization, as guided by a culture.

A complementary way of conceptualizing the transition from relying on symbolization guided by a culture to relying on a technical and economic rationality is to reflect on what happens when a goal is imposed on a situation. Doing so can have only a very limited meaning in a dialectically enfolded world. In contrast, it can make complete (technical) sense in situations where the technology-based connectedness dominates the culture-based connectedness. The imposition of a goal intellectually and practically rearranges the local fabric of relationships. Those directly relevant to achieving the goal are retained as the focus of attention; those relationships that are indirectly relevant may be retained in the background; while the relationships irrelevant to the accomplishment of the goal are simply ignored. This intellectual rearrangement of the local fabric of interdependencies is then imposed when attempts are made to achieve the goal. It creates all manner of distortions that become externalities to this goal-driven behaviour. Consequently, the local culture-based connectedness of a symbolic universe becomes a significant part of these externalities. So is the local fabric of relationships that constitutes the local ecosystem. These externalities to goal-directed behaviour will be translated into market externalities and via them into Market forces, of which only one instance has generally been recognized: the great invisible hand of the Market.

The approach to the world based on symbolization guided by a culture that we once used to make sense of the world and live in it is thus diametrically opposite to doing so by means of technical and economic rationality. In the case of the former, the meaning and value of anything is derived by symbolizing it relative to everything else in our life and the world, with an open-endedness that must be closed by symbolizing the unknown and absolutizing what we know and live by means of myths. In the case of technical and economic rationality, the significance of anything is established relative to itself according to the internal functioning that delivers something we depend on or desire. The limitations of this orientation must also be protected – by secular myths that "interpolate" and "extrapolate" all these rational technical and economic elements into a way of life that can sustain us. Technical and economic rationality can operate only on elements of our lives and the world that have first been commoditized and reified.

For example, before a part of a life-sustaining ecosystem (such as land) can be reduced to a piece of terrain that can be described on a deed and traded in a real estate market, it must be commoditized and to a considerable degree reified. Before human work, which is integral to a life that helps to make up a community and the local ecosystem, can become labour and be traded in a labour market, it must be commoditized and reified. In both cases, doing so involves the translation from one architecture into another. It intellectually twists and distorts the fabric of reciprocal relationships that sustains all living entities and their mutual interdependence. When this is followed by carrying out the translated activities, the intellectual externalities will translate into economic externalities, for example. In the case of a Market economy, they produce the well-known market externalities. These are the unintended and often unforeseen consequences of a particular market transaction between two parties – consequences that affect third parties in the present or the future who did not participate or agree to the transaction. These can be positive or negative. Millions of such externalities will cumulate into powerful Market forces that drive a society into often unwanted directions. Everything that was once more or less taken care of by a culture now accumulates in and shapes everything through these forces. Economic rationality thus achieves the efficient allocation of goods and services, but it is accomplished by straining everything that sustains life.[65] In our contemporary societies, technical externalities predominate, as we will see in the next section.

The introduction of the technical division of labour and machinery of all kinds also unfolds a society's technology from its way of life. It goes hand in hand with the unfolding of the economy from a society. Most pre-industrial traditional societies did not have a technology the way we understand it. What they did have was a multitude of relatively distinct technological elements that primarily evolved relative to their use, that is, in relation to their immediate sociocultural and ecological contexts. This technological embeddedness led to highly appropriate and generally sustainable technological elements, which archaeology could readily identify as bearing the stamp of a time, place, and culture.[66] To be sure, there were occasional interactions between these technological elements that exercised some influence, but generally speaking, they did not amount to anything as decisive as what we encounter in our technologies today.

The introduction of the technical division of labour and the associated unfolding of the technology-based connectedness from the culture-based connectedness of a society necessitated a greater linking together of industrial technological elements through exchanges of inputs and outputs. This reorganization of technology and the economy would eventually yield a kind of architecture ideally suited to discipline-based approaches. These had been pioneered in the sciences, but this took some time, as we will see in the next section. Initially, technological knowing and doing had been based on symbolization, experience, and culture. It was then entrusted to the technical and economic rationality that eventually led to a complete reorganization by means of disciplined-based approaches. It allowed more and more technological knowing and doing to be expressed in the architecture of mathematics – and thus in an architecture of non-life. Every area of specialization now began to act as a transmitter and receiver of technical information through abstracts, journal articles, conferences, collaboration through invisible colleges, and so on, thereby creating a global system. This system continued to respond to human needs and desires, but to a far greater extent to this new internal mechanism of development that produced millions and millions of permutations and combinations resulting from this global information flow and that triggered all manner of technical advances.[67] Technology thus began to take on the character of a universal system with respect to all societies that participated in its application and development. Its evolution became non-linear and impossible to predict by technological forecasting.[68] As we will examine in some detail, the domains associated with the disciplines that jointly constitute a technological specialty were separated from experience and culture and thus could be entered into only by means of a technical and mathematical imagination.[69] This access would be accomplished by extending the way we apprehend the world by seeing, at the expense of doing so by symbolization guided by a culture.[70] These transformations had many consequences, including the creation of a great many technological externalities associated with the way technological knowing and doing was reorganized on the basis of disciplines.

As a consequence of the unfolding of an industry and the economy from the remainder of society, society became essentially divided against itself. One "part" was increasingly adapted and evolved by means of technical and economic rationality, while the other "part" continued to rely on symbolization guided by a culture but suffering from extensive desymbolization.[71] Why did this not translate into a kind of cultural schizophrenia in people's lives? The answer is the same as always: we need to look for what, in the experience of the members of these societies, was limitless, autonomous, and thus could not be shaken by anything they knew or lived. We should not jump to conclusions and look for the kind of myths that gave rise to the traditional moralities and religions. If we did so, the behaviour of entrepreneurs or wage earners would not make any sense. For example, if an entrepreneur, based on his convictions, decided to pay his workers a living wage, doing so would have to be funded by charging higher prices – in which case he would be competed out of the markets for his products and he would cease to be an entrepreneur. Alternatively, he could finance higher wages from his personal wealth. Sooner or later it would be depleted, and the entrepreneur would either have to cease doing business or continue his original behaviour. In other words, the technical and social roles of entrepreneurs were determined by the economic system. Their choice was between serving it regardless of their moral and religious convictions, and withdrawing from their technical and economic role as entrepreneurs. Much the same argument can be made for the economic role of the wage earners. The (cultural) morality and religion of their society could not continue to play a role unless they were assimilated by that system. This happened by the splitting of Christianity into liberal (love of neighbour) and conservative (love of God) branches that could readily be assimilated into the system to serve it rather well.[72] Similarly, the common law systems could no longer be entirely rooted in the metaconscious values dialectically enfolded in the organizations of the brain-minds of the members of the industrializing societies. Law had to take on an increasingly organizational character.[73]

It is for these and other reasons that the behaviour of entrepreneurs and wage earners could be approximated by what later, in the discipline of economics, was referred to as *homo economicus*, a kind of return on investment, or utility-maximizing algorithm. The collective output of these algorithms, which maximized return on investment, determined the supply of goods and services; while the collective output of the algorithms maximizing the utility that could be derived from a wage determined the demands for goods and services. Next, the "mechanism of supply and demand" mediated between these supplies and demands one good or service at a time. The very possibility of modelling this economic system in terms of an architecture of non-life should be interpreted as a sign of the enslavement of people to this system. Legitimating all this in terms of self-interested behaviour is to disregard the fact that all human communities throughout history have always greatly constrained this kind of behaviour by their traditional moralities and religions.[74]

To make this new way of life meaningful and purposeful required the kinds of myths that would somehow prioritize economic goal-driven behaviour over the kind of culture-driven behaviour characteristic of all of humanity until then. In other words, the claim of self-interested behaviour must be complemented by the commonplace incomplete descriptions of a Market economy. In the short term, it would produce the best possible world for most people in the industrialized world, as Adam Smith predicted.[75] However, all this would be undermined by what was left out of his description: the accumulation of many and diverse market externalities combining into Market forces that would threaten the fabric of all life.[76]

It was Max Weber who warned humanity against the rise of the phenomenon of rationality, which would "shut it into an iron cage."[77] As noted, a dialectically enfolded architecture of human life makes the imposition of a cultural goal somewhat meaningless. Everything is related to and evolves in relation to everything else on many different levels of metaconscious depth at the same time. To how many levels should a cultural goal apply in order to be meaningful? What will be the negative consequences of ignoring the levels of metaconscious depth of life not taken into account by a cultural goal? In other words, the living of a life cannot be replaced by a hierarchy of goals as long as that life remains highly dialectically enfolded. In many languages, concepts such as efficiency could have no meaning until a high level of unfolding set in. Why measure anything on its own terms rather than in terms of the contribution it makes to the overall interconnectedness of human life and society, of which it is an integral part? Doing so could not have any meaning until there was a great deal of unfolding of human life, which began with the introduction of the technical division of labour and the translation of human work into the architecture of non-life. In turn, the more the dynamic equilibrium of the technology-based connectedness was disturbed by the introduction of ever faster and more efficient machines and processes, the more economic rationality was necessary to ensure the economic viability of each and every change. The associated imposition of economic goals produced spectacular economic growth but also an unprecedented straining of the fabric of relationships to which these economic goals were applied. Economic goals, unlike their cultural counterparts, measure everything on their own terms by means of output-input ratios (ratio-nality).

To sustain human life in these divided societies with contradictory architectures, and to protect that life from cultural chaos, required non-traditional myths that were unprecedented in human history. What we are looking for are those entities in individual and collective human life at that time that were so self-evident that anything else was rendered unliveable. Individual diversity could thus be constrained by this new cultural unity.

It was self-evident to people of that time that industrialization brought material progress. Each and every year, the members of the industrializing societies experienced the growing use of machines and thus their increasing dependence on them. Every year these machines were becoming larger, faster, and capable of producing a greater output. Every year there were more factories full of these machines, with the result that their collective output rose steadily. There was nothing on the horizon of people's experience to indicate that these trends were temporary and would soon be halted by entirely different and more influential ones. On the contrary, this material progress seemed to be unstoppable.

It was equally self-evident that this ever greater material output could satisfy the needs of ever more people. It could therefore be anticipated that poverty would decline and eventually be wiped out altogether. The social scourges that poverty produced would also disappear: the inability to feed and take care of one's children; the need to steal to feed them; the moral dilemmas this produced that so often led to a sense of hopelessness and defeat; turning to alcohol in despair; the widespread reliance on alcohol leading to family violence, breakdown, and the erosion of community; children born out of wedlock as couples saw no way of making a family work; families losing breadwinners as they were deported or hanged for as little as stealing a loaf of bread; and so on. Once everyone's basic needs were met, the reasons for all this destructive behaviour would disappear, and it could reasonably be expected that socially things would improve a great deal. Besides, the human ingenuity that had, for the first time in human history, eliminated poverty could surely be diverted to address whatever social problems remained. As a result of these developments, there was no reason to suppose that, once everyone had enough of everything they needed and then a little surplus to enjoy life and, still later, achieve a measure of affluence, all this would not make them happy and content.

It may be supposed, therefore, that the symbolic "interpolation" and "extrapolation" of the experiences of that time by the organizations of the brain-minds of the members of these industrializing societies would become their deepest metaconscious knowledge. It created the first generation of secular myths. Material progress would extend into social progress, and social progress would extend into happiness. Progress would be everywhere and touch everything in a limitless fashion. All a person had to do to contribute to this progress was to work hard to contribute to the growing economic output, and thus help to achieve everything else that humanity had always hoped for. The happiness that could thus be achieved was potentially limitless.

The first central secular myth, or secular sacred, thus comes into focus. There was plenty of land and labour, thanks largely to a preceding agricultural revolution that permitted the production of more food with fewer people. What was the scarcest "input" into this new economic system was capital – the lifeblood without which nothing could be accomplished. As industrialization advanced, everything in industry required an ever greater investment, and so did the infrastructure to support it. If the renewal and accumulation of capital could be ensured, everything was possible. Consequently, the cultural unities of the first generation of industrial societies were constituted from the first secular sacred of capital along with the sustaining secular myths of progress, work, and happiness.[78]

The new cultural unities of the first generation of industrializing societies were thus constituted of the first secular sacred sustained by the first secular myths in human history. They represented the deepest metaconscious knowledge implied in the daily-life experiences of the members of these societies. Similar profound transformations of metaconscious knowledge occurred on all lesser depths. For example, it led to the formation of much more materialistically oriented metaconscious images of people's sociocultural selves. It was as if a new "human nature" had been formed to match the new "cultural DNA." Similar transformations on many other levels of metaconscious depth led to intuitions that were aesthetically expressed in art and literature.[79] The meanings of many words in the English language changed.[80] Despite significant socioeconomic class differences, there was nevertheless a deeper underlying cultural unity.[81] It was all part of the process of industrialization with its dual components of "people changing technology" and "technology changing people," through the internalization of many different daily-life experiences that jointly and metaconsciously built the organizations of people's brain-minds.[82]

The emergence of these cultural unities and the transformation of people's sociocultural selves implied that humanity had always gone about its business in the wrong way. It had sought to oblige its gods by (individually and collectively) living good moral and religious lives according to a culture and attending to its material necessities within these constraints. All this had been a huge and tragic mistake. What humanity should have done, and what it was now demonstrating, was that the reverse strategy was required to achieve everything people had always wanted. Concentrate on satisfying society's material needs, and everything else will be granted in terms of social improvements and spiritual happiness. This challenged every traditional moral, political, and religious tradition. The technology-based connectedness of human life was to be improved at all costs because it was only in this way that the culture-based connectedness could be improved beyond what earlier civilizations had been able to accomplish. The first secular cultural unities thus implied a hierarchy between the emerging economic approaches and whatever remained organized by means of a culture. Set your economic goals and do not worry about what may be externalized in the process, because in the end everything will fall into place. Do not be bothered by the self-interested behaviour that the system encourages, because this apparently amoral and a-religious behaviour will in the end produce a new kind of secular paradise. The "entrance fee" to this utopia remained well hidden: people would have to give up their lives so that they could be turned into non-lives as "resources" for the economic system.[83] The term is entirely appropriate because human beings would increasingly surrender part of their persons and their lives – such as their "hands," their "brains," or their ability to coordinate and manage these two by suppressing the others. These would be "mined" from them, not unlike the way other resources are mined.

In the United States of America during the second decade of the twentieth century, the technical division of labour began to take on the forms that we now easily recognize. Its application to the "collective hand" (its horizontal dimension) took on the form of the assembly line based on interchangeable parts. Its application to the "collective brain" (its vertical dimension) led to scientific management and the emergence of industrial engineering. Jointly, these developments created the

Fordist-Taylorist system of production that helped make America the leading industrial power of that period.

Max Weber provided the classic description of the collective brain as a bureaucracy.[84] In order to carry out its mission, an organization identified the primary supporting functions that would be required. These were generally allocated to its departments. In turn, the departmental functions needed to be sustained by a diversity of secondary functions, which were allocated to various sections. In turn, the functions performed by each section were grouped in job descriptions as the "organizational niches" to be filled by the members of the organization. Having thus differentiated all the functions and subfunctions of the organization, their coordination, integration, supervision, and management had to be ensured by a hierarchy of positions that each more or less corresponded to the overall responsibility for a subfunction, function, or group of functions. The extent of this vertical division of labour was indicated by the number of levels of supervision, middle management, and senior management.

The results can be represented in the form of an organization chart that maps the "wiring" of the collective brain, much like the way one would engineer a complex integrated circuit on a chip. All the dialectically enfolded relationships between the members of such an organization that might stem from their personal lives and could give rise to informal communications were strongly discouraged, even though their occasional usefulness in a blocked situation was certainly recognized. The architecture of the organization was thus unfolded from the culture-based connectedness of a society. Many of the early bureaucracies in industry were closely tied to the local technology-based connectedness within the organization.

These bureaucratic organizations underwent two subsequent major transformations. The first one occurred around the time of the Second World War as a consequence of a growing reliance of the organization on highly specialized discipline-based approaches to knowing and doing that needed to be synthesized and brought to bear on complex tasks and situations faced by the organization. The classical description of this transformation was provided by John Kenneth Galbraith, who distinguished bureaucratic organizations from what he referred to as technostructures: as the "collective brains" of large corporations, government ministries, or other large organizations engaged in the application of highly specialized knowledge to complex tasks.[85]

A second major transformation soon followed. We have nearly forgotten the difficult lessons that were learned when the first word processors and then computers were introduced into the above kinds of organizations. Those corporations that treated the word processor as an advanced typewriter requiring minimal organizational changes were unable to gain the productivity increases that others were able to achieve by recognizing that it was not merely a tool in the hands of an organization. The same lessons had to be learned again when organizations were unable to gain the expected productivity increases through the applications of computers. All this changed when it was gradually realized that everything had to be reconceptualized in terms of the equivalent of information assembly lines with interchangeable "information parts." All business processes had to be re-engineered in this manner. Next, they had to be integrated into large databases whose information flows represented the integration of all the functions of an organization. The corporation was thus re-engineered in the image of the computer.[86] This eliminated the need to re-enter information into all the computers spread around an organization, so that the transfer of information between them was greatly facilitated. There was a kind of "just in time" production of information throughout the entire enterprise. It produced the so-called enterprise systems used by any large institution today.

The architecture of these enterprise system-based organizations may be conceptualized as transforming the local technology-based connectedness into a local technique-based connectedness comprising the storage, transmission, application, development, supervision, and management of all information. This includes the flows of materials, energy, their composites (such as parts, subassemblies, and final products), human resources (from unskilled to highly skilled), specialized scientific and technical knowledge, capital, buildings, public relations, advertising, research and development, marketing, land, and all other relevant resources. It is important to stress that we are dealing with flows of information and not with symbolic representations of these flows. This technique-based connectedness, represented by the enterprise system of the organization, is thus unfolded from the culture-based connectedness within and beyond it. The organization now has the architecture of non-life built up with rules, algorithms, and the like that can be coded as mathematical operations of the system.

This simple complexity allows for the running of very large organizations whose size far exceeds what might have been accomplished by symbolization guided by a culture. The principal reason is that this simple complexity can be made more manageable by dividing it into relatively distinct and separate divisions conceptualized as domains connected by inputs and outputs, a method that was first pioneered by General Motors. The internal workings of each of these domains can then be organized in terms of networks of smaller domains again connected by inputs and outputs. It greatly facilitates extending the scope as well as the level of local detail.

The limits of these enterprise systems are now becoming evident. Before any human activity can be incorporated into them, it must first be translated from a dialectically enfolded character into that of non-life. It is from this perspective and with a great deal of hindsight that all previous developments related to industrialization, urbanization, rationalization, and secularization can be interpreted as preparing the way for an almost unlimited use of information technology and everything built up with it.[87]

It is in this context that we can best understand the development of the new artificial intelligence based on neural nets and advanced statistical procedures that permit deep learning from big data, as mentioned in the previous section. It has little to do with living intelligence, despite all thinking to the contrary. This development is a big step forward from the first generation of artificial intelligence because it is much more successful at translating complex human activities into an architecture of non-life. The true significance of these systems of artificial intelligence is thus in extending the capabilities of the enterprise systems of the leading "information utilities" of our global economy. Like their electrical utility precursors, they represent a tremendous force for a new kind of centralization that goes hand in hand with a new kind of decentralization. As their architectures of non-life engulf more and more of our daily lives, these systems have been justifiably referred to as "weapons of math destruction."[88] This analysis takes on its full significance in the context of the rise of the phenomenon of technique as examined by Jacques Ellul.[89]

Technique and the Architecture of Non-Life

What is commonly referred to as industrialization, urbanization, rationalization, and secularization involved a great deal of unfolding in a diversity of elements from a dialectically enfolded way of life and the symbolic universe of a society's culture.[90] This unfolding meant a full or partial translation from the architecture of this culture into another architecture. The effects were twofold. The way of life and symbolic universe of the culture were desymbolized as the adaptation and evolution of some relationships and webs of relationships were withdrawn from them. At the same time, these withdrawn relationships gradually began to link together into what eventually became what we commonly refer to as *reality*, with an architecture based on the principles of non-contradiction, separability, closed definitions, measurability (including quantification and possible mathematical representation), and a simple complexity. This dual development increasingly led to human life being placed within the context of, and evolving in relation to, this reality and relying less and less on culture.

For example: Imagine the way in which the newscasts on the traditional and the "new media" provide us with the information we need to maintain and evolve what is left of our symbolic universes. It becomes immediately evident how these newscasts provide us with the most discontinuous, incoherent, and piecemeal accounts of what is happening to human life and the world. Events come out of nowhere, receive a few minutes of attention that may be distributed over several days, and then disappear, likely never to be heard of again. Even if our symbolic universes have not been extensively desymbolized, they would soon lack any coherence were it not for extensive public relations efforts supported by a variety of techniques, but that transform these updates into "event-dots" to be integrated by the program into mythological patterns.[91] These patterns are interpreted by their audiences as commonplaces that take on their meaning in the context of the deepest metaconscious knowledge (cultural unity) that reigns over and orders people's lives. The newscasts contribute to a bath of images (accompanied by image-words) into which the media immerse people's lives, which jointly tell people everything they need to know for the conduct of their daily lives. This bath of images is a compensation for a highly desymbolized way of life and culture.[92] It is an integral part of a social integration propaganda that makes life in contemporary mass societies with their highly desymbolized cultures liveable.[93] This bath of images accompanied by image-words thus performs what tradition and culture took care of in earlier societies, except that it is now an extension of our visual apprehension of our lives and the world – our reality.

In the same way, presentations of complex issues are made by means of the powerpoints of this world, which use distinct and separate bullets, flowcharts, decision trees, and so on. This method presupposes that we are dealing with a reality built up with the architecture described above. In this reality, it is possible that discipline-based approaches to knowing and doing have no limits, and facts are no longer relative to the conceptual framework in relation to which they can have a sense. It is as if human life, society, and the biosphere can be represented as complex sets of domain-life entities connected by flows of resources of all kinds.

The more a reality was unfolded from people's ways of life and symbolic universes in industrial civilization, especially during the last half of the twentieth century, the more symbolization guided by a culture ran into its inherent limitations. In some areas of society, these limits were obvious from the very beginning. For example, in the chemical and electrical industries, what was available to the senses was of little or no use in understanding chemical reactions or the behaviour of electrical circuits. Similarly, as the unfolded complexity of a way of life increased beyond a certain point, it became very difficult to make sense of it. Even technical and economic rationality were beginning to encounter some limits.

These kinds of limits to what could be accomplished by civilization, experience, and culture were overcome by reorganizing all human knowing, doing, and organizing by means of discipline-based approaches. Such approaches had been developed in the physical sciences for some time, but their broader implications did not become apparent until the transformations associated with the unfolding of human life and society were well advanced. This gave rise to what has been studied as the separation of knowing and doing from experience and culture, of which we will present a brief overview.[94]

Contemporary societies are in part characterized by many of their members making use of two parallel modes of knowing and doing, with one embedded in, and the other separated from, experience and culture. For example, before entering an elementary school, children have a metaconscious knowledge of the grammar of their mother tongue, as shown by their relative competence in the language. At some point, they will acquire an explicit knowledge of this grammar, but this development does not displace the previous one. That this is the case is evident when we volunteer to correct the grammar of a paper written by a foreign student. There will be instances where our metaconscious knowledge of the grammar allows us to say that, although we cannot explain exactly what is grammatically wrong with a sentence, people here would never write it this way.

In growing up, we also acquire a great deal of metaconscious knowledge about the physical behaviour of our bodies and the world as we learn to sit, crawl, walk, bike, play ball, climb trees, and so on. When we enter a physics class in high school, however, all this metaconscious knowledge is set aside as we (through our imaginations) enter the domain of the discipline of physics, in which we learn to solve a variety of problems. In experiments where students were given physics problems described in terms of daily-life situations they gave one answer, while if the same problems were cast in the form of physics, they gave a different answer. This has given rise to the distinction between intuitive physics and school physics.[95]

A construction worker who has been erecting structural steel for high-rise buildings for a long time will acquire a metaconscious knowledge of the strength of beams and columns. When one day the construction crane brings up a beam, he may look at the span it is to bridge and immediately know that it just does not look right. An engineering student who has taken a number of stress analysis courses can only make this kind of judgment after she has completed the necessary calculations that she has learned in classrooms and from textbooks. However, the metaconscious knowledge of the strength of materials will never grow into a knowledge of stress analysis regardless of how much experience a person has, and the reverse is also the case. In other words, the construction workers mediate the job situations they encounter through symbolization, experience, and culture. In contrast, engineering students and graduate engineers mediate the same experiences through the domain of the discipline of stress analysis. Examples of this kind can be multiplied almost indefinitely, as a university education allows people to make sense of the output of a variety of technologies such as X-ray machines, electron microscopes, ultrasounds, and many other devices, where before they had this education they saw nothing meaningful in these images.

All discipline-based approaches to knowing and doing mediate any situation through the domain of a discipline. The domain contains theories of the category of phenomena dealt with by the discipline, expressed as much as possible in mathematical form. For example, when we began our physics classes in high school, we learned to imagine a domain in which everything that happened could be expressed in Newton's laws of motion. These laws related entities such as forces and masses to explain the effects the former had on the latter. These effects were constrained by boundary conditions, as in the case of a force accelerating a mass along a frictionless horizontal plane that constrained the motion. We then learned that solving this problem was the exemplar for dealing with free-falling bodies, by imagining them moving along a frictionless vertical plane accelerated by a gravitational force. The sliding of a block up an inclined plane by the pull of a string passed over a pulley to a falling block could be solved if we recognized it as a combination of two motions, which could be solved if we treated them as one being essentially similar to the first kind of problem, and the other to the second kind of problem. These two earlier problems and the variations on them that we had solved thus acted as exemplars.[96] It was in this way that we learned to populate the domain of physics with an ever greater diversity of exemplars. At the same time, these exemplars gradually began to approximate some of the situations dominated by physical phenomena in our world, as we learned to compensate for friction, inertia, air resistance, and more. However, there was a limit that this domain could not transcend: unlike the world around us, it always remained a mathematical domain containing exclusively physical phenomena. To make all this possible we were told about the scientific mythology comparing a detached observer, a scientific method, and objective facts, which are at best worthy ideals but ultimately unachievable by the members of a symbolic species. Moreover, any mediation through the domain of physics has made it impossible to deal with any situations in our world in which phenomena other than physical ones have to be taken into account.

We can take this one step further by interpreting the discipline of physics as an exemplar for calculating the stress distribution in a solid material that sustains certain loads and is supported in a particular way. Alternatively, it can be conceptualized as an exemplar to investigate the flow of a liquid or a gas. We create a mathematical model of a solid, a liquid, or a gas. A continuum is such a model of a solid material whose properties have been uniformly distributed throughout space. A fluid is a mathematical model of a liquid or a gas with its properties so distributed. These models have turned out to be extremely useful under a wide range of conditions.

The discipline of stress analysis sets out what may be regarded as the laws describing the way a continuum deals with forces and moments exerted on a material, including the way they are transmitted throughout the continuum. Boundary conditions can then be imposed on this continuum to give it the same geometry as a beam or column in our world that is loaded and supported in a particular manner. Stress analysis may thus be regarded as a theory in mathematical form that sets out the fundamental processes that interpose themselves between the constraints set out by the boundary conditions. It allows for the systematization of all our experiences of beams and columns: carrying loads in a variety of ways, being made up of many different kinds of materials, and having a variety of cross-sectional shapes. Every situation is now characterized by the same continuum transmitting the same fundamental processes described by the theories, while any differences can be taken care of as particular geometrical details, local configurations, and the properties of the continuum representing those of real materials (such as a modulus of elasticity). There is thus a complete reversal in perspective. Whereas technical and economic rationality dealt with a particular situation at hand, the discipline-based approach to stress analysis does no such thing.

Stress analysis begins by unfolding a particular category of phenomena from human life and the world, replacing it with a theory of the underlying fundamental processes that cannot be observed, symbolized, or experienced in the usual manner, and representing what cannot be observed in the domain of its discipline. This domain is thus separated from experience and culture and regards the situation from the "inside out" as it were, to encounter the boundary conditions. In contrast, both technical and economic rationality regard a situation from the "outside in" and then rearrange it according to the relevance of the goal imposed on it. In the case of this discipline-based approach of stress analysis, the situation is unfolded from human life and the world, translated into the architecture of reality, and replaced by the fundamental processes that occur in the domain of the discipline. In contrast, the technical and economic rationality rearrange the connectedness of anything according to a goal and thus remain more or less on the level of experience and culture.

Initially, discipline-based approaches made perfect sense. They extended symbolization, experience, and culture beyond the limits they could go by substituting the domain of a discipline "between" the boundary conditions that could be experienced or measured. The domain thus became a distinct and separate "world." Hence, such approaches initially operated entirely within symbolization, experience, and culture as a kind of cultural framing of the application of a discipline.

As discipline-based approaches began to spread throughout the way of life of a society, the balance was completely reversed. Instead of each discipline being embedded in human life and society, at some point so little remained of a way of life and symbolic universe that they could no longer frame anything. Especially the efforts made by the United States during the Second World War gave these developments a tremendous boost. The reasons are clear: in a war it is a question of killing or being killed, with the result that performance and efficiency are all that matter. This encouraged the representation of war-related operations as relatively separate and distinct domains, whose desired outputs obtained from the necessary inputs had to be increased to the greatest extent possible. Following the war, these approaches spread throughout human life and society. It led to a complete reversal between the interlinked discipline-based approaches and the role an increasingly desymbolized culture continued to play.

For example, engineering design operates on the level of design exemplars in dealing with how a particular technological entity, process, or system functions in human life and the world through all its connections with them. If technology is to serve us, surely design should have remained the master activity in order to ensure that, on the level of symbolization, experience, and culture, a particular technological advance makes a genuine contribution. Instead, the entire curriculum was handed over to discipline-based approaches and their analytical exemplars, with the result that design became an embarrassing addition that no one any longer knew how to teach.[97] The same development occurred in the business schools.[98] In the end, not a single faculty within the contemporary university remained unaffected.[99] It is at this tipping point that technique, built up from discipline-based approaches, became a phenomenon in itself. Technique eventually became a system that all but took over from the role of culture in the ways of life of the societies that increasingly contributed to the building of a global civilization.[100] The framework within which discipline-based approaches played an important role and could act within their limits had all but disappeared. This represented the fundamental crisis of our current global civilization, whose universal means have lost touch with the living of human life. It continues to be lived relative to unique local situations but now only by highly desymbolized cultures.

It is important to stress that our present situation is not the result of "bad" discipline-based approaches and everything built up with them, but from ignoring their limits, with the result that our universities and other institutions have all but banished all other approaches by which such limits could be transcended. The phenomenon of technique has now all but displaced culture from the ways of life of contemporary societies in our global civilization. It may be conceptualized as the system of means largely arrived at through discipline-based approaches, for the purpose of making anything to which they are applied as efficient as it can be on its own terms, and doing so in every area of human life.[101] The evidence of this is all around us. If we examine any aspect of our contemporary ways of life, it becomes immediately evident that most adaptations, reorganizations, and improvements are based on scientific and technical experts of all kinds applying the discipline-based approaches to knowing, doing, and organizing that they have acquired in university. These approaches are used to endlessly improve the pedagogy and operations of our schools, the running of factories and offices, the delivery of healthcare through hospitals, the improvement of the performance of athletes and sports teams, the retaining and expanding of audiences for shows diffused by the media, the conduct of military operations and wars, the public relations efforts of any large institution, the execution of complex policies by governments through their departments or ministries, the improvement of the impact of religious broadcasts by megachurches, the running of all financial institutions, the provision of services ranging from insurance to tourism, the delivery of information services custom-tailored to our preferences, and just about anything else. Jointly, these discipline-based approaches and everything built up with them have all but replaced what, in traditional pre-industrial societies, would have been accomplished by their cultures.

The implications of all this are thus tightly bound up with the substantial differences between culturally and technically mediating our relationships with others and the world. Unfolding almost everything into a domain of one kind of discipline or another, or into a mathematical model of their interactions, involves a translation into the architecture of non-life. Individually and collectively, we are behaving as if what we refer to as our way of life is nothing more than a vast collection of relatively distinct operations by which we interpret and manipulate human life and the world, but in the image of non-life. A simple complexity thus emerges that can be expressed in terms of autonomous facts amenable to a variety of techniques to undertake what such facts all but compel us to do in our "improvement" of everything. It turns everything into universally efficient entities that no longer fit into and are compatible with their surroundings, as these are increasingly made up from similar "foreign bodies." It imposes enormous psychological, social, economic, political, moral, and religious stresses within human life, and impairs the functioning of the biosphere. As noted, we now live in the world on the basis of what we visually perceive and imagine, and suppress what we have learned about it by listening to language in order to enter a symbolic universe. As toddlers and children are increasingly dealing with an entirely new stream of experiences constituted of images and image-words coming from screen-based devices of all kinds, they experience a very different world – a world constituted as a reality containing a highly desymbolized cultural sphere limited to intimate relationships.

We have become so accustomed to human knowing and doing being organized by means of discipline-based approaches that we do well to remind ourselves that for all but the last hundred years or so, humanity lived without them. It amounts to making sense of and dealing with the world one category of phenomena at a time. Physics deals with physical phenomena, chemistry with chemical phenomena, biology with biological phenomena, economics with economic phenomena, sociology with social phenomena, political science with political phenomena, and so on.[102] Each discipline is associated with a domain in which the theories of the category of phenomena it examines set out the basic processes in mathematical forms to the greatest extent possible. Consequently, each discipline translates a category of phenomena abstracted from an enfolded biosphere, and a dialectically enfolded human life within it, into an architecture of non-life. All our doing has been similarly organized by disciplines that operate in much the same way.[103] The names of the departments of any faculty of engineering are but one example: metallurgy and materials, chemical engineering, mechanical engineering, electrical engineering, computer engineering. The subsections of other departments such as civil engineering, industrial engineering, and aerospace reveal the same pattern. Consequently, "behind" the way of life of our society, we encounter a great many domains that mediate the adaptation, organization, and evolution of every aspect. Nevertheless, such domains do not collectively form the scientific and technical equivalent of a symbolic universe, nor is one required. The role of culture in evolving the way of life of a society has all but disappeared. Nevertheless, a highly desymbolized culture remains in people's personal lives and in getting babies, toddlers, and young children launched as members of a symbolic species by listening to language.

Returning to the organizational structure of any faculty and university, it quickly becomes obvious that almost all knowing and doing is in the grip of discipline-based approaches. Apparently, we see no need for alternative ones. Since the graduates of these universities tend to hold key positions in all of the organizations required to maintain and evolve the equivalent of a way of life and institutions of a society, we may assume that during the past two hundred years, humanity as a symbolic species has gone from a primary reliance on culture for making sense of and living in the world to a primary reliance on technique and a secondary reliance on symbolization and culture.

In the spheres of human knowing and doing it is as if all our needs can be satisfied by discipline-based approaches, either because they have no limits, or, if they do have limits, because they are not serious enough for us to concern ourselves with them. Otherwise, our universities would be scrambling to research such limits in order to invent other approaches for transcending them. Given that we have surrendered ourselves and our institutions to these approaches, almost everything is potentially at stake.

It would appear, therefore, that individually and collectively we live with our discipline-based approaches the way earlier societies lived with their traditional gods. The latter were the only entities in all of human history that were autonomous, without limits, and omnipotent in their jurisdictions. If this were not the case, it would be impossible to explain how the discipline-based approaches of science have led to a kind of secular religious scientism, of which one of the best known forms is the new atheism. It can be legitimated only by an implicit or explicit belief in the autonomy and omnipotence of science in the kingdom of human knowing. The new authority and legitimacy of discipline-based approaches also challenged all traditional religious knowledge. For example, within the universities they transformed Christian theology and religious studies into disciplines, thereby opening the door to these approaches. This resulted in an inevitable confrontation between the authority of the knowledge that had traditionally been derived from the Bible and that of the knowledge gained by discipline-based approaches. The clash took one form in conservative Christianity and another in liberal Christianity. The consequences were so extensive that their analyses have produced a separate work, which follows the present one.

Similarly, it would be impossible otherwise to understand our widespread technicism, which holds that our discipline-based approaches to problem solving in the service of improving everything are entirely objective, rigorous, and value-free, and thus autonomous and omnipotent in the kingdom of human doing. Perhaps we are not nearly as secular as we imagine ourselves to be. We have plunged human life into an entirely new range of contradictions that, like those who have gone before us, we veil from ourselves most of the time.

In the political domain, we encounter similar contradictions. If everything has become political because the state was increasingly obliged to organize and adapt what had traditionally been accomplished through symbolic cultures, does this mean that everything in individual and collective human life has become the business of the state? Is the state omnipotent and autonomous in the politics of a nation? It was but a few centuries ago that almost every moral, political, and religious tradition was opposed to a large state, on the grounds that it is either the state or the people who decide what is to happen in our lives. Especially when slavery became almost universally regarded as an unacceptable form of human life, it is rather surprising that, other than a few anarchists, every tradition has more or less learned to live with a large state employing vast organizations making use of the latest discipline-based approaches to legally frame almost every aspect of human life. Promises to downsize the state sound a little hollow, since they are not accompanied by ways of dealing with the root causes. What has compelled the state to grow so large over the objections of virtually every tradition? Any society with a highly desymbolized culture and thus faced with the loss of its largely self-regulating character has no alternative. It is true that most of us work for the state more than two-thirds of each year, and that this state is not very effective at doing what culture used to accomplish, but without it a society can no longer survive.

During the twentieth century, we saw the rise of the first three secular political religions, namely, communism, national socialism, and democracy (especially in the United States and Canada). All three serve the state.[104] From a historical, social, and cultural perspective, these new secular political religions performed exactly the same roles in the industrialized societies that traditional religions had performed in their precursors.[105] The results have been, and continue to be, equally bloody and a justification for the almost limitless expansion of the military-industrial complex, thus diverting resources away from preventing the very conditions that could make war necessary.[106]

At least in North America, politicians rarely speak of a public good as what lies behind their motives and actions. It has become all too common to hear them talk about "moving forward," as if anyone has ever moved backwards. This would appear to indicate that these politicians have essentially bowed to the necessities imposed by the global technical order that we are building with the universal means of discipline-based science, technique, and the state. It has created a political illusion.[107] Since we now regard ourselves as secular and rational beings, it has also endowed human history with the ability expressed as "time will tell" or "history will judge" – implying that we have surrendered these collective human discernments, increasingly rendered impossible by means of highly desymbolized cultures, to a play of forces dominated by universal means. There is no longer any thought of traditional gods other than the one some Americans believe stands right behind their president and their way of life.[108]

The most essential features of this thumbnail sketch of our global civilization's predicament can be readily verified by inquiring into the limits of discipline-based approaches to knowing, doing, and political organizing, and assessing their significance. Their limitations become apparent when we recognize that these approaches are responsible for our spectacular success at improving everything on their own terms, and our equally spectacular failure to ensure that all these "improved" elements of our lives and our world fit together to constitute lives and a world that are improved as a whole. Once again, discipline-based approaches are an extremely useful and powerful intellectual and practical invention, but, like all human creations, they are very good for certain things, not so good for others, and completely irrelevant to still others. When used in an unlimited manner, they will therefore produce a great many undesired and unforeseen difficulties. We now come back to the question we posed at the beginning of this Introduction. Do we live with these discipline-based approaches as our new secular gods, that is, as if they have the attributes associated with the traditional gods of the past? To settle this question, we need to inquire into their limits.

The current intellectual and professional division of labour within our universities is built up with autonomous disciplines that each transmit, advance, and apply the theoretical knowledge of one category of phenomena built up in its own domain, but without any meaningful references to the categories of phenomena dealt with by other disciplines. Consequently, such a division of labour is well suited to dealing with situations characterized by the influence of one category of phenomena dwarfing the influences of all other categories of phenomena contributing to the situation, to the point that their influences may be neglected. It also deals with situations where all but one category of phenomena are essentially static. The limits to discipline-based approaches to knowing and doing are thus embarrassingly obvious. Such approaches cannot adequately deal with situations that involve multiple categories of phenomena making significant contributions, to the point that they cannot be neglected without undermining our scientific understanding or our technical reorganizing without significant undesired effects stemming from what has been unscientifically neglected or technically externalized.

It so happens that these latter kinds of situations are typical of a living world, while the first two categories of situations where discipline-based approaches can be applied within their limits are almost exclusively encountered in the "world" of technology. As noted, this world is deliberately built up from sequences of domains where the inputs into any domain are received from the previous one, transformed into intermediary desired outputs, and then passed onto the next domain. The transformations of inputs into intermediary desired outputs are organized by endlessly repeating a cyclical process carried out by a particular instant of a single category of phenomena. All other categories of phenomena are deliberately kept at bay by a variety of means that are integral to the design and organization, in order not to perturb the efficiency and productivity of the transformations. In contrast, in a living world everything tends to be related to and evolves in relation to everything else, with the consequence that separate and distinct domains almost never occur: and when they do, they are never autonomous. The biological wholes are enfolded and the cultural wholes are dialectically enfolded, in each and every part. Consequently, each "part" is internally connected to the whole and externally connected to it via its relationships with all the other "parts." No such relationships occur in an architecture of non-life.

It is now clear why physics, chemistry, and molecular biology became the "model" disciplines. The situations they deal with are all characterized by one category of phenomena being so dominant that all other categories of phenomena can be neglected. Examples include the Big Bang and the subatomic and molecular "worlds." These are the kinds of situations to which discipline-based approaches can be scientifically applied. The above interpretation also explains why the disciplines of the social sciences were much less successful, at least until they came into the grip of survey methodology and statistical approaches for dealing with the results. They deal with dialectically enfolded situations which discipline-based approaches translate into an architecture of non-life.

Despite these obvious limitations, when we ask a professional audience what science will never be able to know, or what technology will never be able to accomplish, the strangest answers will be forthcoming. For the greater part, the usual answers manifest our widespread scientism and technicism. It would appear, therefore, that individually and collectively we live with our discipline-based approaches to knowing and doing as if they are without limits, autonomous and omnipotent in their domains. We have thus created a global civilization that acts as if everything living can be understood in terms of the kinds of situations that almost exclusively occur in non-living entities, processes, and systems, just as in the "world" of technology. If our understanding of everything includes a reimagining of everything living in terms of an architecture of non-life, we are denying ourselves the kinds of insights indispensable for human life. Moreover, if we act on this understanding, we remake the architecture of what is living in the image of what is non-living. We are essentially busy killing ourselves and everything else that lives on our planet.

Our spectacular successes and accomplishments as well as our equally monumental shortcomings and failures thus appear to share a common denominator. The improvement of whatever has an architecture built up from domains is successful in most cases. However, improving something whose architecture cannot be broken down into such domains, as in all enfolded or dialectically enfolded living entities and systems, will distort the internal integrality and external compatibility with their contexts to create the "foreign bodies" that contribute to our human, social, economic, political, and environmental crises. It is as if our global civilization has a collective wish to live as the living dead. Moreover, our secular religious attitude toward our most powerful creations has made it almost impossible to be "truly realistic" by not distorting what is happening to our lives and our planet.

These secular religious attitudes are sustained by new cultural unities constituted of the secular sacred of technique and the state-nation(s), with the sustaining secular myths of science and history.[109] As the first generation of secular cultural unities, these also create a hierarchy prioritizing the discipline-based approaches to knowing, doing, and political organizing at the expense of their symbolic and cultural alternatives. However, these alternatives have now been so highly desymbolized as to be incapable of providing an ultimate meaning such as human happiness. The secular myth of history now limits everything to what can be achieved by discipline-based approaches – only time will tell, and history will judge the results.

In a living world, each entity helps to constitute the niche and ecosystem for all the others, as they also do for it. Similarly, our lives help to constitute the social equivalent for the other members of our community and society, as they do for us. It is this sustaining reciprocal interdependence of a living world that is being externalized in our knowing, doing, and political organizing through our secular sacred and its sustaining myths. Further evidence of this comes to light when we examine the introductory textbooks to a wide variety of disciplines. Each textbook constructs a theoretical domain into which are admitted only those situations dominated by the category of phenomena assigned to the discipline. Human life and the world are thus unfolded into a corresponding array of domains.

For example, let us begin by examining the index of an introductory textbook to economics for entries about natural resources, possible resource crises, global warming, an environmental crisis, or any such reference in acknowledgment of the dependence of an economy on the biosphere. According to the first law of thermodynamics, no economy can either create or destroy the matter and energy on which it depends, thus making it fundamentally dependent on the biosphere. As noted, it must borrow matter and energy from the biosphere, transform them for our use, and eventually return them to the biosphere. We will certainly find a few such references, but when we check the pages indicated by the index, it becomes immediately evident that, even when the interdependencies between the economy and the biosphere are acknowledged, they play no role whatsoever in the building up of the theoretical domain. Even environmental economics (in contrast with ecological economics) is essentially an end-of-pipe approach to economically dealing with the undesired outputs of an economy.

All this ignores the fact that it has been estimated that 93 per cent of what we borrow from the biosphere does not end up in saleable products.[110] Consequently, pollutants are the main output of an economy – things that are produced like all the desired goods, but that cannot be sold. The costs of dealing with these pollutants is beginning to catch up with the wealth earned through industrial production, unless the facilities are moved to places with limited environmental standards.

Another consequence of the discipline-based approach to economics is that schools, centres, and institutes for the environment in our universities rarely collaborate with engineering and management faculties to examine in a meaningful way how our ways of life "produce" the environmental crisis along with the goods and services that we desire, or are persuaded to desire. In the same vein, if we search the index of an introductory economics text for entries on science and technology, we discover that it is apparently possible to describe a modern economy with polite acknowledgments of their economic importance and value, but never are highly specialized scientific and technical knowledge incorporated as a significant, and possibly the most significant, factor of production.[111] Overall, the discipline of economics differs radically from its precursor, political economy, in that the latter examined economic phenomena in the context of and in relation to other relevant categories of phenomena, much as we do in daily life when we focus our attention on something and everything else remains in the background. Economics is now totally in the grip of discipline-based approaches, just as political economy was in the grip of symbolic approaches.

As a second example, consider introductory texts to the discipline of sociology. Once again, it appears to be entirely possible to theoretically understand social relationships, families, groups, communities, and societies with no essential references to science and technology. This is hard to understand, considering that almost every daily-life activity directly or indirectly depends on a technology of one kind or another. Moreover, we now live in an urban habitat that is a technological creation. Little attention is being paid to the founding thinkers of sociology, who attempted an interpretation similar to those of political economy by examining social phenomena against the background of all the other categories of relevant phenomena that help to constitute human life and society. Instead, students are encouraged to master survey methodology to permit a more effective intervention in social behaviour, but without an overall comprehensive theoretical understanding of what is happening, they lack the prerequisites for asking decisive questions on a survey. As a consequence, social engineering based on survey instruments can frequently be summed up as "garbage in, garbage out."

In the engineering disciplines, we encounter the mirror-image problem, as indicated by a detailed examination of undergraduate engineering education.[112] It looked at what future engineers are taught about how technology interacts with human life, society, and the biosphere as well as the extent to which they learn to make use of this understanding to do their work, while at the same time adjusting design and decision making to prevent or greatly minimize "collisions" with everything around technology. Given that the curriculum is almost entirely given over to disciplines, it ought to come as no surprise that the answer was: almost nothing. The study was comprehensive, assessing almost every course offered by a typical faculty of engineering by means of a composite score derived from the rated scores of evaluating all components (textbooks, student notes, quizzes, exams, laboratories, handouts, and so on) according to the estimated time students devote to them. The so-called complementary studies courses that make up 12.5 per cent of the curriculum and must be of a non-technical nature were also scored to determine whether or not future engineers might profit from them and learn to adjust design and decision making in a preventive manner. These courses showed the same low scores because the theoretical domains of the social science disciplines were entirely void of technology. As a result, they could not contribute to preparing future engineers for their professional obligation to protect the public interest above all in exchange for professional autonomy in the form of self-regulation.[113]

This diagnosis led to a prescription for preventive approaches that made use of the findings of relevant disciplines by interpreting them in a non-disciplinary fashion by means of resymbolization. Although this accomplishment was widely recognized,[114] it appeared unthinkable and undoable in the eyes of university colleagues, chairs, and deans.

What this cursory glance at some introductory textbooks to various disciplines confirms is that our present intellectual and professional division of labour within the university is built up from autonomous disciplines that transmit, advance, and apply the theoretical knowledge of a single category of phenomena built up in each of their domains without any meaningful references to the categories of phenomena examined by other disciplines. It is thus ideally suited to the "world" of technology with its architecture of non-life. When it is used beyond these limits, it systematically externalizes all enfolded and dialectically enfolded relationships, including the way living entities have their constituent elements both internally and externally connected to one another, as well as these entities themselves being internally and externally connected to their surroundings. There is thus a reciprocal interdependence of everything on everything and simple causal relationships are impossible. Because of their enfolding, the constituent elements cannot be separated the way they are in classical or information machines. Measurements are much more difficult to undertake, and quantification and mathematical representation can play only a very limited role. There is an enfolded complexity which implies that no entity can be understood without its evolution or history as the description of how it helped to constitute the milieu of all the others, as they did for it. The many extinct and endangered species, as well as our mental health crisis, are but a few symptoms of our externalizing something that appears to be fundamental for the life of our biosphere, our communities, and ourselves. However, this must not be interpreted as a kind of metaphysics, but an attempt at resymbolizing what is happening all around us. Our discipline-based organization of science has turned it into what, for the greater part, is an unscientific co-phenomenon of technique.

In conclusion, we will attempt to come to grips with what this means for our daily lives. Our present situation is the result of a growing portion of humanity having given up its primary reliance on symbolic approaches, which gave rise to the earlier tradition-based ways of life and symbolic cultures, in favour of primarily relying on universal alternatives. These alternatives all share a common architecture with how we visually see and deal with human life and the world. There are situations in our lives where this makes a great deal of sense.

For example, when we are suddenly confronted with a dangerous animal, surviving the encounter becomes paramount and every other aspect of our life and the world must be discounted in relation to it. Our attention thus becomes focused on those details relevant to adopting and acting on a strategy for survival. Without such a threat, we would have symbolized the situation in a manner that would have been a great deal more respectful of its interrelatedness, including its enfolded and dialectically enfolded character. Such a situation would then have been dealt with on many levels at the same time, including the contradictory aspects of many of these levels. In other words, our response to it somewhat resembles a discipline-based approach: we take into account only those details that negatively or positively affect our survival and neglect everything else in order to improve our chance to live. The same kind of strategy ought to govern our behaviour when riding a bike in a busy metropolitan downtown area. We are always one step away from a life-threatening situation, against which we can best defend ourselves by dealing with what we see in relation to where we wish to go by avoiding the usual symbolization.

In the same vein, our global civilization increasingly compels each society, community, workplace, and individual person to join the global race. It is no longer simply the arms race of the past, life-threatening as it was. It is now a much more comprehensive race to squeeze every possible economic advantage out of the discipline-based approaches to improve the efficiency, productivity, and power of everything. Individually and collectively, we have become like an elderly couple living in a modest home on a fixed pension that appears every year a little less adequate. Their home is heated by an ancient forced-air furnace that runs almost constantly during a severe winter. Rightly or wrongly, the couple believes that, sooner or later, the furnace will break down because of overheating some part or another. They worry about it a great deal and decide that, by lowering the set point on their thermostat a little bit at a time, they will reach a setting that will allow the furnace to periodically shut down and cool its components. The condition of the furnace now helps to determine the set point, which ought to be governed only by their desired level of comfort. In the same way, our surrender to global markets in the name of "free" trade is levelling the playing field to the lowest common denominator with minimal or no labour, health, social security, and environmental standards. We are thus making the world as safe as possible for technique while making it increasingly unsafe for people. All this is done in the name of being "realistic." After all, if we do not do it, someone else will, and we may then condemn ourselves to the new secular hell of underdevelopment relative to the global technical order. In this way, our global technical civilization makes us behave like the elderly couple – adjusting our lives, our values, and our expectations to what is "realistic" in accordance with our latest secular myths. We imagine ourselves as constantly confronting the danger of being left behind in the self-imposed global race to squeeze everything possible out of our discipline-based approaches. There appears to be little that can stand in our way. Our cultures have been extensively desymbolized, and whatever metaconscious values they still embody are so weak that they appear to be incapable of sustaining any resistance. We have become so accustomed to the spiralling accomplishments of our most powerful creations that we more or less accept that we, individually and collectively, must spiral down.

This outlook also permeates our daily lives. In societies dominated by the traditional and the new media, we have become so accustomed to our dependence on images and image-words that we are in danger of losing our ability to know when we ought to pay a lot more attention to what lies behind all these appearances. For example: on a date we may be somewhat bewitched by the looks of the other person and discount some of the things about his or her life "behind" the appearances that we like less. These can be the sources of future difficulties in the relationship. If we gradually become more committed to the other person, the exact opposite needs to happen: such details must receive a great deal more attention in order to work them out before they begin to strain the relationship.

When we make use of the discipline-based approaches in our personal and professional lives, we separate ourselves from experience and culture to suspend ourselves in a triple abstraction. Suppose we were participating in improving the running of a factory, office, school, university, government ministry, or hospital by means of discipline-based approaches. We must first abstract this entity from the world by replacing the latter with the inputs received from it, and the outputs returned to it. Next, from the process that transforms these inputs into the outputs, it is necessary to abstract whatever aspects are commensurate with the disciplines we have mastered and to repeat this process until all the important aspects of the process are covered. The process of transformation is thus reduced to its constituent aspects, with all enfolding and dialectical enfolding eliminated from consideration. Finally, within the domain of any discipline utilized in this endeavour, any promising alternatives as sources of possible improvements or possible ways of eliminating certain difficulties cannot be distinguished in terms of cultural values expressing relationships with human life, society, and the biosphere. The previous two abstractions have made this impossible. Consequently, the decision process must be abstracted from all human values and considerations. It can be dealt with only by criteria applicable within the domain of a discipline: the ratio of the outputs produced obtained from the requisite inputs to yield "rational" behaviour. Everything is thus considered and treated on its own terms, as if it was not dependent on and integral to a complex fabric of relationships without which it could not exist.

In other words, to apply discipline-based approaches to anything in human life or society is to "explode" the interconnectedness of everything into isolated elements by means of this triple abstraction. It is the diametrical opposite of any approach based on symbolization and culture, which seeks to understand the contribution of anything in terms of its effects on the interconnectedness of which it is a part, and without which it cannot exist. In contrast, optimization begins by isolating something from this connectedness in order to understand and improve it on its own terms in isolation from everything else. Doing so will make it into a foreign body within this connectedness, whose internal integrality and external compatibility with regard to it have been weakened or destroyed. It is as if discipline-based approaches intellectually and practically deal with human life and the world through the mediation of this triple abstraction, essentially creating a constellation of isolated elements that can only be improved by means of ratio-nality.

This rationality has permeated almost every aspect of our contemporary ways of life. Many people feel that common sense has disappeared. They are entirely correct if by common sense they mean the body of shared meanings and values that used to constitute a culture. For example, the trades that repair our homes work in much the same way: each trade deals with what is an issue in its own domain, and frequently creates several problems in the domains of other trades. What a trade does is rarely framed by a context-rich interpretation of the situation it deals with, thus signifying an inadequate symbolization of it. Generally speaking: we lack a shared understanding and a way of dealing with everything in a broad context according to what used to be a way of life and culture of a society. Everything appears to be understood and dealt with as if it was isolated from everything else, and needs only to be measured and improved on its own terms. Its internal workings are modified by means of rationality.

Putting it more strongly: if we physically isolated ourselves from the fabric of relationships within which we live, we would immediately die from asphyxiation. If we did not die from asphyxiation, we would soon die from dehydration. If we did not die from dehydration, we would soon starve. In the same vein, if we could not get rid of our waste products, we would rapidly poison ourselves. Nevertheless, the intellectual and practical use of discipline-based approaches beyond their limits does much the same thing eventually, since all the essential interconnectedness that sustains all life becomes a mere externality.

All practitioners of discipline-based approaches are suspended in a triple abstraction that isolates them from anything they might have known or done based on symbolic approaches. Their humanity becomes externalized and reified in their involvements. Again, the fault does not lie with discipline-based approaches themselves. Our problems stem from our secular religious attitudes toward them, which make it impossible to restrict their use and keep them within the limits of what they are capable of doing. Within their limits, these approaches can achieve genuine accomplishments. However, beyond their limits they are almost entirely destructive of all life. We resemble a highly inexperienced contractor who has not yet learned that everything in our home cannot be fixed with his favourite hammer because of a secular religious attitude toward it. There is no problem with a hammer, provided that it is used within the limits of its capabilities and traded for another tool to go beyond these limits.

It may be objected that when operation research-like approaches are utilized, the discipline-based approaches applied to particular aspects of a situation can each yield an equation. These equations can then form a set that can be solved simultaneously. Such approaches thus allow for a more comprehensive use of a set of discipline-based approaches. However, doing so does not eliminate the fundamental issues we have been explaining. Even if we possessed an almost unlimited computing power to solve very large sets of such equations, the above analysis would still apply. Each equation is established in the domain of a discipline and consequently necessitates a translation from the architecture of a living biosphere and symbolic universe into an architecture of non-life. This applies to all our computer simulations and software packages through which we mediate more and more interventions.

We are back to the same root problem. Everything in human life and the world is no more and no less than what it is. Treating it as anything more by means of religious or secular religious attitudes will not endow it with some kind of magical powers to deliver anything more than what it is capable of delivering while participating in an interconnected fabric of relationships. To return to the example of the hammer: it is very good for driving nails, but do not expect it to cut wood or drill a hole. It will simply produce a big mess. This brings us to the topic of this work: in the days of our youth, will we be able to learn symbolic approaches that will enable us to resymbolize our cultures within their limits and, correspondingly, restrict our discipline-based approaches to function within their own limits?

For example: when elementary school children are encouraged to do their homework with the powerpoints of this world, they are implicitly told that anything in human life and the world can be reduced to separate and distinct bullets. They are thus being socialized into the ways that will aggravate our current predicament. This is but another way of externalizing the enfolded and dialectically enfolded complexity of anything living as separate and distinct points. It encourages the destruction of all true understanding of complex issues involving living entities. Our teachers appear to be asleep at the switch, and are behaving as secular priests and priestesses who believe that anything can be accomplished by the kind of education they deliver. Since they themselves are in the grip of several pedagogical, psychological, and cognitivist disciplines, they too are suspended in a triple abstraction and have little awareness of what is truly happening to themselves and others. They are but one example of how all of us are stuck in an indiscriminate use of discipline-based approaches because of our secular religious attitudes.

Technique, the State, and the Law

In the context of the developments just described, we can begin to appreciate the achievement of Roman law as one of the pillars of Western civilization.[115] It departed sharply from the Greek approach, which had essentially attempted to "ground" the laws of their society by means of philosophical foundationalism. The results were so asocial and ahistorical that no human community had even come close to practising anything of this kind. By eliminating all dialectical enfolding in the relationships between a ruler and a people, parents and children, people and slaves, and so on, they developed proposals that no community in human history has ever attempted to put into practice. In contrast, the Romans sought to establish an overriding political orientation in their culture(s), one that was not enslaved to philosophical foundationalism but was guided by practical considerations. Over time, they forged a concept of the state that was to take charge of, and evolve, a legal framework for human life that would be tested in juridical trials based on evidence brought forth by witnesses. Their legal system was thus rooted in experience and legal reasoning. It also by and large respected the self-regulating character of the cultures whose societies they had compelled to enter the Roman Empire. Because the cultures of those days were not yet desymbolized, except during the transitions from one historical epoch to another, the full significance of this achievement did not become apparent until the emergence of technique and its desymbolization of all cultures.

The relationship between the culture of a society and its legal framework critically depends on metaconscious knowledge, especially the values and myths implied in the organizations of the brain-minds of the members of a society.[116] This dependence may be illuminated by asking why we obey many laws of our society without most of us ever having read them. In a democratic society, there needs to be a relatively pragmatic and prudent approach that carefully "stretches" the metaconscious values of the people in a desired direction. If a particular law strays too far from these metaconscious values, it risks being widely disobeyed, which, in a democratic society where no excessive force can be applied, leaves a judge with no alternative but to declare the law inapplicable. Consequently, to regard the law as the founding "social contract" of a society ignores its relationship to culture.

Since traditional societies symbolically grounded themselves by means of their cultural unities, common law systems were very closely tied to the metaconscious values enfolded in the organizations of the brain-minds of their members. Moreover, these metaconscious values reflected a deeper metaconscious knowledge of the symbolic grounding of a culture by its myths. Hence, these laws could be nothing else but what was thinkable and doable within a cultural unity unless brutal and repressive totalitarian measures were taken by a ruler.

The rootedness of a common law system in the cultural unity of a society became greatly weakened as a consequence of the introduction of the technical division of labour, followed by mechanization and industrialization. The economy was split off from the remainder of society by taking on a diametrically opposite architecture that could not be organized and evolved on the basis of a culture. There was a corresponding desymbolization of the culture, creating the need for a compensating intervention of the state to extend the legal system to encompass this new economy. Whether this took the form of imposing a codified legal system or of assuming a growing role in what was less and less of a common law system made ultimately very little difference. As the legal system was less and less rooted in the increasingly weaker symbolic grounding of the society, the law took on a more organizational character that could only be ensured by the state.

With the emergence of technique and the parallel desymbolization of all cultures to unprecedented levels, the need for the state to take charge of the legal system increased all the more. Nevertheless, there is significant evidence that what is legally thinkable and doable remains tied to our second generation of secular myths.[117] Technique took on an ever stronger organizational character that needed to be legitimated by the state. This legitimation was greatly facilitated by the second generation of secular myths, with the state-nation(s) and technique as two poles of the secular sacred. This enormously facilitated the state's growing dependence on all manner of legal techniques. Law had also become a discipline, with all the consequences we have noted.[118] This appears to be borne out by some preliminary research that showed significant biases of the courts in favour of technique.[119] It is also in this light that we need to reinterpret the greatest challenge to all democratic societies, which came from a court decision declaring corporations to be legal persons.[120] There is little question that this organizational character of contemporary law needs a great deal more attention.[121] We are now as close as we have ever been to the Roman conception of the role of the state in creating and adapting a legal framework for human life. The long-term implications are bound to be enormous. Our cultures continue to be desymbolized as the formation of secular myths continues to depend on the formation of metaconscious knowledge. It may put even greater pressure on the state to legitimate itself and its uses of the law. There also appears to be little question that our courts are increasingly in the grip of all manner of legal techniques that provide enormous advantages to those parties who can afford the latest ones. Justice is now in the grip of a technical battle.

Growing Up and Living with Technique

The above overview of my previous work would suggest that the growing up of babies and children has also undergone a fundamental transformation. Each new generation continues to begin their development as members of a symbolic species by acquiring what has become a highly desymbolized culture, in order to make some initial sense of their lives and their world relative to the lives and the world of those who nurture them. Initially, their encounters with reality are made sense of, and related to, as if nothing radically different had been inserted into their world, because the organizations of their brain-minds are not sufficiently developed to make the essential distinctions. For example, their encounters with screen-based devices may make little sense to them. However, their ongoing exposure to this reality gradually creates a stream of experiences whose architecture is the diametrical opposite of the architecture of the experiences corresponding to their acquisition of a language and a desymbolized culture. Making sense of and relating to the latter requires the development of first a primary frame of reference, followed by a secondary frame of reference. As this secondary frame of reference develops further, it will gradually begin to differentiate the experiences of reality from all others, and this leads to an entirely new development that will be referred to as a tertiary frame of reference.

The secondary frame of reference develops to symbolize and differentiate everything from everything else in people's lives. However, the child's encounter with screen-based devices and the eventual learning of disciplines involves domain-like structures as well as the domains of disciplines. These are separated from experience and culture and thus cannot be made sense of in the context of the life and world of the members of each new generation. The foregrounds of such experiences must be directly differentiated from each other in order to make sense of reality and of the domains of disciplines. Once the secondary frame of reference begins to differentiate the two bodies of experiences, it is busy breaking itself up into a limited modification of itself and the tertiary frame of reference. At this point, children begin to learn something about their highly desymbolized culture relative to technique and something about technique relative to their highly desymbolized culture. In the case of the latter, they have learned that relating everything to everything else through symbolization was limited to the construct or domain in question as a consequence of its separation from experience and culture as well as its autonomy from all other disciplines. The process of symbolization thus falls far short of its potential and we experience this in our daily lives as desymbolization. At the same time, what is learned about our highly desymbolized culture is that it is excluded from all the constructs and domains that helped to constitute reality and the technical order. Once again, we experience this as desymbolization. We will examine these developments in babies and children in the following chapters.

Since the relationships between babies and children and the anti-society in which they grow up remain dialectically enfolded despite high levels of desymbolization, a complementary perspective is required that examines how their growing up corresponds simultaneously to their embodying it. They become individually unique expressions of it. In the third chapter, we will examine some relevant aspects of these anti-societies as forms of collective human life characterized by the dominance of discipline-based approaches and a highly restricted use of a desymbolized culture.

As before, the intellectual perspective and theoretical approach will be that all human knowing, doing, and organizing is relative. Nothing we can do is absolute. We can only create religious or secular religious attitudes toward something and thus insert it into a religious or secular religious sphere. What this means is that there can be no understanding of technique other than by relating it to culture, and vice versa. The two are each other's diametrical opposites, and yet our civilization has permitted technique to almost completely take over from culture. However, all our knowing, doing, and organizing are also relative to a vantage point created by absolutizing what we know and live by our secular myths. Babies and children growing up in our societies will have to acquire the necessary secular religious attitudes in order to make sense of their lives and their world. Doing so has become a great deal more complex since they must deal with both what we think of as reality and what remains of a cultural order in our non-technically mediated relationships with others.

Particular attention will be paid to this transmission in the learning of mathematics and the discipline-based approaches of science and technique. In the Epilogue, we will show that Judaism and Christianity (as the dominant religions of the civilizations that gave birth to our global one) might have played very different roles in these developments if they had understood the danger of treating anything as god-like, that is, as autonomous and limitless. In such a case, children growing up within these communities would have faced a difficult task in the days of their youth.

Before we begin our inquiry, a note of caution may be helpful. When complex issues in human life and society are being interpreted and dealt with as if they had the architecture of reality, these issues become so highly simplified that there is little middle ground based on a knowledge of how, in a dialectically enfolded human life in the world, the positive aspects are indissociably linked to the negative ones, thus making such issues extremely complex. In a middle ground of less complexity, there is a greater likelihood of disagreement, but this is accompanied by an appreciation of the complexity and why others may take a different position by giving different meanings and values to significant details. It opens us up to appreciating that possibly we underestimated the importance of certain details and thus encourages a measure of respect for those with whom we disagree. It is even possible that others could convince us that some aspects were more meaningful and valuable than we had previously appreciated. The effects of technique on our highly desymbolized culture have created a kind of "cultural soil" in which neo-fascistic tendencies can flourish and where people can support politicians who in another age would have been dismissed as too simplistic, divisive, and one-dimensional. An enormous amount of time is spent on the media in North America discussing a few politicians who are masters at translating everything into a simple complexity that allows them to dismiss almost everyone else. It is mostly a complete waste of time. We had better take a good look at ourselves and our dependence on our "cultural soil." Every age has its share of people with extreme views and highly simplistic solutions to everything. There is little that can be done about this. However, when such people can get themselves elected, we ought to interpret this as a symptom of deeper issues undermining what little remains of our communities and democracies.

1 Our Physical Embodiment within the Relativity of Life and the World

Can We Escape Our Embodiment?

There are those among us who believe that we can reasonably expect to live eternally on the Net in the near future. There are others who think that we are rapidly moving to a point where the quantity of information stored on the World Wide Web will soon cross a threshold amounting to a qualitative transformation of all this information into a kind of collective global intelligence. All arguments of this kind are inevitably circular. The facts on which they are based are gathered by and take on their significance relative to an interpretive framework, and the significance of this framework is relative to the facts it has interpreted and with which it has been built up. This circularity is embedded into that of our lives. We live each experience relative to all other experiences and thus as a moment of our lives, while the significance of our lives is relative to all the experiences that have been interpreted and lived. Integral to this living of our lives is our physical, social, and cultural embodiment in the world, because it is the vantage point from which they are lived. This circularity in turn is embedded in our lives having been lived with others in a community that we help make for each other and through which we make history. The culture of this community grounds all these nested circularities by symbolizing everything that potentially can be known and lived as what has become known and lived, which will continuously be extended by what will be discovered and lived in the future. In this manner, the lives of the members of a community become suspended in myths. It is now widely accepted in disciplines such as cultural anthropology, the sociology of religion, and the history of religion that the presence of such myths is manifested by each and every society, without any exceptions, creating a morality and a religion in order to serve these myths and be served by them.

Our present global civilization has obscured this general relativity of human knowing and doing by means of an intellectual and practical division of labour that is based on autonomous disciplines. In the Introduction, we noted that these disciplines each create their own unique context in a manner that is the diametrical opposite to how symbolic cultural approaches "map" the general relativity of human life and the world. First, they each represent this general relativity one category of phenomena at a time. Doing so is scientifically and technically valid only if the architecture of this general relativity has a simple complexity. Second, whatever situation a particular discipline scientifically studies or technically "improves" must be dominated by one instance of the category of phenomena assigned to the discipline, to a degree that all other categories of phenomena can be neglected. Proceeding in this way attributes either a relative autonomy to this category of phenomena because it is temporary, or a permanent autonomy that negates the general relativity of human life and the world. For example, studying the universe immediately following the so-called Big Bang scientifically legitimates the relative autonomy of physical phenomena, but this autonomy vanishes in daily-life situations because these involve multiple categories of phenomena making non-negligible contributions. In the same vein, scientifically examining molecular and subatomic phenomena can at best provide them with a relative autonomy that they will almost certainly lose near black holes. The general theory of relativity in physics cannot account for all branches of physics, let alone the relativity of physical phenomena to all other categories of phenomena in our daily lives. As a result, the discipline of physics can provide us with ever more accurate mathematical descriptions of a physical universe, but can never give us a deeper insight into their meaning and value relative to those of all other categories of phenomena integral to the general relativity of human life and the world. It will require this entire work to somewhat develop what may be regarded as humanity's struggle with the general relativity of its life and the universe.

Because of the massive scale on which we are now dealing with ourselves and the world by means of discipline-based approaches rather than their symbolic counterparts, we have failed to respect this general relativity. Instead, we have overvalued our discipline-based approaches to the absolute, as a consequence of what we will discover to be our secular religious attitudes towards them. This makes it impossible to recognize their limitations, which is why our universities and other institutions are entirely given over to them. Consequently, their global and universal use is making as big a mess as a carpenter's hammer would if she used it as a universal tool.

Any context on which all human knowing and doing ultimately depend is necessarily an integral part of the general relativity of human life and the world, with the result that such knowing and doing are vulnerable to cultural chaos. This has been confirmed by the failure of philosophical foundationalism and by the impossibility of finding independent foundations for our most successful and powerful creations such as mathematics, science, and technique. Nevertheless, these human endeavours did not succumb to cultural chaos any more than did human life itself. In our present understanding, their survival depends on their being grounded in myths, as uncomfortable as this may make us. In this chapter, we will begin to explore this grounding by examining how utterly dependent we are, in all our activities, on our physical embodiment in the world. In the next chapter, we will turn our attention to our social and cultural embodiment.

Within the general relativity of human life and the world, the role of science is the great intellectual rearranger by entrusting everything to disciplines, and the role of technique is the great rebuilder of this simple unfolded complexity in order to make it as good as it can be by improving it one element at a time. They are thereby necessarily imposing a technical rationality that can achieve nothing else but the improvement of the internal functioning of these elements to obtain the greatest possible efficiency. Technique also demands an unlimited confidence in the state to ensure by political means that all these more efficient elements are organized in such a way as to produce the greatest possible benefits, with little regard for any harmful consequences. It amounts to recognizing what our global civilization is busy accomplishing with its universal means: intellectually, practically, and politically having our cake and eating it too. We continue to bring up each new generation (at least initially) to become members of a symbolic species; but then we concentrate on preparing them to play their role in expanding a global order that shuts the door to the threats of cultural chaos by means of secular myths that prioritize universal means over those that have made us a symbolic species.

What is happening to our lives and the world may thus be interpreted as an irreconcilable conflict between culture and what has been introduced as technique. Both these human "inventions" deny the relativity of human life and the world. Nothing in our lives and the world is what it is in itself. Everything is what it is relative to something else. Anything that is what it is in itself becomes an entity without limits, having lost its dependence on everything else, and thus autonomous from everything else because any contact with it will have no effect. However, in human life and the world, where everything is related to and evolves in relation to everything else, such entities result only when our secular religious attitudes are directed toward them, thus turning them into something equivalent to the traditional gods of the past.

These observations are not at all intended as an introduction to a kind of metaphysics of life, but a summing up of what we learn when we attempt to make an intellectual map of the interrelatedness of human life and the world in their enfolded and dialectically enfolded complexities. Everything is both internally and externally connected to everything else in its surroundings because of this enfolding. This is the case within any living entity as well as in its relationships with everything else on which it depends. It gives rise to very different kinds of relationships from the ones we encounter in the simple complexity of what we have referred to as reality, which has an architecture of non-life. For example, reality can exhibit simple causal relationships, which are impossible in an enfolded or dialectically enfolded architecture.

We have noted that cultures deny the relativity of a community by symbolizing the unknown as more of what its members individually and collectively know and live, thereby absolutizing the collective body of experience as the only possible vantage point as well as a body of common sense for living in the world. The price to be paid for being rooted in an ultimately unknowable universe, however, is an enslavement to traditional myths represented as gods and served through a religion and a morality. An enslavement to the gods is the price that must be paid for making life liveable by closing the door to cultural chaos.

In contrast, a discipline-based science, technique, and politics sweep the skies clean of traditional gods by means of secular myths that translate the architecture of all life into the architecture of non-life by universal approaches to knowing, doing, and organizing. All life is thus reified and out of reach of the threat of cultural chaos.

Understanding the interconnectedness of human life and the world by means of intellectual mapmaking is diametrically opposite to doing so by a discipline-based approach. Each of these has its unique strengths and weaknesses. They could potentially complement one another, provided that each one is used within its limits and their relationship is not distorted by secular myths sustaining secular religious attitudes toward the discipline-based ones. The potential complementarity is destroyed as each approach is pushed toward its asymptote: intellectual mapmaking distributes our available attention over the broadest context possible, while a discipline-based approach concentrates all available attention on a single category of phenomena that dominates the situation being studied. Hence, intellectual mapmaking ultimately becomes knowing less and less about more and more and, in the extreme, tends toward knowing nothing about everything. In contrast, a discipline-based approach, pushed within its limits, evolves through ever-greater specialization, and thus a knowing of more and more about less and less, and in the extreme tends toward knowing everything about nothing. Such is the relativity of our understanding.

There is a matching relativity of everything we seek to understand. It is only the intellectual mapmaking approach that is able to respect the relativity of what we seek to understand, because it is designed to deal with a living world. It may be interpreted as the conscious or explicit counterpart of the way we appear to metaconsciously make sense of everything by means of symbolization, experience, and culture. In contrast, a discipline-based approach to understanding or intervening in a situation is necessarily limited to those situations that are dominated by a single category of phenomena, to the point that the contributions of all other categories can be neglected. Excluded is the way a situation under study (even when it is dominated by a single category of phenomena) is integral to a great deal else that may make this dominance transitory by depending on changes in its broader context. A discipline-based approach may thus provide a temporary "snapshot" of the situation under study in a manner that could require other investigations. For example, in physics a unified theory for integrating its major branches has thus far escaped us, and this may or may not be illustrative. The fact that the disciplines of the natural sciences are able to make extensive use of mathematics may or may not imply a translation from an enfolded architecture to that of reality as suggested by the physicist David Bohm.[1]

When we intellectually attempt to restructure something to match it to something we believe we do understand, we are vulnerable to what Devereux has referred to countertransference reactions, which "adjust" the process of understanding in order to reduce anxiety and tensions to bring it in line with what we already understand.[2] Given the first and second generation of secular myths, such countertransference reactions are bound to be common in our global civilization. What individual and collective human life has maintained, evolved, and transmitted from one generation to the next as a body of common sense (the shared meanings, values, and everything cultural) no longer has any credibility, having been devalued to pure subjectivity. It has thus become impossible to recognize what intellectual mapmaking (as the conscious or explicit critical equivalent of symbolization) could possibly contribute to the modern university, and via it to all other institutions of society in order to overcome the lacunae in discipline-based knowing and doing. As we have briefly attempted to sketch in the Introduction, the discipline-based approach to knowing essentially "explodes" human life and the world into autonomous categories of phenomena by sweeping under the intellectual rug all situations in which they deeply affect one another. Similarly, the discipline-based approach to doing superimposes the manipulation of everything as if the domains of disciplines are always autonomous from each other.

There is a fundamental relationship between our finitude, the relativity of our being in the world, and our embodiment in everything. Our finitude involves all manner of limits, including our reciprocal relatedness to our surroundings, without which we cannot maintain our physical, social, and cultural embodiment. Our physical embodiment can largely be understood in terms of our inability to create or destroy the matter and energy involved in any human activity and in the constant repair and replacement of cells in our bodies. Similarly, our social embodiment involves a constant reciprocal exchange with others in order to maintain and evolve our lives and our social selves enfolded in them. We only need to remind ourselves of situations such as a child living alone in the woods and then returned to a human community, of people held in solitary confinement for long periods of time, or of people participating in sensory deprivation experiments to realize how fundamental and sustaining our social embodiment is. Finally, despite two generations of secular myths and the accompanying deep symbolization of our cultures, we continue to rely on the latter to keep the door shut to cultural chaos. How well we are succeeding – or failing – may become evident from the epidemic of anxiety, depression, and mental illness, as any first responder to emergencies can testify.

As we prepare to shed further light on the fundamental importance of our physical embodiment in the world, we can use two different intellectual frameworks that will gather two incompatible sets of "facts" for interpretation. We will adopt the framework arrived at through intellectual mapmaking as opposed to the discipline-based approaches, developed to interpret what has been happening to human life and the world since industrialization, urbanization, rationalization, and secularization. In the Introduction, we provided a brief overview of the research that led to this intellectual map. Such a map is a theoretical elaboration of what would have been our symbolic universe if such a universe had not been displaced by the reality built up by the original and new media as well as the discipline-based approaches to knowing, doing, and organizing. The making of this map requires "translation" of the many facts gathered by the disciplines in our universities from the domains relative to which they were gathered into this map. In this translation, careful attention has been paid to the way the "facts" gathered by disciplines have been sheltered from secular religious attitudes rooted in our secular myths. We are not suggesting that an intellectual mapmaking approach is exempt from such secular religious attitudes, but that it engages in a struggle to arrive at an interpretation of human life and the world that negates as much as possible anything sacred, autonomous, or limitless either in a traditional or contemporary secular form. Despite its obvious limitations, discipline-based science is no longer self-critical, as is painfully evident from the many "science worship hours" on the media.

The Great Cultural Divide in the Relativity of Human Life

As noted in the Introduction, from the perspective of our history as a symbolic species, a cultural divide began to occur with the introduction of a technological and economic order into the cultural orders of the first generation of industrializing societies. It began to divide individual and collective human life into two spheres with diametrically opposite architectures. The dominant symptom was the occurrence of separate and distinct economies within these societies. Within the architectures of the technological and economic orders, the symbolic approaches on which every culture had relied until then began to reach their limitations, which led to the eventual separation of knowing and doing from experience and culture by their reorganization in the form of disciplines.[3] These disciplines enormously restricted the scope of symbolization, which resulted in high levels of desymbolization of this discipline-based knowing and doing. At the same time, the accompanying displacement of cultural approaches in daily life to the benefit of their discipline-based counterparts also resulted in the desymbolization of all cultures. In other words, this cultural divide was between a humanity living as a symbolic species and a "re-engineered" humanity as a highly desymbolized species relying less and less on symbolization, experience, language, and culture to the benefit of technique.

Before this great cultural divide in human history, babies and children experienced a cultural order gradually enveloping them through the intermediary of the lives of others. Following the great cultural divide, these cultural orders have become so highly desymbolized by technique that babies and children encounter only highly desymbolized remnants embedded in an order of non-sense.[4] These orders of non-sense are organized by a technique-based connectedness that envelops and dominates what little remains of the sense of the culture-based connectedness of human life and society. The ongoing adaptation and development of this technique-based connectedness requires no reference whatsoever to what anything and everything means for individual and collective human life, or to the value it contributes to that life, thereby shutting out all sense. Jacques Ellul has demonstrated this through five primary characteristics of technique.[5]

The rationality of technique transfers everything belonging to human life and the world into the domains of disciplines whose architecture shuts out any need for referring to what makes sense. Another characteristic, that of the necessary linking together of techniques, is an extension of what I have examined in earlier works,[6] as the chain-reaction-like character of industrialization was followed by technicization, which engulfed everything in their paths with little or no dependence on sense. The automatism of technical choices is rooted in the necessary choice of the highest ratio of desired outputs to the requisite inputs (such as efficiency, productivity, and profitability) imposed by practitioners suspended in a triple abstraction. No reference to what this means for human life and society is either possible or required. The self-augmentation of technique produces a technical development that does not depend on human goals or aspirations. Its primary component results from each technical advancement producing a diversity of incompatibilities with human life, society, and the biosphere. These combine in a range of problems that must be addressed if technical development is not to be undermined. Hence, technique feeds on the problems it creates. A secondary component of this self-augmentation results from the flows of technical information within technique. These transform each discipline into a receiver and transmitter of information in a global system of information flows regarding technical advances that, when received by a particular area of specialization, will frequently and almost automatically trigger further developments. As a result, technical developments continue to be driven by traditional inventions and innovations, but their effects on the global information system greatly enhance their technical significance. All these permutations and combinations of technical information lead to a component of technical development that is completely at arm's length from human wishes and desires. It has made the global technical system exceedingly non-linear, with technical breakthroughs next to impossible to forecast. Finally, another major characteristic of technique is its universality. Because the domains of all disciplines are separated from experience and culture, they are independent from any local context. Their mediation thus produces results that are technically valid all around the globe, even though they may be incompatible with local conditions, but since technical validity does not refer to any human or social contexts, these incompatibilities are inevitable.

In sum, no decisive human intervention is required in the global system of technique. The human participants have an education in one or more disciplines, and this suspends them in a triple abstraction that makes any reference to sense next to impossible. Consequently, their behaviour conforms only to the requirements and opportunities of the system.

Before the great cultural divide, humanity lived in a relativity that potentially would have made sense were it not for each group and society having to deal with the threat of cultural chaos. Consequently, the "sense" of traditional cultures was alienated by their need to ground this relativity in myths. Following the great cultural divide, humanity gradually began to live in a relativity of non-sense, which also needed to be defended against cultural chaos. This was done by grounding the now highly desymbolized cultures in secular myths. These myths legitimated these highly desymbolized cultures by directing their members toward discipline-based knowing, doing, and organizing, which no longer had any limits according to these myths and thus were not to be interfered with by cultures. However, babies and children now growing up in this relativity of non-sense cannot do so directly. They first have to acquire the desymbolized language and culture of their society in order to make sense of their lives relative to the lives of the others who nurture and sustain them. In other words, their early development begins in the highly limited relativity of desymbolized sense, and only when it is sufficiently advanced can the desymbolizing influences of a relativity of non-sense begin to have an effect. There is as yet no indication whatsoever that babies and children can directly enter a relativity of non-sense, but there is a great deal of evidence to suggest that their symbolic development can be seriously undermined, with detrimental effects. Situating human life in time may be among the most difficult tasks. These possibilities will become evident later in this work.

In the beginning of our global civilization, humanity began to build a very different relativity of lived relationships, which needed to be grounded in new ways. This manifested itself through the emergence of new secular political religions and thus the transformation of all traditional religions, beginning with those of the industrializing societies of western Europe. The new emerging secular political religions were sustained by the metaconscious knowledge that the new relativity of relationships was "saving" society by gradually permitting it to materially sustain all its members. How this would be accomplished varied according to the particular secular political religion. At the same time, the traditional religions, whose function had been to bind the members of any society together, had to relinquish the saving of a people and concentrate on the "saving" of individual members by ignoring their alienation and reification. To understand this, let us briefly return to the genesis of our global civilization as outlined in the Introduction.

In the historical accounts of the so-called Industrial Revolution, what is generally overlooked is that, when "people changed technology," there was a simultaneous "technology changing people" through the vast influences that industrialization, urbanization, and rationalization were having on people and the social fabric. The people living in those days built up the organizations of their brain-minds with experiences that jointly made it unthinkable that the growing trends all around them would not go on forever. Consequently, ever more machines housed in ever more factories would produce an ever-growing output that would soon make poverty unthinkable. Along with it would disappear the scourges produced and sustained by this poverty. The viability of individual and collective human life would thus steadily improve along with its quality, and this could only make everyone happier. All society had to do was to ensure that sufficient capital would be renewed and accumulated in order to make the required investments. It was this investment of capital that could produce all this material, social, and spiritual progress. With a great deal of hindsight, this brief description makes explicit the deepest metaconscious knowledge that grounded individual and collective human life in the first generation of industrializing societies in western Europe.

Intuitions of these secular myths were developed by many thinkers as different as Adam Smith and Karl Marx.[7] Again, with a great deal of hindsight, it is evident that, without the myths of their days, their thinking would collapse like a house of cards. For the followers of Adam Smith, society would be "saved" provided that everyone acted in their own self-interest (the contrary of the value system of all traditional cultures), and free markets would take care of the rest to produce the best possible world for most people; but this omits the market externalities that cumulate into market forces. In the case of the followers of Karl Marx, the Communist Party, endowed with the absolute truth of how human history would move through five progressive stages, had to act on this knowledge for the benefit of all humanity and transform capitalism into the communism of the fifth and final stage of human history. Alienation would cease, and the freedom that had been lost at the end of the first stage of human history would be regained because people would no longer exploit people and people would no longer exploit nature. In both secular religious visions, humanity would be "saved" from its terrible past, but the ways of salvation according to democracy and communism were radically different.

With the salvation of humanity at stake, neither democracy nor communism shied away from a little collateral damage. The democratic West had little regard for all the people who were being left behind, or for the people it colonized in order to obtain required resources or expand markets. In the communist West, anyone who disagreed with the Communist Party, which acted on the truth of human history, had to be re-educated or eliminated. With industrial civilization bogged down in the scourges of unemployment and inflation, National Socialism proclaimed another path for saving humanity, and thus the stage was set for an extraordinarily bloody twentieth century. All three political religions directly and indirectly sacrificed more human lives on their altars than possibly all traditional religions did collectively during any equivalent time span. The irony is that as a consequence of "technology changing people," *homo economicus* became transformed into a consuming humanity with insatiable needs that were open to almost all technical possibilities (both material and information), enslaved to them as far as financial resources allowed, but committed to them only until something more performing was produced by the system of technique.

It may well be objected that democracy was essentially compelled to sacrifice many lives to defend, maintain, and expand itself. Such a view cannot stand any careful scrutiny. Have we forgotten the Korean War, the Vietnam War, and the invasion of Iraq, to mention only a few examples, which many people with a great deal of hindsight now regard as terrible mistakes? For many years, the annual meeting of the American Association for the Advancement of Science included a symposium dealing with the possibility of diverting the American economy away from its overwhelming dependence on the military-industrial complex, against which President Eisenhower had already warned the nation in his farewell address.[8] John Kenneth Galbraith and others have pointed out that the dependence of the public sector in the American economy on war and space is deeply disturbing.[9] The relationships between democracy and war are complex and cannot be underestimated.[10]

As noted, the emergence of a so-called secular humanity put all traditional religions into a terrible predicament. As the dominant religious influences on Western civilization, Judaism and Christianity were not exempt, since they served the moral and religious needs of the societies involved in the pioneering of this beginning. The moral and religious needs of the new secular societies were being met by the new secular political religions. What was left for the traditional religions were the "private" and "spiritual" needs of those who continued to cling to the moral and religious traditions of the past. In the case of Christianity, the images of God that various Christian traditions had made for themselves under the influence of Greek philosophy, Roman law, and Islam[11] were entirely foreign to the new time, place, and technical order.

In the eyes of many people, Christianity was doomed unless it engaged in one of the two following strategies. One involved playing down the importance of God with the guidance of science, since all traditional religious needs were in decline. The focus had to be shifted to the vast needs the new emerging world had for a greater love for one's neighbour, given the upheavals and suffering involved in the bringing in of the new order. The alternative strategy was in a large part a reaction to the first. It deemed that it was essential to assert the ultimate religious significance of God, but to do so not in the context of collective life but in the context of individual life – even though the Jewish and Christian Bibles focus on God's people and the community of all believers, and individuals are always addressed as members of these communities. The above reinterpretations perfectly matched the gradual transformation of traditional societies into mass societies with unprecedented individualistic orientations. These orientations were in part the result of the desymbolization of all cultures, which greatly diminished the abilities of such cultures to sustain social relations, bonds, and communities. The social and historical stage was set for Christianity to be transformed into the business of my God, my conversion, my salvation, my sins, my church, and my going to heaven. Of course, this description of these two strategies is highly simplistic, but it is only intended to capture the new orientations of the emerging liberal and conservative forms of Christianity. The one emphasized the love of one's neighbour at the expense of the love for God, while the other did the reverse.

Both these new forms of Christianity were well adapted to the new so-called secular age that was emerging. The conservative form continued to cling to Greek philosophy through its use of universal theological knowledge. Such knowledge appeared valid everywhere and thus was readily confused with what is true. However, something that is valid everywhere cannot be lived anywhere, because the relativity of human life is lived in relation to a time, a place, a highly desymbolized culture, and a global technical order. Christianity was replaced by statements of belief, rules, and doctrines that were difficult, if not impossible, to live relative to a culture. It might have been helpful if some theologians had remembered that the Greeks attempted to save their culture from relativism by grounding it in universal knowledge, which most likely accelerated the decline of Greek culture. Conservative theologians have often engaged in a somewhat similar strategy, which also resulted in making Christianity more difficult to live. In contrast, the liberal form of Christianity has benefited enormously from the collapse of foundationalism, which created a greater flexibility driven by a concern for local suffering and the need to save people from their misery and restore them to a more meaningful and purposeful life.

Following the great cultural divide, Christianity was thus effectively assimilated into the world by breaking the unity of the gospel, always summed up as an indivisible love for God and for one's neighbour, whom God also loves. There were criticisms to the effect that it was but the highest form religion had achieved (given the myth of progress) and a false awareness of human life and the world as set out by Karl Marx;[12] and that it could be explained from a sociological and historical perspective as satisfying the moral and religious need of a society, according to Max Weber.[13] These criticisms, and the many that followed, were mostly valid. If the Christian community had thought of itself as following a way that was entirely other than that of a morality and a religion prescribed by its God, such criticisms might have provoked a fundamental theological rethinking, if not a repentance. Nothing of this kind happened. For example, in North America after the Second World War, an individualistic Christianity was blended into the American way of life and the secular political religion of democracy.[14]

The salvation of the new secular humanity took a somewhat different course during the second half of the twentieth century. A cultural realignment became necessary in response to intuitions of new forms of the deepest metaconscious knowledge, which were displacing the older forms in the organizations of the brain-minds of the members of the so-called industrially advanced societies. A few secular "heretics" critiqued the older forms of the two remaining secular political religions. Communist orthodoxy was challenged by the communist government of Czechoslovakia. Possibly with the benefit of the most extensive experience with industrialization in eastern Europe, this nation officially questioned what it regarded as communism's only marginal advantages over Western capitalism. A commission was charged with investigating the situation. It re-examined the theories of Karl Marx and found that highly specialized scientific and technical knowledge now played the role capital had played in the previous century.[15] Everything had changed, with the result that entirely new economic, social, and political approaches needed to be invented. Russian tanks soon rolled in to stamp out this secular heresy with regard to orthodox communism. In China, Mao Zedong also undermined Marxist orthodoxy by claiming that China could directly pass into the fifth stage of human history without going through capitalism.[16] In the end, communist orthodoxy essentially vanished along with the former Soviet Union.

In the West, John Kenneth Galbraith showed that the new and decisive role played by highly specialized knowledge in industry, the economy, and the state demanded significant shifts in economic and political thinking.[17] Although he acknowledged that these developments were integral to what Max Weber referred to as the rise of rationality, Galbraith essentially limited his analysis to the above spheres, and his work failed to come to grips with rationality and technique.[18] Nevertheless, Galbraith came to the secular heretical conclusion that we were now serving the economy rather than its serving us.[19] Max Weber had gone much further by warning humanity that it was shutting itself into an iron cage of rationality.[20] A little later, Jacques Ellul warned against the rise of autonomous technique, by which he meant that its social and historical influence on humanity had become much greater than the influence humanity appeared to be able to exert on it.[21] The result was the enslavement of humanity to technique. Almost without exception, the philosophers of technology regarded this conclusion as a philosophical statement. In any case, many people failed to recognize that our conformity to technique as a consequence of technique's changing people represented the latest form of the alienation of humanity.[22] All public discourse about a common public good vanished, only to be replaced by a discourse about how to best "move forward." I am not aware of any society ever deliberately moving backwards; and I interpret these kinds of statements about moving forward as a public acknowledgment that we no longer have any choices: either we take advantage of the latest techniques or we will condemn ourselves to the secular hell of underdevelopment. What is good for science, technique, and the nation-state (or state-nation) will produce the only history that can "move us forward."[23] Of course, no one paid any more attention to the new secular heretics than they did to the religious heretics of the past. Humanity remained on the same page of having made a new beginning for itself, and it failed to understand how anything else was possible. Our secular salvation carried the day.

Christianity also had its share of new heretics. Some were shaken in their liberal beliefs by the terrible upheavals of the twentieth century. Others saw the contradictions created by an individualistic conservative Christianity attempting to serve the moral and spiritual needs of its adherents. There were very few true heretics who saw what had happened to Christianity. My favourite one is Jacques Ellul, whose thinking is neither liberal nor conservative because it is based on an attempt to reunify the gospel as an inseparable love for God and one's neighbour. His model for doing so was Qohelet (commonly translated as Ecclesiastes).[24] I have attempted to explain his approach in a previous work.[25]

The Relativity of Our Lives before Screen-Based Devices

When we seek to understand how babies and children build the relativity of their lives, it is essential to recognize that all this is deeply influenced by much of what has happened since the great cultural divide. This relativity is built up in relation to and within the relativity of the lives of everyone around them, which are embedded in the relativity of their society grounded in its secular myths. In other words, babies and children are gradually being confronted with the "silences" of our highly desymbolized cultures, where certain thoughts and actions are essentially blocked by these myths. As we will see, toddlers may initially be rather baffled by the way we interact with our screen-based devices. In the opposite direction, most adults will have little appreciation for the importance of the way babies' and children's lives are being embodied in the world, given that adults are immersed in technically mediated relationships that require only a minimal embodiment, participation, commitment, and freedom. In the same vein, we have so taken for granted the distinction between public schools and private religious schools that we are generally unaware of the secular religious attitudes that all schools help to transmit in order to protect our society from cultural chaos. Public schools are not supposed to transmit traditional religious attitudes while so-called religious schools are required to do so, but they all transmit the secular religious attitudes necessary to sustain our society. Similarly, what we regard as the moral and religious aspects of the development of babies and children is almost entirely determined by the position we take relative to the moral and religious traditions that have served our civilization in the past. All we need to do is to embark on a comparative approach between how babies and children developed the relativity of their lives in our highly desymbolized cultures and how they did so in the traditional cultures of the past; and we will discover the "silences" of our own culture. Symbolizing the unknown as more of what is known and lived clearly rules out all kinds of possible thought and actions. A social and historical understanding is essential; without it, the discipline-based approaches of philosophy or theology will completely distort the situation.

In order to develop a sense of the relativity of human life along with the theoretical and practical consequences, we will endeavour to briefly sketch the beginnings of an intellectual map of that relativity. The beginning of every person's life is relative to a society whose way of life and culture influenced how our parents met, developed a relationship, and produced a child. This sexual reproduction and the emergence of oneself, from an embryo to a child, was relative to many conditions that actually or potentially influenced our development. For example, the society of our parents may have declared a war on drugs or launched educational programs regarding the use of alcohol and smoking during pregnancy, but all this fails to consider why people face the risks associated with their use. Society loves to imagine that people simply do these things out of ignorance, with the result that a little education is expected to go a long way toward bringing about fundamental changes. Rarely is a community willing to acknowledge that the consumption of drugs and alcohol amounts to a societal illness stemming from a failure to sustain its members. For example, its way of life may be such that many of its members lack access to appropriate employment and thus to meaningful participation in their society. For those who do find employment, social epidemiology shows that our workplaces are a primary producer of physical and mental illnesses that are largely preventable by redesigning work.[26] The correlations between unhealthy work and smoking, alcohol consumption, drug abuse, anxiety, depression, or aggressive behaviour are well known but ignored.[27]

Similarly, a way of life can create the kinds of conditions under which it is very difficult to sustain the lives of others, and for them to sustain our lives. When breakdowns of relationships become frequent, a variety of consequences occur, from an increase in sexually transmitted diseases to many children's lacking a stable nurturing family. To turn all this into moral judgments and condemnations of others is to completely miss the point. What we are discussing are, first and foremost, structural issues related to the capacity of a way of life to adequately sustain the lives of its members and our responsibility for it as we help to sustain that way of life. Surely, no one would argue that highly individualistic mass societies are well equipped to sustain the lives of their members. This lack of support produces incalculable harm that may or may not have affected children's early development in relation to their society.

It must be remembered that at any time, our relationships with our physical and social surroundings are reciprocal in character: we change our society and our society changes us by our internalization of our experiences, which are symbolized by neural and synaptic changes to the organization of our brain-minds and thus to our lives and our persons. If we can influence our society more than it influences us, we are relatively free. However, if our society influences us a great deal more than we can influence it, our life is alienated by it. Similarly, our health is determined by this reciprocal relationship, in which our society imposes certain demands on our immunological, psychological, and social resources. If these demands can readily be handled by them, we remain healthy. If, these demands begin to tax, strain, or overload our resources, we will become ill. In other words, we cannot dissociate our collective responsibilities from our individual ones, or vice versa, as is so frequently done to justify ourselves or our society. Initially, there is always a stigma against people who are unable to cope with the demands placed on their resources by a new development in their society. As more and more people experience the same problem, a point is soon reached where many are either directly affected in their own lives or indirectly via the lives of those dear to them, and it becomes easier to feel sympathetic than to be judgmental. The stigma weakens and eventually disappears as the issue has more or less become everyone's affair. It has happened with divorce, remarriage, birth control, abortion, gender issues, gay marriage, mental illness, suicide, euthanasia – and surely will be repeated for future issues. It happens to everyone who has an opinion about an anonymous social issue and then needs to confront it in a child or in a person dear to them.

The relative character of our life thus began with our embodiment in the world as an embryo. Its development cannot be understood apart from situating an egg and a sperm in the surroundings relative to which the embryo was formed and began to develop. We will not deal with this development, other than that it involves the transformation of stem cells relative to the whole of an emerging human body as it is influenced by the DNA and the surroundings. This process turns these stem cells into expressions of that whole as member cells of specific tissues, organs, and the larger entities built up with them. There is no part-whole distinction, since the DNA is enfolded into each cell; that cell in turn becomes enfolded into particular expressions of that whole essential for sustaining a life. There is thus no creation of "parts" that are then "assembled" into a whole, as occurs within the architecture of anything technical.

Instead, during the development of an embryo, every cell is internally connected to the whole present in the form of the DNA, and externally connected to all the other cells of the entity it helps to constitute, and via it, to the remainder of the emerging body. All such relationships involve molecular and cell interactions constrained by the expression of the whole. Everything depends on inputs of matter, energy, and microorganisms, from which is extracted whatever is necessary for the formation and maintenance of the stem cells, their differentiation, and their building the larger expressions of the whole. Any part of these streams of matter and energy that cannot be utilized, or that constitute the waste products of such utilization, must be removed; all these necessary functions can take place only if the surroundings of the emerging body are capable of transmitting these flows – flows that ultimately connect this emerging body to the biosphere, since matter and energy can neither be created nor destroyed, in the biosphere and beyond. At the same time, these flows of matter and energy into and out of the emerging body can contain microorganisms that may or may not contribute to these processes. The mother's body directly sustains this development, and it in turn is sustained by the biosphere supporting all life.

Although we take this for granted, we are able to express these developments and make them intelligible by means of our language. The reason is in part that its architecture has evolved to mirror that of all life. Both architectures can materially and culturally deal with human life and a world in which everything is related to and evolves in relation to everything else, because nothing is what it is in itself.

As the embryo unfolds the complexity of the human body biologically enfolded in the DNA, unborn babies begin to express themselves in relation to the womb, and via it, relative to the bodies and lives of their mothers. They begin to move about and use their limbs, of which the strongest expression may be experienced by their mothers as if they are kicking. It may therefore be supposed that, well before the formation of the cortex as the completion of the organ of the brain, unborn babies begin to include some self-expression in their lives as soon as it can be sustained by a rudimentary capacity of the organization of their brains. Doing so becomes greatly enhanced by the formation of the cortex just prior to birth. Unborn babies continually unfold their lives to the level that can be sustained by their bodies.

The formation of the cortex thus represents a further aspect of the embodiment of unborn babies. Its development adds to the complexity of the embodiment of babies in the womb by permitting the senses to contribute to whatever embryonic experiences they had up to that point. These embryonic experiences and their growing bodies jointly prepare unborn babies for the new kind of embodiment they enter into following their birth. In any case, it is likely that, before their birth, babies may experience the rhythms of the lungs, hearts, and stomachs of their mothers and possibly some sounds from beyond her body. At any time, such experiences are an expression of and thus enfold the entire development that has taken place up to that point. Of course, such an enfolding remains unconscious.

What then, if anything, is added to the development of unborn babies by the cortex? It certainly is not the "stuff" of the mind elaborated by Western philosophy for centuries. All this has turned out to be mostly nonsensical relative to what we now know, but it made sense in terms of what was thinkable within the architecture of philosophy. What the cortex adds is an expansion of the organization of the brain to potentially fully coordinate the embodiment of babies in the world as members of a symbolic species. There is no need whatsoever of the "stuff of the mind" if we interpret what is happening from this perspective, that is, from within an architecture that anticipates, and prepares for, the acquisition of a language and a culture. The symbolic organization of the complex array of neurons and synapses may be interpreted as another step in the ongoing process of babies developing as members of a symbolic species. This potential is no afterthought, nor a bursting on the scene of an embryonic consciousness, but something that was unfolded from the DNA from the very beginning. Each detail of this development is a particular unfolding of this whole and an enfolding in its unique expression of it. All this far transcends the notion of brain plasticity, which is a necessary characteristic of the brain of a member of a symbolic species. It provides the brain with the ability to symbolically coordinate the lives of babies in the world by continuously building these lives with each and every new experience, symbolizing it as one of the moments of a life.

A symbolic memory has nothing in common with a machine memory. It does not store information. It symbolically builds up the lives of babies with new experiences. Doing so is possible because each experience of their lives results in a modification of the brain by means of neural and synaptic changes that symbolize it. Brain plasticity is both a consequence of and a prerequisite for babies learning to live their lives in the world. What we refer to as the mind is essentially a continuation of the kinds of developments we have been describing. The only reason for introducing a distinction between the brain and the mind is that the organization of the brain is the consequence of the unfolding of the complexity of the DNA of a symbolic species. The mind is the result of the development of this organization, attributable to babies' gradually emerging awareness of their lives in the world, interpreted as symbolic experiences of those lives. We cannot radically distinguish the brain from the mind, any more than we can radically distinguish nature and nurture. Hence, we will adopt the interpretation of the organization of the brain-mind emerging from the organization of the brain as the consequence of symbolic experiences beginning to play a role in these developments.

Since each and every experience is symbolized by neural and synaptic modifications of the organization of the brain-mind, this organization also begins to enfold a great deal of metaconscious knowledge of babies' lives in the world. It derives from the way symbolized experiences participate in the dynamic organization of the brain-mind, to which babies have no conscious access. However, as members of a symbolic species, babies and children learn a great deal through the development of metaconscious knowledge implied in the organizations of their brain-minds, and they express this metaconscious knowledge in their behaviour.[28] All such possibilities are a manifestation of the organizations of their brain-minds symbolically working in the background. Each experience becomes an expression of the life these organizations symbolically map, with the result that they are all lived as moments of a life.

A great deal of babies' learning to live their lives can be understood if we assume that these lives are built up from differentiated experiences. Each experience is directly differentiated from those that are the least different, and via them from all the others that are similarly differentiated from their neighbouring experiences, thus ordering an entire life. As a result, we may imagine relatively distinct clusters of differentiated symbolic experiences corresponding to daily instances of activities such as being fed, burped, washed, diapered, and played with, as well as being put back in the crib to explore the surroundings when awake and sleeping the rest of the time. Each cluster of these activities begins to build up metaconscious knowledge implied in the differentiated cluster. Jointly, these differentiated clusters that have been put into the organizations of babies' brain-minds begin to symbolically map their lives. Additional "deeper" metaconscious knowledge is generated, but there is as yet no metaconscious knowledge of a physical, social, or cultural map reader or anything else that would require them. What is happening from the very beginning is that the organizations of the brain-minds of babies symbolically interpolate and extrapolate all the experiences of a baby into a life, rudimentary as this life may be.

Using our imaginations, we may attempt to dialectically interpret our experiences as babies and enrich this interpretation as well as test it by observations and by the findings of a great many disciplines. As unborn babies we must have been essentially one with the world, especially prior to the development of the cortex. Nevertheless, each of us was a unique expression of everything that had been unfolded from our DNA, but this expression was still dominated by the ongoing unfolding of this potential relative to our interacting with our physical and biological surroundings via our umbilical cord and the womb of our mother. It was as if we were still hidden within all this relatedness.

Once our cortex began to form, this situation gradually began to change. Our senses became connected to and coordinated by this more complex organization of our brain. As noted, it may well be that we gradually gained some awareness of the rhythms of our mother's lungs, heart, and stomach; and some sounds from beyond our mother's body may have been "experienced" as well. This more complex organization of the brain also facilitated a variety of movements, some of which our mother may have made intelligible as "kicking." We had entered the final preparations for our birth.

Following our birth, we lost our physical and biological integrality with our mother's body and life. We now faced more complex embryonic experiences, which can perhaps be imagined and made intelligible as a hyphenated word indicative of as yet undifferentiated constituent elements: subject-verb-indirect object-object-surroundings. It is the result of the organizations of our brain-minds further developing their symbolic potential by mapping our experiences into our life, while our expression of that life was still limited by the absence of a physical, social, and cultural self acting as a map reader and user to find our way in the world. For example, our being held and fed would be constituted of a simultaneity of sensations derived from our senses and our body that had as yet no meaningful pattern or order, other than what we now imagine by means of a string of hyphenated words. Even these imagined strings of hyphenated words would be different for each of the activities that made up our daily routine. As a baby we would have had no idea whatsoever of the meaning and value of each activity, which we have designated as a string of hyphenated words to make it somewhat intelligible. Nevertheless, as a baby we were able to symbolically differentiate them from each other. We learned something about what we would later call eating relative to what we would later call being washed and diapered, and about being washed and diapered relative to eating. At the same time, we would learn something about eating relative to being burped, played with, and lying in our crib exploring our surroundings or sleeping, and about each of these activities relative to eating. It was radically impossible to learn anything about eating in an absolute sense, that is, as something autonomous from everything else in our life and our surroundings. Of course, this included each eating experience being directly differentiated from all the others, and via them from the experiences of all other activities. As a result, each experience, no matter how embryonic it may have been, could begin to take on a rudimentary meaning and value relative to all the others in our life, as they did to it.

It is important to emphasize that it is a lived meaning and value that symbolizes something of what all the other lived experiences are not, and vice versa. As such, an experience becomes a unique way of living a particular moment in our life relative to all the other lived moments. It is not a question of an analysis having been performed as to what constitutes a set of shared characteristics of all eating experiences, for example. What an experience is in our life is what all the others are not, and vice versa. Consequently, the meanings and values of all the experiences are dialectically enfolded into each other.

Enfolded into the organization of our brain-mind as a symbolic map of our life was thus an embryonic self that lived each experience as a moment of our life. However, a physical, social, and cultural manifestation of that self was absent as yet. Nevertheless, a specific daily-life activity that in retrospect we have interpreted as a string of hyphenated words was beginning to take on some kind of order, as expressed in its relative meaning and value.

It would appear impossible, therefore, to deduce that our life was living itself. Metaconsciously, we were learning a great deal about our life by living it. Doing so was entirely relative: how we lived each and every activity was how we did not live all the others, and vice versa. How we lived particular incidents of that activity was how we did not live all the other incidents of it, and vice versa. All this may appear purely speculative, were it not for what we know about our dependence on the development of metaconscious knowledge in growing the organization of our brain-mind. This metaconscious knowledge is confirmed by our expressing it in our behaviour, which enfolds it. For example, as adult members of a cultural community we express a metaconscious knowledge of a unique conversation distance, eye etiquette, and body rhythms when we talk to others face to face. Similarly, we express a metaconscious knowledge of how our way of life and culture arrange individual and collective human life in time, space, matter, and the social. Ultimately, we express the deepest metaconscious knowledge when we participate in the history of our community as persons of a time, place, and culture.[29] All our adult skills fundamentally depend on metaconscious knowledge, which is why we did not succeed in knowledge engineering and the building of expert systems except in circumstances that approach a relatively closed micro-world.[30]

At no point can it be said that our lives live themselves. There is always an embryonic intervention in each and every elementary experience by that life itself as symbolized in the organization of the brain-mind – an expression of the unfolding of our physical and biological selves increasingly complemented by the enfolding of expressions of this through our lived activities. This unfolding first gave rise to the development of a metaconscious physical self. In turn, this metaconscious physical self opened the door to the formation of a metaconscious social self, which in turn paved the way for the development of a metaconscious cultural self.

Imagine the beginning of all this for many of us. We were lying in our crib playfully exploring our surroundings. Initially we were moving our limbs for no particular reason other than from time to time encountering something interesting. For example, we might accidentally touch a rattle suspended above our crib and experience a tactile and aural sensation. We might kick against the cloth wall of the crib and experience a soft tactile sensation. If we wiggled our limbs with all our might, we may have been able to shake the crib. All such discoveries rewarded our efforts, but if we desired to experience more of something, it may have been somewhat frustrating because of our limited ability to replicate anything at will.

We were as yet incapable of distinguishing between sensations originating within our body, sensations derived from a more or less voluntary or involuntary use of our body, and sensations of our body encountering others or something in our physical surroundings. Initially, we were unable to focus our eyes, since we did not have any awareness that there was something meaningful to focus on. Nor did we have any awareness that moving our limbs was partly under our control and not an external event in our surroundings. We can thus imagine experiences involving sets of sensations whose only connection was their simultaneous occurrence as detected by the organization of our brain-mind. As babies, we learned something about any such set relative to other ones and about other ones relative to it as a result of their being differentiated from each other in the context of the organization of our brain-mind. As this differentiation became endlessly refined, one such cluster of the embryonic experiences of lying in our crib during which we touched parts of ourselves would gradually become differentiated from another sub-cluster where we touched something else in our crib such as a blanket or a suspended toy. The basis for the differentiation of the two clusters is that in the case of the former there were two tactile sensations, as when one of our hands touched another part of our body, while in the latter there was only one such sensation.

At the same time, such developments occurred in all the other clusters associated with the other activities of our life. For example: when we were held to be fed, there would be one set of associated experiences with only one tactile sensation, such as touching our mother with our hand (and this would often evoke a response), and another subset where we touched ourselves (which would often go unnoticed by our mother). The same kinds of distinctions may have developed in all the other clusters of symbolized experiences associated with other activities. All this contributed to a growing metaconscious knowledge of our body being distinct from our surroundings. Some bodily expressions stood out because we received an emotional response from our mother or from others who made the situation intelligible, as if we were deliberately reaching out to them. Such responses in turn encouraged other developments. Our body was not merely distinct from our surroundings, but a way of getting one kind of response from our mother or other people and quite another from objects in our surroundings.

Of course, all these developments are hardly limited to tactile sensations and the possible emotional responses associated with them. They may have encouraged us to more carefully follow the visual "blurs" of our mother, and in the process we gradually learned to focus our eyes on her. It opened the door to our responding to sensations of being reached out to with hugs, kisses, and gentle stroking. In parallel, our ears began to add aural sensations to our life that could take on some meaning in the context of other sensations, such as those associated with our mother approaching our crib to pick us up while talking to us. Similarly, we may have begun to detect that certain activities were accompanied by reactions from others about unpleasant odours in which we were enveloped from time to time. These reactions may eventually become associated with the tactile sensations of soiling our diaper. The more our embryonic experiences became enriched in an ever-increasing diversity of ways, the more comprehensively our experiences could be differentiated from each other, and the more their meanings and values were enriched. So was the metaconscious knowledge built up in the corresponding areas of our brain-mind.

It is through these kinds of developments that we began to intuit our being a physical self distinct from the physical selves of others, pets, plants, and objects. The more we intuitively externalized this metaconscious knowledge of our physical self in our behaviour, the more we developed the ability to make a deliberate use of our limbs and our senses and to make a growing range of sounds in an effort to elicit responses. As our metaconscious knowledge of our physical self was thus increasingly expressed and was met by more specific responses, the door gradually began to open to non-verbal communication. We learned to differentiate our crying to indicate feeling hungry, wet, cold, bored, or tired, and so on. Depending on the response received, such non-verbal communication developed in accordance with how our mother's language and culture were able to make them intelligible to herself and to others. It was thus relative to her and the cultural context she embodied that much of the above developments took place. All this was further reinforced by contacts with significant others who shared that same cultural context. We were beginning to learn something about the way our life was physically embodied in the world relative to our ways of expressing ourselves, and about our expressions relative to our physical embodiment. At the same time, these interactions were implicitly and explicitly guided by a shared context: the language and culture of our significant others.

The growing physical awareness of ourselves led to more and more of our experiences taking on a kind of internal order, one that arranged the experienced sensations around the ever more intentional physical expressions of our self. We can make this somewhat intelligible by assuming that an activity that we have thus far represented as a string of hyphenated words is beginning to take on an order of a physical self expressing itself or responding to something, and in the process encountering a growing variety of entities in our world. In other words, the string of hyphenated words would gradually be differentiated toward a form of subject-verb-indirect object-object-surroundings according to a particular cultural context. As babies, our ability to make a greater mark on this interconnectedness further increased as we learned to sit, crawl, and eventually walk.

It is important to keep in mind what Merleau-Ponty, Hubert Dreyfus, Samuel Todes, and others have so convincingly shown, namely, that the physical embodiment of our life in the world puts its stamp on all our experiences and self-expressions.[31] The way our senses are located in our bodies, the way this favours certain ways of moving about, the way our physical embodiment expresses itself in a gravitational field, and much more – all this makes our lived experience so unique that any forms of deep learning and artificial intelligence simulated on digital computers (which are not embodied) can at best imitate only tiny fragments of human intelligence.[32] Our physical embodiment in the world constitutes our vantage point from which we experience everything and initiate our responses and interventions. Simulating this by universal approaches that have no vantage point of any kind is impossible.

Furthermore, meaning, imagination, and reason also have a bodily basis, as pointed out by Mark Johnson.[33] He argued that without our imagination we cannot make sense of our experiences, find anything meaningful in them, or see any reason to seek a deeper knowledge of our life in the world.[34] In a different context, much the same point has been made by Castoriadis.[35] Like other philosophers, Johnson combats what he refers to as the objectivism common in Western philosophy and in the commonplaces of our Western cultures.[36] Objectivism holds that our world is what it is regardless of what we know or think of it. Consequently, there is a rational structure to this world that we are able to mirror intellectually through reason by means of the rules of logic. It is a kind of mechanistic interpretation by which logical algorithms produce knowledge without any essential dependence on our physical, social, and cultural embodiment in the world. It is the asocial and ahistorical view that requires no life, depending entirely on (dead) rules. However, our life in the world fundamentally depends on the use of metaphors that arise from our imagination and which deeply influence the meaning of our experiences and constrain our inferences. Since everything is related to and evolves in relation to everything else to give human life its relative character, our vantage points constantly evolve and nothing can ever repeat itself in quite the same way. Nothing is objective; everything is relative and open-ended. We are constantly depending on imaginatively extending a previous experience in order to interpret another. It is in our fundamental dependence on our imagination for making sense of and living in the world that we are individually unique and yet culturally typical.

For the same reasons, all human languages and cultures cannot be reduced to a universal set of concepts that mirror the structure of an objective world. Many languages and cultures are incommensurate with each other in a variety of ways. Consequently, the human subject cannot be eliminated; it is the dialectical tension between his or her uniqueness and a cultural unity that is at the core of all human understanding of the past, present, and future. We are, above all, social and historical beings. Also in science, we have discovered that there are no experimental data independent from theory. Knowing more is not necessarily knowing better because of the non-cumulative growth of discipline-based scientific knowledge.[37] In sum: according to Mark Johnson, we depend on embodied and imaginative structures of understanding by which we make our lives in the world intelligible.[38] No objective knowledge can exist independent from our human vantage point. We frequently use our imaginations to project patterns of understanding of one area of human life in the world on another, particularly when this other area is less tangible and more abstract. We thus imaginatively tend to build our understanding from the concrete to the abstract. Consequently, the body is enfolded into the brain-mind and the brain-mind is enfolded (as well as dialectically enfolded) into the body, in part through our understanding of our life in the world. Our brain-mind is internally and externally connected to our body, and vice versa. Following Plato and Aristotle, Western philosophy was unique in its belief that it could call on a transcendent, universal, and autonomous rationality that was completely separate from all experience and culture.

A complementary way of approaching the essential role of our imagination and creativity in (individually and collectively) living our lives in the world builds on what we explained earlier. When babies learn to sit, crawl, and eventually walk, they begin to actively and playfully explore their surroundings. They begin to manipulate objects, to look at them from all angles, to investigate what they can discover by exposing them to various senses, and so on. As they learn something about themselves relative to these objects, and something about these objects relative to themselves, there is a kind of reciprocal interplay until whatever new is being discovered becomes less and less significant. At this point, there is a kind of fit between what they have learned about themselves relative to these objects and what they have learned about the objects relative to themselves. When little new is discovered, their curiosity is no longer rewarded. This appears to be the general pattern underlying the developments described above. During the emerging awareness of their physical embodiment, babies learn something about themselves relative to everyone and everything they encounter, while they learn something about these encounters relative to their physical embodiment until a kind of fit or dynamic equilibrium is achieved between these two linked aspects of their development. At this point, they will have developed a metaconscious awareness of having a good grip on the situation. A body-world distinction is thus imposed, when they fit into their encounters as these fit into their lives. This going back and forth in order to get a grip on everything requires a great deal of imaginative and playful embodiment that is always open-ended and relative until it eventually becomes constrained by a culture and, in the end, by its myths.

As noted, throughout these developments our lives were always working in the background. With the emergence of a metaconscious knowledge of the physical self, the influence of our lives working in the background became increasingly important. As babies, we learned to focus our attention and thus overlook certain things in preference to others. This affected each and every experience by what eventually would be made intelligible through a foreground-background distinction. Our emotions played a very decisive role in these developments as they helped us to respond to the love of others. These significant others helped to draw us into our world and into their lives in a manner that was increasingly full of meaning, purpose, and direction. If we had to randomly explore every possibility, as if we had to invent our own culture, our development would have been greatly retarded, if it could have taken place at all.

The General Relativity of Human Life and the World before Screen-Based Devices

Additional insights into these developments may flow from asking the question: if there is no objective world "out there," and if in our emerging global civilization our relationships with that world are primarily mediated by discipline-based approaches and only secondarily by symbolic and cultural approaches, is there anything we can know about that world that may affect our interpretation? Possibly the only thing we can know about that world is that it appears to be a relative one. Nothing within it can be understood independently from its participation in and dependence on a highly interrelated complexity. For example, what we refer to as an object is the result of our intervening in the way matter participates in the biosphere by means of closed cycles and our tapping into the way energy powers that biosphere through linear chains of successive transformations. Since we can neither create nor destroy the matter and energy on which the manufacture, use, and disposal of an object depends, we rearrange this interconnectedness to produce an object intended to be stable during its use. Following this use, it may be disassembled, with some components and materials retrieved for remanufacturing and reprocessing for reuse. The remainder is left to deteriorate in a landfill by participating in the relationships within that landfill and beyond. The object is also designed to rearrange something in the way of life of a community, for some purpose; but following its manufacture and use these rearrangements often turn out to be different from those anticipated. It is for these kinds of reasons that archaeologists have been able to do their work. The objects they retrieve bear the stamp of a particular time, place, and culture; while today they bear the stamp of the discipline-based approaches brought to bear on them. Consequently, there is nothing objective about an object. It is a temporary rearrangement of the interconnectedness of the world to enable its design, manufacture, use, and disposal. Instead of thinking about objects spatially, it may be more fruitful to think of them relationally and temporally; that is, extending what we can access by listening, language, and culture rather than by seeing. As such, the design, manufacture, use, and disposal of an object represent successive temporal events in what is commonly referred to as its life-cycle, as if objects had lives. This once again highlights our confusion and the lack of appreciation of the differences between the architectures of our physical artifacts and living entities.

In contrast, all plants and trees are living entities resulting from seed-producing pollination and the unfolding of the potential of their DNA. All their cells have the biological whole enfolded into them, with the result that they are internally and externally connected to each other. It is this enfolded relativity that participates in and depends on the relativity of the connectedness of the biosphere. The same is true for all animal life, derived from embryos resulting from sexual reproduction and capable of unfolding the physical and biological potential of the DNA embedded in them. All this life may also be regarded as temporal events of an enfolded relativity interacting with and depending on the sustaining relativity of the connectedness of the biosphere. In these cases, these events can be arranged in a life-cycle.

Humanity shares its physical and biological makeup with all animals as an enfolded relativity of relationships that interact with and depend on the larger relativity of relationships within the biosphere. However, human beings are also radically different in their culture-based artificiality, which has no corresponding niche in a local ecosystem or the biosphere. We become human by listening to language and acquiring a culture. We thus become dialectically enfolded living entities sustained by a biological enfoldedness, and these jointly constitute our relativity.

The relativity of non-living and living entities jointly constitutes the relativity of our world. This relativity cannot be understood in terms of extending the architecture of what is visually apprehended, which has led to the philosophical concept of an objective world. Instead, an emphasis on the relativity of that world extends what has made us human until now by listening to language and acquiring a culture that has dealt with this relativity by *living* it. The universality of moralities and religions (either in traditional or secular forms) may be interpreted as this relativity posing a threat to individual and collective human life. Each and every culture has dealt with it by suspending human life in myths, thereby creating for itself gods of a traditional or secular kind that needed to be served by their members' conformation to a morality and a religion. In other words, the relativity of human life in the world was anchored by symbolizing the unknown and the future as "interpolations" and "extrapolations" of what the members of that culture knew and lived and thus its myths became the only ultimate vantage point and direction for human life in that relativity. Every culture thus closed the door to cultural chaos during the epochs in its historical journey – epochs between which it had to transition from one set of myths to another that was proving itself to be more liveable. Our history may also be interpreted as indicating that culture was able to interact with that relativity in very different ways, thereby interposing a unique symbolic universe between its members and this relativity. The symbolic universe of a traditional society thus represented a *relative* objectivity of a shared body of experience acting as a body of common sense that implied a vantage point and a historical direction. Consequently, from a temporal and historical perspective associated with language and culture, everything may be regarded as an "event" in a natural history interacting with "events" in a cultural human history.

It is difficult to exaggerate the tremendous benefits our emerging global civilization would derive by shifting its prioritizing of extending the architecture of what is visually apprehended over the architecture of what can only be apprehended through language and culture. A shift to a more temporal rather than spatial interpretation would, for example, revolutionize the engineering and management of technology in a beneficial manner. It would then be much easier to appreciate the limits of discipline-based approaches and would encourage universities to research alternatives by which we can transcend these limits. This would undoubtedly create more liveable and sustainable ways of life. I have essentially spent my entire career demonstrating this possibility, as implied in the Introduction.

The difficulties that each and every culture appears to have encountered in confronting a temporal and historical relativity (even the Greeks could not *live* what Socrates, Plato, and Aristotle advocated) have resulted in the relativity of human life in the world being lived with as *sameness*. By symbolizing the unknown as more of what was already known and lived, the past and the future were symbolically dominated by making them essentially the same as the present, which thus became the absolute vantage point and direction. Babies, toddlers, and children thus exhibit a transition: first encountering the relativity of their surroundings by means of playfulness and then gradually transcending those surroundings to enter a symbolic universe, or a world of sameness. Initially, this playfulness is received with a great deal of openness and love and is encouraged by the child's significant others making sense of it as attempts to respond to them and their surroundings. Doing so enhances babies' growing awareness of their physical selves distinct from their surroundings, and with that growing awareness the door opens to non-verbal communication. Given the limited character of this non-verbal communication, the sameness of the culture of the significant others may be expected to only lightly impinge on the playfulness of babies and toddlers. However, with the emergence of language, all this changes, as we will examine in the next chapter.

The interpretation of human history as a rich diversity of responses to the relativity of human life and the world by means of a culture must not be turned into yet another metaphysics. It is impossible to construct a general sociocultural theory of the relativity of human life and the world, since life cannot be represented in the architecture of non-life. All that we can know is that, regardless of whether we mediate our relationships with the relativity of the world by means of symbolic approaches or by means of discipline-based approaches, this mediation is not neutral. One culture, by means of its symbolic approaches, may filter out certain aspects of the relativity of the world, modify others, and arrange them in a manner that is entirely different from the way the symbolic approaches of another culture do so. In the same vein, our discipline-based approaches filter and modify our interactions with the relativity of the world, beginning with their limitations, as outlined in the Introduction. What they all share is the need to close the door to the threat of cultural chaos by symbolically grounding the relativity of human life and the world. In one way or another, all cultures find it necessary to introduce sacred elements that they treat in the same way as were the traditional gods of the past, now replaced by their secular equivalents.

Does this interpretation imply that babies and toddlers may also experience the threat of the relativity of their lives and the world? It would appear highly unlikely because of their limited self-awareness and their equally limited symbolic mapping of their surroundings, both of which are undergoing rapid development, however, as one discovery after another enters their lives. There is always more to come, and this understanding sustains their playful attitude. At the same time, the unfolding of their physical, biological, and symbolic potential occurs within the "social womb" of the cultural community of their significant others. These members have learned to deal with relativity according to their culture, and they begin to transmit some of this every step of the way. The relativity of their life and the world is thus never fully available to babies and toddlers, in part because of their limited development in their social womb.

All this confirms what we know from a historical and sociological perspective. We have all grown up with the moral and religious commitments of our significant others, whose culture has structured our social wombs. In other words, if we had grown up in a different time, place, and culture, we would have very different commitments. To acknowledge this is not a relativization of all moralities and religions. We do not have any conscious access to the traditional or secular myths that symbolically ground these commitments. In North America, this brings us face to face with a common misunderstanding regarding the "cultural melting-pot" approach to building a society from immigrants, or doing so by a multicultural approach. As noted, both the first generation of industrial societies and the second generation of so-called industrially advanced societies that succeeded the first had cultures symbolically grounded in secular myths. Each of them implied a hierarchy between the increasingly universal approaches for building their ways of life and what remained of their symbolic approaches for conducting people's personal lives. As a result, the dominant Christian and Jewish traditions had to be transformed – from being a people charged with a unique presence in the world to something that was mostly individual, personal, and private. This transformation greatly weakened the relevance of these traditions for what was happening to society and, via it, to these faith communities.

From a historical and sociological perspective, children growing up in the Jewish and Christian communities, in their acquiring a culture to deal with the relativity of their life and the world, became first and foremost committed to the secular myths that symbolically grounded their relativity. Many of them began to question and judge their traditions by means of these secular commitments. Generally speaking, this led to two responses. Some young people left, as they could not reconcile the two. Others stayed by essentially learning to split these traditions into a vertical and a horizontal dimension so as to give priority to one or the other. The resulting conservative-liberal split made the coexistence of their secular commitments and these traditions much easier. The secular religious commitments to our contemporary ways of life built up with universal approaches to knowing, doing, and organizing are thus served in public life, and anything symbolic and cultural (and this includes the cultural roots of the community into which they are born) is served in their personal lives. It is thus entirely possible to have a melting-pot approach or a multicultural approach without plunging a society into a relativism, as would have been the case if this had occurred in traditional societies. For example, it was found that during the decades following the Second World War there coexisted Protestant, Catholic, and Jewish embodiments of the American way of life.[39] When, decades later, many young people abandoned these now highly assimilated traditions, their secular religious commitments to their society were not in the least affected.

It is important not to misunderstand this interpretation of what has been happening. It does not confirm that the Jewish and Christian traditions were, like all other moral and religious traditions, created by a culture to deal with the relativity of human life and the world. Nor does it confirm that the people who abandoned these traditions were simply shedding one kind of moral and religious commitment (commonly misinterpreted as a loss of "faith") for the secular commitments of "the world." The Jewish and Christian Bibles exhibit a much more radical approach, as shown by the book of Ecclesiastes, which, surveying the history of the Jewish people and its accomplishments, declares that all is vanity. Surely this is impossible if people's lives were not also suspended in myths.[40] The resulting contradictions in our lives are becoming more than a little embarrassing, especially in politics. As noted, following the great cultural divide, especially during the last century, everything became political because the state needed to intervene in one area of human life after another as a consequence of a weakening culture-based self-regulation, as well as other factors. A limitless politics presupposes an omnipotent state, and such a state in turn presupposes secular religious attitudes toward it. The developments were amply confirmed by the rise of secular political religions in the twentieth century. If organized Christianity had understood this situation, it might have recognized that both the conservative and liberal ways of engaging in politics amounted to the service of the secular successors to the traditional gods and religions of the past. For the reader of the opening chapters of the Jewish Bible around the time these texts were written down and received by the people, the serpent was a common deity in the religions of the surrounding peoples. They would have also known that the service of other gods was forbidden. Religion was thus portrayed as the instrument by which humanity broke with God. Consequently, the engagement of organized Christianity in secular political religions is more than a little problematic, regardless of whether this takes a conservative or liberal form. It points to another major contradiction at the heart of modern Christianity, with implications that are so far-reaching that they have been dealt with in a separate volume.

The playful attitude babies, toddlers, and young children have toward the world may now be interpreted as a reciprocal interaction between the relativity of their lives and that of their world. It requires a symbolic approach that attempts to make sense of a situation by metaconsciously differentiating it from all prior symbolized experiences, retaining the one that appears to be a suitable paradigm, exploring the situation accordingly, observing what happens, and repeating the cycle in a playful manner until a good fit is achieved because nothing much new can be discovered. There is thus a playful adaptation of their relativity to the world somewhat guided by the non-relativity of others. They gradually learn to deliberately direct their attention toward certain aspects while glossing over others, while a great deal of the interpretation and response is left to routine made possible by their lives working symbolically in the background. All such developments are self-reinforcing due to their reciprocal character: as babies and toddlers change a situation by their presence, the situation simultaneously changes them. The organization of their brain-minds gradually begins to function as symbolic mental maps. However, this may well change later on depending on the kind of society in which they grow up. The organization of the brain-mind functioning as a symbolic mental map began to break down under the pressures of industrialization, urbanization, and desymbolization, eventually weakening it to the point that only its most stable and thus deepest metaconscious knowledge could act as a kind of compass or gyroscope.[41] This situation could not be sustained for many generations, resulting in another mutation of the organization of the brain-mind. It began to function as a kind of radar in the lives of the members of the so-called industrially advanced mass societies.[42] Further significant changes continue to occur under the ongoing pressure of desymbolization, as we will examine in some detail in later chapters.

When we critically analyse what babies and toddlers are able to accomplish as they learn to make some sense of and live in the world as members of a symbolic species, a significant role must be ascribed to the enfolded as well as the dialectically enfolded character of the symbolic processes involved. There are no simple causal relationships, since the organization of the brain-mind is internally and externally connected to the body (in part through a growing metaconscious knowledge of a physical self), and to the external world. The mind is enfolded and dialectically enfolded into the brain; the body is enfolded and dialectically enfolded into the organization of the brain-mind; this organization dialectically enfolds the life and self of a person; and these in turn dialectically enfold all the experiences of a person's encounters with others and the world.

This complexity may be illustrated by our examining the "structure of experience" of a person's life. Each experience symbolically integrates the five dimensions (corresponding to the five senses) directed toward the external world as well as the dimensions corresponding to what can be detected within the body and the organization of the brain-mind. Depending on how our attention is directed, each experience symbolizes what is directly relevant to it in a foreground, and everything else in a background. This distinction similarly organizes what all the dimensions of experience contribute to the experience as a whole by organizing all the lived sensations into a corresponding Gestalt that interprets all these sensations within a dimension of experience. Each lived sensation reflects a set of stimuli that has taken on a meaning in a person's life, to be distinguished from all the sensory "noise" that cannot as yet take on a meaning in that life.[43]

Before a new experience was symbolized to allow it to take its place in this structure of experience, the situation it symbolizes was first anticipated by means of our life symbolically working in the background. It prepared us for what would routinely happen. We took in the situation at a glance based on this anticipation and immediately began to become aware of what was not routine or for whatever reason attracted our attention. We then became fully involved in the situation, thus imposing a foreground-background distinction on it according to our focus of attention, leaving a great deal to be dealt with as a matter of routine by symbolically differentiating it from all previous symbolized experiences and thus giving it a place in our life by integrating it into the structure of experience. Consequently, insofar as our life is working in the background, the new experience dialectically enfolds our life. As we direct our attention in a certain way, something non-routine as to our person and our life is dialectically enfolded as well. It is not a question of a passive "information processing" but an interactive living of a situation. The resulting neural and synaptic changes to the organization of our brain-mind simultaneously modify our life in response to that situation, with the result that, next time we approach a similar situation, we do it in a manner that includes the way our person and our life have adapted and evolved. Our involvement in a situation is such that we dialectically enfold our experience of it by means of symbolization, differentiation, and integration relative to all prior experiences.[44]

The complexity of our adult behaviour owes a great deal to delegating as much as possible to our lives routinely working in the background, thereby leaving our conscious attention to be directed according to our interests, concerns, hopes, fears, or any other expression of ourselves. There can be no symbolic mental map, compass, or radar without our physical, social, and cultural self enfolded in the organization of the brain-mind acting as its reader and user for finding our way in the world. Any attempt at removing our involvement in a moment of our life results in defence mechanisms such as getting drowsy or daydreaming, as was demonstrated in sensory deprivation experiments or in the performance of highly repetitive work. Even our most routine activities can be interrupted at any time, with the result that it is difficult to imagine anything being replicated by means of algorithms or neural nets. These always translate any human activity from the architecture of human life in the world into an architecture of non-life, following which it can be simulated on a machine. However, we must not imagine that this has anything in common with how we live it.

These reflections on our adult abilities to make sense of and respond to a situation provide us with another window into the way babies and toddlers begin to embryonically learn by means of symbolization, differentiation, and integration sustained by the organizations of their brain-minds. However, care must be taken that with a great deal of hindsight we do not reconstruct what is happening in the lives of babies and toddlers by means of a different architecture. For example, their development must not be reinterpreted in terms of stages in the development of logical thinking. Similarly, language acquisition must not be reduced to its rational dimension. This may be illustrated with Ludwig Wittgenstein's reflections on what the entities in our world have in common.[45] Our use of language shows that we do not group them into tables, chairs, beds, cabinets, and the like according to their sharing a set of characteristics, but according to a network of crisscrossing and overlapping family resemblances. In the case of tables, some may share a round, oval, square, or rectangular top. Some of these, but not necessarily all of them, may also share having legs, but others may share a pedestal base or sawhorses. They may not all share the same height, as in the case of dining room tables, coffee tables, and end tables. Consequently, babies and children cannot be discovering a list of the characteristics of tables. What they are learning is to live with tables according to their meaning and value for a variety of activities and how these activities involve tables. These activities may be differentiated from others that do not involve tables, and so on. Children learn to live with tables as entities that bring into easy reach their food, for example, and everything they need for eating it in accordance with the customs of their culture. They will also be watching how adults act around tables when, for example, they are talking to each other in the living room with drinks and snacks within easy reach thanks to coffee tables and end tables. Similarly, they may watch their parents do some office work at the dining room table and notice that everything they use is within easy reach. Reconstructing all this by means of a theory of language unavoidably involves the objectivization and reification of what babies, toddlers, and children experience in an after-the-fact manner based on a different architecture: an architecture of enfolding and dialectical enfolding encountered only in lived activities. It permits levels of integrality and integrity in our symbolic development that would be unimaginable in an architecture of non-life.

We must not imagine, however, that babies and toddlers repress all their early experiences. In the days of the emergence of psychoanalysis, the cultures of West European societies were still working largely unnoticed in the background, despite significant desymbolization producing a variety of psychic difficulties, some of which have led to new forms of mental illness.[46] Consequently, it was as yet unthinkable that this inability to remember our earliest experiences was the result, not of repressing them, but of our inability to remember them since it is impossible to make our life intelligible without our having an awareness of our physical self distinct from everything else. We cannot relive experiences without an awareness of a self. For the same reason, it is not necessary to hypothesize that our episodic memory develops at a later stage.[47] Our memories are an integral dimension of our being sustained by the organization of the brain-mind, where the symbolization of all experiences has produced neural and synaptic modifications to this organization, thus interacting with all other "memories." When the organization of the brain-mind undergoes major changes as a result of the emergence of a body-world distinction and an accompanying awareness of being a physical self, older "memories" cannot be relived and thus remembered. We must not confuse our memories with computer memories, since they have entirely incompatible architectures. For example, our earlier memories can be affected by later ones, which is impossible in a computer memory.

Similarly, the growing abilities of babies and children to make sense of and live in the world are also deeply dependent on their emotional development. This development in turn is closely associated with that of subconscious and metaconscious knowledge. All too soon, this will also be deeply influenced by toys such as dolls with "artificial intelligence" and screen-based devices that populate our world with life reconstructed in the architecture of non-life.[48]

Possibly one of the greatest obstacles standing in the way of appreciating the development of babies and children is our lack of appreciation of how deeply and completely we *are* our bodies in our daily life activities. I will simply recall a few activities that helped me become aware of this and trust my readers can respond with their own. Eating a "simple" French breakfast in a small family-run hotel in downtown Bordeaux with fresh baguette, croissant, butter, compote, and French roast coffee made me mindful of a long tradition that has perfected this food and drink – something that simply cannot be mechanized, automated, and computerized, as any comparison makes plain. The bodies of the craftspeople who developed this tradition were essential in all this, as is my own body's capability of appreciating and enjoying it. The pure physicality of sailing a small boat into a very strong wind with the spray flying in all directions is surely exhilarating to many of us. So is the great calm of paddling a canoe on a northern lake with not a human sound to be heard. Many of the things we love most are testimonies to our bodies, both in producing the experience and in receiving it: the music produced by a violin virtuoso, the insights into human life by our greatest writers, the works of great painters, the "professing" of a great intellectual tradition by a master teacher (refusing to use powerpoint and the like), and the great many daily-life skills carried out with ease and perfection. All of these and more involve our bodies, into which our brains are dialectically enfolded and thus our lives and our selves. They are symbolically represented in the organizations of our brain-minds, making us one seamless whole as members of a symbolic species. Nevertheless, our contemporary ways of life have reduced our awareness of our bodies by minimally involving them as we do more and more things on screens with little embodiment, participation, commitment, and freedom.[49] It raises many questions about life's possibly being much more enjoyable if there was a greater physicality in our experiences. Yes, despite Hollywood's aberrations of it, when we give ourselves to our favourite person in life and make passionate love, we surely cannot imagine it without the complete and total embodiment of this love. We are losing our ability to distinguish this total giving in love from our being seduced by someone or something into a situation that resembles it but that always, in one way or another, limits the participation of our entire being.

2 Our Social and Cultural Embodiment in the Relativity of Human Life in the World

A Hidden Discontinuity

Implicit in the previous chapter is the existence of a hidden discontinuity that babies and toddlers begin to transcend as a result of their acquiring a language. It is the gateway into the culture of the world of their significant others. The surroundings that babies and toddlers playfully and relationally explore do not expand into the symbolic universe of a traditional society, nor into the reality of a contemporary society belonging to our global civilization. Children's initial surroundings constitute an open-ended relativity enabling a playful interaction under the watchful eye of a parent; while living in a symbolic universe or a reality is marked by a conformity to a sameness that has made their relativity liveable by shielding it from cultural chaos.

This discontinuity remains hidden for some time as toddlers continue their playful interaction with the relativity of their surroundings. Their non-verbal communication is enhanced as they learn some vocal signs, and as these vocal signs gradually mutate into the language of their community. During this transition, the significance of anything in their lives reflects how that something is lived relative to everything else, and vice versa. The acquisition of vocal signs changes nothing essential in this regard. However, with the beginning of language acquisition, everything gradually changes. The acquired metaconscious knowledge increases in depth and thus tends either toward the myths of a traditional society or toward the secular myths of a contemporary society. Nevertheless, it is not until the teenage years that this metaconscious knowledge begins to approach the deepest level, on which the myths of a society can form.

In order to explore this discontinuity, we will return to what lies behind the universality of moralities and religions developed, without exception, by all nomadic groups and settled societies throughout human history. As noted, this universality can be explained by the necessity to transform children's playful interaction with an open-ended relativity into a moral and religious life, whose past, present, and future are constituted by the sameness established by symbolizing the unknown as more of what a community knows and lives. Doing so transforms what was merely a body of shared experiences into an absolutized body of common sense. It implies a shared vantage point and direction for the community's historical journey. The door to cultural chaos is thus temporarily closed until many new experiences begin to undermine the established sameness, eventually forming a different body of experience that begins to make more sense and thus better sustains the individual and collective life of the community. No nomadic group or settled society has been able to ignore the necessity to close the door to the threat posed by this relativity. Of course, no nomadic group or settled society is able to experience this transformation, since everything is symbolically mediated as belonging to the one and only possible order by which it lives. The only way a community can make itself aware of having symbolically grounded the relativity it has experienced, and continues to experience, is by means of an intellectual comparative study of a range of cultures, including its own. Such a study needs to pay particular attention to those entities in any one of these cultures that its members experience as being limitless, all-powerful, or self-evident, which are the characteristics of the traditional gods and their secular successors. These characteristics are possible in a general relativity of human life and the world only by bestowing traditional or religious attitudes on these entities. Such attitudes are always associated with the deepest metaconscious knowledge of the members of a culture because they are unable to go beyond it during an epoch in their history. This situation is made liveable by absolutizing it on its own terms, which is achieved by what are referred to as myths. It is only under these conditions that the meanings and values of any entities can appear as evident in themselves and thus as absolute.

Although such comparative studies of cultures are intellectually possible, they are practically extremely difficult for the members of a culture whose myths are being examined. It will inevitably open the doors to higher levels of relativism, nihilism, and anomie in their lives. It is much less threatening, and thus far more common, to simply declare that our societies have become secular. Nevertheless, it remains possible for the members of a culture to follow their conviction that life must be liberated from its alienation by these myths. This would lead them to resymbolize these myths by refusing their absolute character as best as they can in their daily-life relationships. Doing so would require a non-religious basis for their lives. Once again, this has such far-reaching implications that their analysis has constituted a separate work.

In other words, the myths of a nomadic group or a society have nothing in common with a worldview, belief system, ideology, theory, or philosophy of life. We all implicitly acquire the myths of our society by growing up in it as our symbolized experiences build the organization of our brain-mind and all the metaconscious knowledge implied in it. As for our metaconscious knowledge, we can only live our myths and embody them in our thoughts and actions in ways of which we have no awareness. When we do become aware of these myths, they have already been replaced by others, which makes it possible for the previous myths to be intuited and worked out into a mythology.

The universality of moralities and religions in all traditional and secular cultures is a necessary but insufficient response to an open-ended relativity. It is for this reason that Jacques Ellul argued that the universality of legal institutions appears to correspond to a necessity imposed on all individual and collective human life. Unfortunately, he only published a few aspects of this historical and cultural-anthropological approach to the universality of legal institutions, but he gave me permission to summarize his doctoral-level course on this subject in an earlier work.[1] Complementing his explanation with that of the universality of moral and religious institutions as well as the common occurrence of magic, it can be argued that moral and religious institutions were lived relative to legal ones, and vice versa; and that at the same time, each of these was lived relative to magic, and vice versa.[2] Confronting the open-ended relativity of human life and the world needed to be complemented by confronting the experience of the open-ended relativity of time, space, the land, and the social in human life. Since it is a matter of how a nomadic group or settled society lived this, it can be explained only by examples and not by theories. Depending on the primary life-milieu in relation to which individual and collective human life evolved, all this took on many different forms. We will therefore need to make a brief intellectual detour in order to better understand our present situation and how it affects the socialization of children.

The Artificiality of a Culture

Since humanity is a symbolic species, every food-gathering and hunting group and society has established an artificial cultural order within a natural order, thus embarking it on a cultural rather than a natural history.[3] At the centre of this cultural order are the traditional or secular myths that symbolically ground it. This suspension of individual and collective human life in myths could not directly rule all of daily life. The sameness established by absolutizing a shared body of experience would be unliveable in relation to the unrestrained turmoil and unpredictability in the daily life of a people, and vice versa. Consequently, every group and society has experienced the necessity of setting up a system of juridically founded institutions. These essentially create an order of artificial relationships, as if nothing changes, and sanctions for those who transgress it. For example: during a marriage relationship, both parties change in unpredictable ways; but once they enter an institution of marriage, they know what they may expect from one another and what their community or society expects from them. Similarly, several parties can enter into a legal contract that protects each one from the chance that the others do not meet their obligations on the grounds, for example, that circumstances have changed their lives or that their social situation has been disrupted by developments in their community. Legal institutions thus permit relationships to take on a measure of predictability in the face of time by preventing their domination by circumstances. The laws limit how legally bound relationships can evolve, thus making the future much more predictable, and sanction those who disturb this artificially imposed order by causing a divorce, breaking a contract, and so on.

In the same way, legal institutions stabilize relationships in space. A marriage or a business contract is valid within the territory over which a community has established its cultural order, and thus cannot be violated without sanctions simply by one party's moving to another location within the territory. In our present situation, state-nations exercise sovereignty over a national territory within which a legal system orders all relationships essential to its way of life. It furnishes a system of models of behaviour that permits everyone to know what they can expect from all other members of the society, and what other members can expect in return. In this way, a legal system permits a society to master time and space. It also orders and makes more predictable its relationships with local ecosystems and the biosphere in a variety of ways, including property rights.

The particularities of how the legal system of a society helps to mediate the relationship between the sameness of a cultural order and the open relativity of the land (the life-sustaining capacities of ecosystems and the biosphere) vary greatly with the life-milieu in relation to which human life is lived and evolved. The ways of life based on food gathering and hunting adopted by nomadic groups and totemic societies are an example of human life immersed in a natural life-milieu. There was a complete and total dependence on that life-milieu for everything necessary to sustain human life, and thus all the most important threats came from the withdrawal of its support as a result of droughts, floods, hurricanes, earthquakes, and other such disturbances. The ultimate power of life and death came from this life-milieu, which people sacralized by suspending the life of the group or society in myths that symbolically grounded it in the natural life-milieu. The lives of its members expressed this through a natural morality, religion, magic, and related institutions.

When we speak of a natural life-milieu, implying a concept of nature, we are attempting to make the situation of these groups and societies intelligible from our vantage point, but it may not be concluded that they had such a concept. On the contrary, the considerable literature on the cultures of these groups and societies can perhaps be generalized by their not metaconsciously differentiating between the enfolded architecture of all natural life and the dialectically enfolded architecture of their life as a symbolic species. This is confirmed by the fact that these groups and societies symbolized themselves and their natural surroundings as one vast family of life without any essential differences. Since the human groups or societies within it dominated everything through a language and culture, their natural myths extended the symbolized sameness to everything around them. It was entirely self-evident that every element in their life-milieu was open to human communication by virtue of possessing a spirit or a soul. The thoughts and actions that embodied a natural sacred and sustaining myths are commonly referred to as animism.

Human life symbolically grounded in natural myths thus inadvertently opened the doors wide to experiencing the full relativity of the family of life of which it was a part. The concept of a natural order and natural laws or science as we know it were completely unthinkable and unimaginable. Nothing could be counted on, since any or all spirits could change their routine for any reason whatsoever. The sun or the moon might not return each and every day and night. The seasons could not be counted on. The game on which a community depended could change its migration patterns at any time. The food and medicine they derived from their surroundings might not have the same nourishing effects each time. Anything and everything depended on the goodwill and cooperation of the spirits or souls that exercised their jurisdictions; and this goodwill and cooperation had to be ensured by juridical-religious-magical rituals.

For example: no matter how skilled a hunting party might be, they would not succeed in returning with some game to feed their people if the spirits set over these animals had not given their consent. In order to obtain it for a particular time within the territory over which they exercised jurisdiction, the equivalent of a legally binding agreement had to be established. This was done by enacting a juridical-religious-magical ritual that would bind all parties within the community, promising moral and religious obedience possibly accompanied by a sacrificial offering. Even then, the outcome could not be fully assured. Possible explanations were plentiful: Was the ritual meticulously carried out in all its details? Did a member of the group or society anger the spirits by committing an immoral or forbidden act? Did another tribe engage in counter-magic? Did the hunters stray into the territory over which unknown spirits had authority, without appeasing them by an appropriate magic? Contrary to what is usually believed, a legal notion of private property was first developed to protect a group or society from the possibility that the spirits that ruled their surroundings would reappropriate what was theirs. Nothing could be borrowed without their consent, and game could be killed only to sustain human life – and even then only with their consent. This legally assured cooperation was necessary for all relationships between a group or society and the family of life to which it belonged.

We are thus very far from the assumption that the food-gathering and hunting groups and totemic societies simply imitated nature. They symbolically imposed something human on it. They sought to dominate the relativity of their lives by symbolizing everything relative to human life, and vice versa, in a manner that endowed all entities with spirits or souls. The result was natural moralities, religions, magical practices, and natural taxonomies of all kinds,[4] but this sprang from the kinds of myths necessary to symbolize the unknown as more of what was known and lived, to create an absolute body of common sense that implied an absolute vantage point and historical direction. In any case, the natural milieu of these groups and societies could not possibly have issued a command to imitate and obey it – it was ultimately a human initiative.

With the emergence of ways of life that depended increasingly on agriculture, civilizations began to emerge, sometimes in the most inhospitable areas such as river deltas. The cooperation of a great many groups or small societies had to be arranged to create, use, and maintain the irrigation systems required for agriculture in such areas. It marked the beginning of an entirely new life-milieu: a society that was large enough to interpose itself between the group (or small society) on the one hand and the natural life-milieu on the other hand. Thus society gradually became the primary life-milieu in relation to which human life was lived and evolved, and the natural life-milieu became the secondary one. These cultures symbolically grounded themselves in a sacred and set of myths that blended a social and a natural sacred with similar sustaining myths, depending on the historical situation. For example, in ancient Egypt there was a coexistence of a divine Pharaoh and sacred natural elements symbolizing the life-sustaining powers of the Nile.

Turning our attention to Western civilization, which gave birth to our present global one, we have already noted that it was founded on the three "perfections": Greek philosophy, Roman law, and Christian revelation. When Christianity was declared the official religion of the Roman Empire, the early church was swamped with new members.[5] It became impossible to cope without ever larger religious organizations, out of which grew the Roman Catholic church headed by a pope regarded as Christ's representative on earth. Its divine authority could thus not be challenged. It socially and historically became the anchoring sacred of the cultural order referred to as Christendom that ruled over West European medieval societies.[6] Christianity thus began to serve the moral and religious necessities of these societies, leading to what some contemporary Christians (with a great deal of hindsight) would regard as a mixture of religion and revelation, or religion and faith.[7] Christian theology acquired a universal character thanks to Greek philosophy; whether this was more Platonic or Aristotelian made no essential difference whatsoever. The faculty of theology became the central faculty of the medieval university as the one charged with applying God's revelation to understanding human life in a creation. However, this creation had become regarded as containing both material and spiritual elements, the latter including a Christianized version of the Greek soul.

The interpretation of humanity living as a creature in a creation meant that a distinct and separate environment or biosphere "out there" was unthinkable and unimaginable at the time. The symbolic differentiation of a humanity separate from the creation came about as a consequence of the failure of Christian philosophy, theology, alchemy, anatomy, and other inquiries to locate the boundaries between the material and the spiritual. These results helped to pave the way for the emergence of a mechanistic worldview that symbolized the creation in terms of the most perfect material object known at that time, namely, the clock. It was thus imagined that the Creator had put this clockwork together and then allowed it to tick along according to its "laws." Possibly the earliest portrayal of this may be found in the painting of the Mona Lisa. It appears to symbolize an "inner world" behind her eyes and an "outer world" composed of the first real landscape behind her.[8] Eventually, when Darwin's theory of evolution explained how the clock may have assembled itself, the "Great Clockmaker" was no longer required.

In the course of these developments there were numerous conflicts with the sacred authority. For example: Could astronomy engage in empirical observations and mathematical representations that were autonomous from the Roman Catholic church's sacred knowledge of the creation? All this must be interpreted as the medieval societies' symbolical grounding of themselves in myths that were now primarily related to the life-milieu of society and secondarily to the life-milieu of creation (in the process of becoming nature). It turned Christianity into a morality and religion for these societies.

In more complex societies with a greater social division of labour, the functioning of legal institutions tended to become more difficult. Generally speaking, the metaconscious values implied in all human activities are related in the organizations of the brain-minds of the members of any society, with the highest metaconscious value corresponding to the sacred or central myth of that society. Consequently, most individual members lived almost spontaneously in accordance with these metaconscious values.[9] Lawgivers took advantage of this situation to embody them in the laws they created in a manner that intuitively "stretched" them in a desired direction. However, there were strict limits to how far such metaconscious values could be stretched through the legal system. If lawgivers stretched the metaconscious values too far, the corresponding laws might not be spontaneously obeyed, which would make them unenforceable except by the application of external force and oppression. All this presupposed that the socialization of most people did not occur too far on the fringes of their society or was not undermined by serious difficulties, otherwise deviant behaviour and juvenile delinquence could be the result. It must also be remembered that, in many traditional societies, the end of the process of socialization was marked by a ritual of initiation. Through it, the youngsters became new persons as it were, equipped with the "human nature" of their culture. They were now able to accept full membership in their society.

Individual diversity, even when it remains within the cultural unity of a society, can produce social tensions that must be addressed in the course of maintaining the cultural order a society has established within a natural order.[10] Hence, the question imposes itself as to how well the organizations of the brain-minds of the members acting as mental maps are able to mediate all relationships in accordance with the cultural order. People can always convince themselves that because of a non-normal component in a situation, a non-normal response to the situation is called for that does not reflect the appropriate metaconscious values. In the case of a business transaction, no matter what has been explicitly agreed upon, there always remains an implicit part to the agreement; this can cause difficulties when it is used by one of the parties to justify non-compliance. The other party may feel cheated, since his or her understanding of the agreement was different. Both members may feel they are acting in good faith according to their interpretation of the explicit and implicit aspects of the agreement. Such conflicts are difficult to resolve, and when they become widespread, the cultural order of a society may be threatened.

There are also non-normal situations in the sense that there are no corresponding metaconscious values for dealing with them. It is then necessary to consciously and metaconsciously apply a number of values none of which has a clear priority. Contradictory interpretations of the situation can give rise to different members engaging in incompatible courses of action, which could lead to conflicts. The outcome may be significantly influenced by the way power is mediated in the society. If these kinds of conflicts become common, they can weaken the cultural order of a society. If this threat is perceived, it may lead to changes being imposed by means of power. When this happens, divisions within the body social will grow, as will the distance between the organizations of the brain-minds of the members acting as symbolic mental maps, thus altering the fabric of daily-life relationships. The usual mediation of these conflicts by means of language may no longer be able to reduce them to a level of disorganization necessary for cultural evolution. Over time, these issues may be aggravated when the deepest levels of metaconscious knowledge evolve more slowly than the metaconscious knowledge implied in how a conflict is dealt with on a daily-life basis. The exercise of power can greatly intensify these dislocations within the cultural order. A legal intervention may be required as an attempt to reorganize the artificiality of the cultural order right down to the level of daily experience. Once again, the metaconscious values of the community may have to be "stretched" in a somewhat different fashion to reduce the threat to the cultural order. Doing so successfully by legal means may require an intuitive grasp of the particular implications of the sacred and sustaining myths without excessively favouring the most powerful sectors of a society. Such an intervention may then be widely recognized as just and appropriate for what ails the body social. If this fails, lawgivers will have no choice but to lean on their authority, rooted in the way the legal system is suspended within the myths of the society. They can warn the people that disobeying the law can anger the gods. As long as the legal system is an extension of the way morality and religion mediate the distance between the symbolic grounding of the cultural order in myths and daily life, people may accept its authority.

At the beginning of a historical epoch of a society, the distance between the newly emerged myths and the daily lives of the members is generally small enough to be bridged by the culture's morality and religion. As the society evolves and individual differences and conflicts grow, the religious mediation tends to increasingly require a complementary legal mediation. However, when this legal mediation reaches its limits, religious authority may have to intervene, since it represents the highest values in a society. Alternatively, a political authority may intervene, but doing so requires a sanctioning by the religious authority. There is thus a tendency for the legal system to become more complex as the society evolves – as it deals with conflicts and difficulties in the scientific, technological, economic, social, political, and other dimensions of mediation, which jointly mediate all relationships within a society and between it and other societies as well as with local ecosystems.[11]

Generally speaking, traditional societies depended on common-law systems that were embedded in and depended on a culture in general, and on its metaconscious values embedded in the organizations of the brain-minds of the people in particular. These systems helped to establish an institutional framework that mediated between the symbolic grounding of the community in its myths and the daily-life experiences of controlling the relativity of human life.

As noted in the Introduction, contemporary societies that help build our global civilization now mediate all the relationships that constitute their ways of life by means of discipline-based approaches. Consequently, this technical mediation created a new primary life-milieu that interposed itself between us and the secondary life-milieu (what remains of a society) and the tertiary life-milieu (an extensively modified biosphere). The new primary life-milieu is organized as a technical order that has nothing in common with the cultural orders that preceded it. Cultural orders were built up with ways of living in the world that made sense to a community and sustained its life by being symbolically grounded in myths. Hence, the "sense" of a symbolic universe alienated human life while preventing it from being plunged into cultural chaos. Individual and collective human life was thus supported within an order of lived sense.

In contrast, the technical order that humanity has been building for some 250 years is an order of non-sense.[12] As briefly sketched in the Introduction, the technical order first emerged out of the technology-based connectedness of the way of life of a society that could not be organized and evolved by means of a culture, and thus by what made sense to people. An alternative technological and economic approach was developed within industry and the economy, causing them to be split off from the remainder of society, which, for a time, continued to be made sense of and evolved by means of culture-based approaches.[13] This course of events uncovered the limits of culture-based approaches, and these were eventually overcome by discipline-based ones.[14] These approaches rapidly spread beyond the economy to take charge of organizing and evolving the society's entire way of life. The technology-based connectedness eventually mutated and expanded into a technique-based connectedness of a technical order of non-sense. As noted, disciplined-based approaches make no reference to sense, thus building a technical order outside of the domain of sense that had constituted the cultural order of society. What remained of it were only those elements that affected people's lives outside their participation in their society's way of life. New secular myths integrated everything in a hierarchy that banished sense from the mainstream of society and delegated it only to people's personal lives.

The organization and adaptation of the technical order by means of discipline-based approaches is primarily guided by decision criteria based on output-input ratios (such as efficiency). As noted, everything is thus improved on its own terms without any consideration being given to how its compatibility with human life, society, and the biosphere is affected. Such considerations are introduced only when legally imposed, and even then only in an end-of-pipe manner.[15] Consequently, the technical order is built up by people who have learned to behave on its terms. Within a cultural order, such behaviour would be regarded as psychopathic and sociopathic, but within the technical order all this has been normalized by our secular myths.

The consequences are far-reaching. All the institutions that participated in the organization and evolution of the technical order are incapable of acting first and foremost in the public interest. Any notion of a common good appears to have vanished. Our politicians speak only of moving forward. In the most fundamental sense, it has made all political differences between the Left and the Right completely irrelevant, even though the common good is assumed to be a structural characteristic of what primarily orders our lives. It has created a political illusion.[16] Of course, I am not suggesting that everyone's participation contributes equally to the psychopathic and sociopathic character of the universal technical order. On the contrary, it is clear that the chief executive officers of large corporations have a great deal more influence than most other people: but if they stopped this kind of behaviour they would soon be fired by their boards of directors. When politically left-leaning people launch accusations against an increasingly wealthy and very small elite, they endow these people with more power over the system than they could possibly exercise within their technical roles dictated by the necessities of the system. Similarly, when the Right accuses the Left, the unions, the environmentalists, or any group concerned with peace and justice of undermining and threatening the economy, they also show that they have understood nothing of the system that constrains all our behaviour and which is legitimated by our secular myths. Politics has become a tragic tale of division, polarization, and self-justification that entirely ignores our living within a universal technical order. We continue to behave as if we live in societies that differ from their precursors only in terms of enjoying much greater scientific, technical, and economic fruits. We forget that as we transform the technical order, it simultaneously transforms everything through its vast influence on human life, society, and the biosphere. This includes the neural and synaptic changes to the organizations of our brain-minds stemming from our participation in and experiencing of the technical order.

Included in these changes were new roles for the state, legal institutions, culture, and much else.[17] As the cultures of the societies actively participating in building the new global technical order were extensively desymbolized in the process, the traditional high levels of self-regulation of the communities that used to live within a cultural order were all but destroyed. Everywhere the state had to intervene and compensate for this loss by dealing with the endless necessity of reorganizing one aspect of human life and society after another. As a result, everything became political because there no longer appeared to be any limits to the role of the state in human life. The new political authority of the state was established over the territory of a nation (or nations) at the expense of all traditional, moral, and religious authorities. This apparently unlimited political authority (the "state-nation") was absolutized by symbolically grounding it as a secular sacred or central myth.[18]

In the same vein, techniques made up from discipline-based approaches were increasingly used everywhere as if they had no limits. Moreover, the improvements kept coming and there appeared to be no limits in sight. This technical development was symbolically grounded in the secular sacred or central myth of technique.[19]

As a consequence of these and other developments, the legal system now had to extend the newly established technical and political sameness of the technical order of non-sense to what remained of sense in people's personal lives, their families, and the remnants of larger social entities. The state had no choice but to intervene in all common-law systems, which were swamped with entirely new situations for which there were no precedents. All legal traditions would have broken down without such interventions. Moreover, the state now had to use the legal system to co-organize, along with technique, the order of non-sense with reference to what still made sense for most people.[20]

The role of the legal system thus gradually became the primary organizational means used by the state as an overriding necessity to fit a society into the emerging global technical order. We have grown so accustomed to these developments that we live with them as "naturally" as with everything else in our world. We take it for granted that lawyers can draw up papers in a secular religious-juridical and magical ritual that creates out of nowhere a university, a corporation, and a great deal else without providing the necessary elements for the functioning of these newly created entities. For example, a university can be created before it has obtained any funds, purchased any buildings, hired faculty, engaged administrative staff, admitted students, or done anything else that universities do. Similarly, a university can let all its people go, stop admitting students, sell all its physical assets, and still exist as a university.

In the case of a corporation, the secular religious-juridical magic went even further. It simply required a judge to declare a corporation to be a legal person, and it was so.[21] This completely upset what happened in the past when a charter was granted on condition that the organization serve a purpose in the public interest: if it failed to do so, the charter could be revoked. To turn a piece of paper into a legal person is a feat that has changed our world forever. Despite the vast implications, no government put it to the vote, no commission investigated all the implications as a basis for an informed policy, no consultations of the public took place – and yet there were no protests and no angry people in the streets. Nevertheless, it has made all of us second-class citizens in our democracies. The new "legal persons" had orders of magnitude more power and influence than any citizen, and thus were directly consulted by the state and had the means to have it do their bidding. The influence of everyone else was so dwarfed that democracy has become a shadow of its (admittedly not perfect) former self.

Within the economic and financial spheres, similar secular religious-juridical magical feats were accomplished. For example, banks were provided with the right essentially equivalent to creating money: they were only required to retain a strong percentage of what their clients deposited, and could utilize the remainder to make loans and even speculate in global markets.[22] Consequently, when depositors lost confidence in the banks and the majority of them wished to withdraw their deposits, the banks failed. This could trigger other failures and cause larger monetary crises.[23] A very simple cure for this juridical-magical creation of fiat money has been proposed by a number of renowned economists, but thus far no one is willing to protect the common good against it.

There are also many contemporary financial techniques that create new instruments of the kind that produced the great crises of 2008.[24] The entire global economy has been dwarfed by a new kind of economic activity capable of "making money from money" without any intermediary traditional economic activities producing goods and services. It is a highly technical form of speculation involving sophisticated computer algorithms that sense "ripples" in markets in order to "bet" against them, which, according to the calculations of one banker, accounted for nearly 97 per cent of all global financial flows.[25] There was a possible solution: a Tobin tax was proposed to counteract this creation of money from money to protect the common good, but again no action was taken.[26]

In North America we have another kind of juridical-magical creation referred to as pension funds. These belong to the people contributing to them – until a corporation goes bankrupt. At this point, it usually becomes evident that the assets of such funds, although belonging to the employees, have been used as if they were owned by the corporation. Again, there is a solution: protecting society from this kind of secular magic is simple and is practised in many more civilized countries than our own: the pension funds should be put in trust for the employees who own them, and employers should be barred from using these assets as if they belonged to them. Despite some high-profile scandals in Canada, the government, not surprisingly, does the bidding of corporate legal "persons."

These are but a few of the extensive secular religious-juridical magical practices that affect all our lives in ways that can at any time pull the rug out from underneath us as employees, bank clients, investors, pensioners, and others. All this secular religious-juridical magic can be put to the test at any time, although such an experiment would immediately plunge us into a crisis that might require decades of recovery. If tomorrow we could persuade a significant proportion of the population of our society to withdraw all their deposits (assuming they have any) from their banks, there would be instant and massive bank failures and the entire financial system would collapse. The deposit insurance of our governments would not be able to cope, which is why, once again, governments back this magic: banks are "too large to fail."

Most of the consequences of these secular religious-juridical magical feats affect us in more subtle ways. For example: when public chartered accountants calculate the costs communities incur as a result of having corporations in their midst, the answers show that the costs exceed their profits.[27] As a result, these corporations are wealth extractors; much of modern competition amounts to a race to find ways and places to externalize as many costs as possible. The psychopathic and sociopathic character of these corporations and the techniques they embody have made national economies wealth extractors as well. We are hypnotized by indicators such as the GDP, which entirely masks the fact that, when we calculate net wealth creation by subtracting the costs, we observe that most segments of society experience this wealth extraction from their lives as declining levels of purchasing power, social security, and healthcare and retirement benefits.[28] Herman Daly has also studied the new economic activity, showing how it is based on making money from money without any intervening economic activity that delivers a good or a service.[29] We would all be out in the streets demonstrating were it not for the powerful social integration provided by the traditional mass media and the new social media, which describe the supposed reality of our lives as the indispensable complement to our highly desymbolized culture.[30] Without it, however, we would be unable to make sense of the technical order within which we live, since we no longer symbolize it in the way societies did in the past.[31]

From this very brief overview of how the universality of the necessity of legal institutions has played itself out within the three primary life-milieus, it becomes apparent again that "there is little new under the sun." Our lives are totally dependent on legal institutions and a wide range of secular religious-magical juridical practices not so different from the way early communities in the natural primary life-milieu depended on other forms of magic. Hence, in attempting to understand the socialization of children into the reality that has replaced the symbolic universes of past societies, we must realize that it involves the same deep, complex, and all-encompassing religious commitments of past societies – with the exception that they now have taken on a secular form.

Screens as Magic Portals

Because of the artificiality of our primary life-milieu of technique, its technical order cannot be directly experienced by toddlers and children, other than via the intermediary of the behaviour of their significant others toward it. Our dependence on screens is possibly the most important sign of it that our children experience. Increasingly, we can no longer function without them – to operate many of our household appliances, communicate with others, inform ourselves about our world, stay abreast of events in our society, make effective use of all modes of transportation including our cars, do our work, entertain ourselves, shop online, do our banking and pay our bills, borrow electronic books from the library, fill in all kinds of forms such as tax returns and health insurance claims, check and respond to our emails, verify our payroll deductions, and apply whatever new services are being added each and every year. We have become so accustomed to the omnipresence of screens in our lives that we hardly give them a second thought. Nevertheless, we should ask ourselves whether we increasingly live our lives in relation to these screens rather than to our social and physical surroundings. The corresponding experiences are almost diametrically opposite in character. Life on a screen involves minimal embodiment, participation, and commitment as compared to living the kinds of daily-life situations involving face-to-face contact with others and our surroundings. We are back to the question as to the importance of living an embodied life through a cultural mediation or attempting to do so through a technical mediation.

If we are as interested as we claim to be in our freedom, and are thus united in our opposition to slavery as an unacceptable form of human life, we cannot brush the above issues aside on the grounds that we cannot turn back the clock. Such a response is a symptom of our enslavement to our secular myths, which makes it impossible to take seriously any radical alternatives to our present way of life. Even if these alternatives were thinkable, they would not yet be liveable because of the constraints imposed by our present order.

We must confront a situation in which almost nothing about the functioning of our society can be understood through our highly desymbolized personal experiences. All our institutions have been organized around large computerized integrated databases (commonly referred to as enterprise systems), which may be imagined as virtual intellectual assembly lines. All the functions of the institution have been captured by algorithms that include our possible interactions with and participation in these institutions. Based on specific deadlines or timeframes, we get on a screen in order to take our place on these intellectual assembly lines to put in whatever information is required for the institution to function and accomplish its missions. If there are problems, we are usually kept waiting for excessively long periods of time if we want to talk to a human being – who in turn is constrained by what can and cannot happen on this intellectual assembly line. We have effectively become unpaid data-entry clerks in return for "saving time" as a result of a supposed greater efficiency. Despite all this saving of time, no one appears to doubt that we are all "running out of time." One of the keys to understanding this contradiction is that the architecture of our way of life and our institutions is increasingly based on the principle of non-contradiction, while the architecture of earlier societies was one of dialectical enfolding. The practical implication of this difference in architectures is that in the case of the latter, almost every daily-life activity enfolded a wide variety of functions, while our current architecture "unfolds" them, with the result that we must carry them out one function at a time. For example: enough physical exercise was generally enfolded into the daily-life activities of the inhabitants of the cities of pre-industrial societies that there was no need for them to spend time driving to and from a gym in order to get a workout.

We are living our lives in relation to a technical order with an architecture based on the principle of non-contradiction to which we cannot relate other than by disciplined-based approaches; and we exist mainly on the receiving end of technically mediated relationships. With what little time is left to us, we can attempt to live with our partners, children, relatives, friends, and neighbours in relationships predominantly mediated by what remains of our cultures, although subjected to an ongoing desymbolization by a growing technical order. We will briefly highlight a few details drawn from a more extensive study of this order.[32] Toddlers and children are no longer being socialized into the kinds of societies that were common throughout most of human history. Nevertheless, in our daily lives we think that we can merely add the use of screens and nothing fundamental will change very much. We are no longer able to symbolize their use in order to understand their meaning and value for our lives. If we could, we would discover that we have had to give up a great deal in order to live with our screens. In the same vein, we live with our anti-economies as if they still produced wealth, with our anti-societies as if they still sustained human life; we continue to socialize toddlers and children as if we lived in a cultural order, and so on. This illusion is possible because we are unable to adequately symbolize our experiences as a result of the reality of the bath of images appearing on our screens as well as our enslavement to our secular myths. Our anti-societies now sustain us in ways that contribute to a great deal of stress, anxiety, and mental disorder.[33] Let us attempt to imagine our origin and our growing up in this historically unique context of a global technical order. We will put ourselves into the picture, which will help clarify our responsibilities for the kind of future we are helping to create.

Based on the reflections presented in the previous chapter, I am now aware that before my birth, I was entirely enfolded in my relationships, elementary as they were. I was unaware of the vast interrelatedness within which I became an embryo and then a baby physically and biologically distinct from my mother. As I gained a little awareness of being integral to, and yet distinct from, all this interrelatedness, I was able to put that awareness to use in asserting myself. I was never lost in this overwhelming complexity – a senseless wandering within it. Instead, this complexity helped to orient me in relation to it in the many expressions of my life. This complexity, of which I am a part, became increasingly more intelligible; and I gradually was drawn into a playful and more deliberate participation in it all.

I find it very difficult to relive this playfulness, since it was so magical. I participated in the lives of the people around me as a child who knew how to play. Doing so produced constant surprises because things never worked out quite the same way as they had the day before, or the day before that, and so on. Remembering more than a week was very difficult because everything changed so much. There was a constant stream of playful surprises that I eagerly anticipated. Only some of it was scary, but I knew very well that I did not know nearly as much about myself and the world as everyone else did. I was constantly discovering new things and there was never an end to them. I knew I could turn to others if something baffled me, undermined me, or brought me to tears. I knew very well that these kinds of events could always be overcome by a hug or other signs of love and assurance that everything was all right. None of these incidents ever changed anything important. Hence, I could stop crying and feel that everything had been put right again, which made it possible for me to go on playing.

Compared to my adult experiences, there was indeed something magical about this playing. I did not need any reference points, any ability to say that this was meaningful but that was not meaningful, or that this was harmful and that was not. Of course, I knew from time to time that I wanted something so badly that I tried every imaginable trick I knew to get it, but this was not what dominated my play. This playful interaction with the world never deteriorated into a kind of Brownian movement in which anything was no different from anything else, and anything I did had no greater significance than all the rest. My playfulness never resembled a random sequence of events, encounters, and actions that added new chaos in my universe.

My playful exploration of what I encountered was greatly enriched when I began to pay more careful attention to what others said. Initially I thought that these sounds were no different from all the other sounds around me, but I soon learned that this was not the case. I began to learn some words and phrases, and this opened up a whole new world to me. For example, when I got tired of playing in the park, I could ask my mother if we could go home and get something to eat or drink. I also learned to ask for stories about people I did not know and who sometimes lived in worlds very different from my own. It greatly extended the range of my playful interaction with everything.

My mother was always pointing things out to me. I remember that in the park she was often pointing to these funny things that scampered around in trees the way we walk on the ground. Each time she pointed to one, she called it a squirrel. Just when I thought I had figured out what she meant and pointed out a squirrel on my own, she heard a beep, grabbed this thing from her pocket and talked to it. It was as if someone else was there and I was no longer there. I decided I might as well run off and find some squirrels on my own.

The same kind of thing happened the other day when I was playing with my dad. He made a buzzing sound and grabbed his thing to talk to it, and he forgot all about me. It also happened when we were eating together, reading a story before bedtime, going somewhere in the car, and shopping in the store. My parents constantly had to look at these things to talk to them. I could not make much sense of this. Did these things create another room that my parents could go into to talk to people that I could not see? I could still hear my parents talk but they did not see me anymore. This other room seemed to follow us everywhere we went and could stop anything we did whenever my parents popped into it. No activity was safe from this. It made me wonder when I would be able to look into that room because it seemed to be much more important than the room in which I was living. I wanted to see who lived in that room and why they were constantly talking to my parents. These people seemed to be as fascinating to my parents as my bedtime stories were to me. The thing seemed to be putting all of them into a magical land full of all kinds of things that I had never encountered by myself where I lived. Maybe what they talked to in their hand was a window into their world, where I could not go.

I suppose that these kinds of experiences undermined my playfulness to some extent. Should I be making an attempt to explore the room my parents constantly went into when they talked to others who were not there? I could not entirely put this out of my mind because of similar situations. At home and in other places, there were these funny windows that you could look through into another world. The only difference was that when I looked out of a window I could try to go outside and find everything I saw through the window. However, when I tried to look behind these other kinds of windows, there was nothing there. It made me wonder where this world was that you were seeing through these funny flat windows with nothing behind them. The behaviour of my parents toward those windows was also very different from their behaviour toward the small things they pulled out of their pockets, which also seemed to have those kinds of windows. These other windows they mostly looked at. Sometimes such a window appeared to make them laugh, but they certainly did not talk to these windows the way they did to the windows on the things in their pockets. All of this did sometimes make me wonder what was really going on. What was I playing with, and what were my parents so busy with? Were there other worlds that you could learn about through stories or watch through these windows? Should I learn to play in these other worlds? If so, I had no idea of how to go about it. Every time I asked my parents if I could look at the windows on these things they pulled out of their pockets, there were pictures but also a whole lot else I could not make any sense of. Was I to believe that what I could play with by touching and manipulating it was my world that my parents entered into to pay attention to me, while the rest of the time they seemed to escape into their own world?

I know I had vague feelings about all of this as a child; and I now know that this came from building up a great deal of metaconscious knowledge from all these different experiences. Nevertheless, it does not alter the confusing situation of my childhood. Now as an adult, I wonder if some children could have developed the beginning of a mental disorder if they could not quite sort all this out in the way most of us appear to have done. After all, we were all trying to learn to live our lives relative to time, space, other people, and everything else. Surely these apparently different worlds must make growing up much more difficult, with more chances of something going very wrong.

One form of metaconscious knowledge that must have developed from all these experiences being symbolized by neural and synaptic modifications to the organization of my brain-mind may have given rise to intuitions that, with hindsight, can perhaps be expressed as my parents constantly being in a kind of entryway to another world. This world appeared to be a lot more important than anything I had access to by listening as they and others talked to me. Beyond what happened in their immediate experience of everything around them, there must have been something much more interesting and exciting, because everyone immediately grabbed this "thing they held in their hand" to talk to every time it made a sound. I now recognize that, in earlier societies, the interactions between parents and their children were also frequently interrupted, but this was done by others who approached and remained in the same world. Consequently, it may have been a little easier for children growing up in these societies to learn to live their lives in relation to time, space, and the social because it was less complex. Nor was it susceptible to the kinds of difficulties children face today as they grow up in "anti-societies" whose characteristics are diametrically opposite to those of traditional societies.[34] Whatever difficulties did occur would have given rise to different kinds of mental disorders.

Let us once more return to our general analysis of children growing up in a contemporary anti-society with a highly desymbolized culture and a way of life dependent on a global technical order. The transition from living playfully in relation to an open-ended relativity of the world to living in relation to the sameness a community now creates by symbolically grounding this technical order involves a double discontinuity over and beyond the one discussed at the beginning of this chapter. First, there is the discontinuity between an open-ended relativity that has as yet not been closed by symbolically grounding it and a distorted world that has been symbolically grounded, thereby alienating its inhabitants. Toddlers and young children relate to this open-ended relativity as if they had metaconsciously symbolized the unknown as always being radically and unpredictably different from the known and lived. This discontinuity had to be faced by children growing up in all earlier societies as well.

Second, there is the additional discontinuity that stems from language having become the necessary but insufficient condition for participation in the technical order, whose architecture is contrary to that of all enfolded and dialectically enfolded life. Whatever fragments of a symbolic universe may remain as a kind of personal "space" shared with significant others through face-to-face contacts continues to have the architecture of life, even though the culture has become highly desymbolized. This change will be the subject of the following chapters, but will be introduced here in general terms.

Growing Up with Symbolization and Desymbolization

As I have shown throughout my previously cited works, much of our being a symbolic species can be explained in terms of three fundamental metaconscious processes: symbolization, differentiation, and integration. Differentiation refers to our metaconscious ability to sustain our life by making each experience into a moment of it as a unique expression of it through a relationship of reciprocal dialectical enfolding that includes its differentiation from all previously lived experiences. Integration refers to our metaconscious ability to sustain our life as a dialectical complement to differentiation, one that constantly builds and adapts our life with all these experiences and permits that life to fully work in the background. In this way, it facilitates extraordinarily complex behaviour by delegating many aspects to routine in order to free up our attention to focus on what matters most, including what previously may have been dealt with routinely. This building and adapting of our life involves a great deal of metaconscious knowledge implicit in the organization of our brain-mind. It includes our self with its physical, social, and cultural dimensions, and the myths by which we symbolically ground our life as an individually unique expression of the culture of our community.

Symbolization refers to our metaconscious ability to sustain our life by constantly evolving and adapting it. Since no situation in an open-ended relativity repeats itself in quite the same way, it requires dimensions of creativity, imagination, and the paradigmatic symbolic utilization of prior experience. Jointly, differentiation, integration, and symbolization may be regarded as what allows a person to live a life relative to a world in which everything is related to and evolves in relation to everything else, and dealing with our finitude by imposing myths, a religion, and a morality by means of a culture that binds us to a community. No community has ever symbolized the unknown other than as temporal extensions of what is known and lived. Not doing so would have eliminated the necessity of imposing something absolute on what is relative, including the distortion of all relationships as well as alienation and, possibly, reification. On the other hand, imposing something absolute makes individual and collective human life liveable in an ultimately unknowable relativity by shutting the door to cultural chaos, which is the secular equivalent of hell: the absence of the possibility of human relationships.

Our present situation is characterized by our having introduced into the relativity of our world an attempt at a new beginning, by first constructing an economic order and then a technical order based on the architecture of non-life. This must not be confused with the universal response of all societies: to create legal institutions that would limit the unpredictability and surprise of the relativity of human life by introducing and imposing legal models for a variety of relationships. The economic and technical orders also imposed such models, with the difference that these were non-cultural and thus universal because they were based not on an architecture of relativity but on an architecture of reality. As noted, this latter architecture can be expressed in terms of five primary characteristics: non-contradiction, separability, closed definitions, openness to measurement (and thus quantification and mathematical representation), and a simple complexity. Jointly, these are the opposite characteristics to those shown by enfolding and dialectical enfolding, without which biological and cultural life cannot be lived, and thus the architecture of reality is an architecture of non-life.

In spite of introducing a sphere built up with the architecture of reality, we remained a symbolic species depending on a metaconscious differentiation, integration, and symbolization for our living in the world as a society. We did not magically mutate into something else. On the contrary, these abilities, by which we sustained individual and collective human life as a symbolic species, now confronted an open-ended complexity along with a growing constructed order having a simple and closed complexity. Initially, we did not see the risks involved. This can be explained by our using our sight, which is well adapted to dealing with an open-ended relativity but treating it as if it were a reality integrated with our hearing, which permits the acquisition of a language, a person's entry into a culture and thus becoming a member of a symbolic species. The coexistence of these two functions was generally well symbolized by traditional cultures whose myths prioritized the latter over the former. For some two hundred years, however, we have been evolving desymbolized cultures that have done the exact opposite: by giving priority to the growing construct of reality as "the way forward." It is both the cause and the effect of highly desymbolized cultures. As noted, the first generation of secular myths implied that all previous cultures had mistakenly gone about their business of making life as good and liveable as they were able by serving their gods through their religions and moralities. Humanity was to live by bread alone, that is, by working hard and carefully allocating capital to raise the material standard of living. Every other kind of progress would follow, bringing about the greatest possible human happiness. This "sameness" was extrapolated to all of human history, which was seen as a struggle for survival that was finally coming to an end thanks to this new technological and economic progress. The leading moral and religious traditions of the industrializing societies were turned inside out and upside down; and they have not recovered to this day because the second generation of secular myths has the same orientation with respect to culture and our being a symbolic species.

In practical terms, what happens to our metaconscious abilities of differentiation, integration, and symbolization when they encounter an order of reality? The simple answer is: desymbolization, but what does this involve? Let us imagine an experiment on the vocabulary of a human language. Suppose we had agreed to shorten the dictionary of this vocabulary by eliminating every twentieth entry. Of course, these words would then also have to be eliminated from the explanations of the remaining words. If a symbolic language (unlike any computer language) is characterized by dialectical enfolding – with the result that the meaning of any word is what all the other words do not mean, and vice versa – then the meanings of all words would be impoverished by this editing project. For the moment, let us not attempt to imagine what kind of sociological and historical conditions might somehow approximate or correspond to this kind of venture.

Next, consider another version of this experiment. Instead of eliminating every twentieth word, imagine restricting the vocabulary to those words required to describe everything in the domain of a discipline such as physics. The result would be that all words related to non-physical phenomena would be eliminated. What we would be left with is the kind of vocabulary to which we were introduced when we took physics in high school: forces, masses, acceleration, momentum, velocity, inertia, frictionless planes, pulleys without inertia, air without friction, and so on. In high school, we were told that physicists are neutral observers who objectively investigate physical phenomena and enter the results into the domain of physics. Nothing of their persons and their lives is supposed to affect this; their involvement would thus be limited to observations, experimental data, experimental design, statistical analysis, and so on. Clearly, whatever remained of the vocabulary of a symbolic language would be a tiny fraction, with the result that the meanings of its constituent words would be greatly simplified. It may be argued that this remaining vocabulary might approximate what physicists would use when they are engaged in theoretical or experimental physics. This interpretation is a simple and incomplete meaning of desymbolization. The design of our experiment has overlooked a rather important detail, namely, that the discipline of physics evolves over time.

Next, let us imagine another version of this experiment, one which takes into account the constant social and historical adaptation and evolution of a symbolic language through the lives of the members of a community who speak it. Now repeating the imaginary experiment would require taking into account the changes in the vocabulary of a language over time and how this would affect the subset used by practising physicists. If we think about this issue a little more deeply and bring in what we know about the differences between the history of a community and the history of a scientific discipline, we would be compelled to conclude that the original experiment was faulty in its design. A high school class in physics provides us with the clue as to why this is the case. As we will show in later chapters in greater detail, by growing up in a society the students entering a high school physics class have already learned a great deal about the physical phenomena involved as they grew up learning to sit, crawl, walk, run, bike, jump, climb trees, play ball, and much, much more. However, their first physics class does not build on the knowledge the students have acquired. The reason is straightforward. They are not learning physics in the context of the world in which they live. They are learning physics in the context of the domain of physics, which is a construct represented by means of free body diagrams, definitions of terms, equations that bind these terms together, and so on. In other words, they enter into a domain that exists nowhere in their world. They can only enter it by means of their mathematical imaginations, because the domain of physics is a theoretical construct that has the architecture of reality, with the result that it is entirely discontinuous with the world in which the teacher and the students live and "do" physics. This is no typical human activity, and to simply call it scientific is akin to attributing some kind of secular magic to it. As we will see, it belongs to a category of human activities that is entirely separated from experience and culture, which is why all earlier knowledge of physics (the kind embedded in experience and culture) must be set aside when entering the physics classroom.

This theoretical construct, which embodies a great deal of experimentation, is built up with the characteristics of non-contradiction, separability, closed definitions, quantification and mathematical representation, and a simple complexity. Relativity theory, quantum mechanics, and all the subsequent attempts at unifying physics are based on mathematics and thus on the architecture of reality. Each term can be defined by means of a closed definition, without which any mathematical representation would be impossible. Consequently, the vocabulary of physics relative to the vocabulary of a symbolic language does not differ in extent as much as it differs in being universal and thus non-cultural, lacking any enfolding or dialectical enfolding, and thus having an incommensurate architecture. Young people who want to become physicists must overcome the double discontinuity to which we referred earlier on. The global technical order is built up with a multitude of these kinds of constructs associated with discipline-based approaches.

In our daily lives we encounter numerous such constructs as well. As an example, consider the newscast on a television network. If we attempt to compare our watching what is happening to human life and the world on such newscasts with how we would experience these reported events if we could have participated in them, we would encounter the same kinds of double discontinuities. We do not need to go into any detail to recognize that what we see has been "shot" by one or more camerapersons who essentially determine our focus of attention with the guidance of a reporter or producer. If we had been present at the "shooting," we could have made our attention wander over the scene to get a good grip on the broader context. From all the digital images that were taken, one of the producers of the program will select various fragments in order to reintegrate them into something that "works" for the program – something that will attract and hold the attention of a large audience in order to ensure sufficient advertising income to pay expenses. Again, as viewers, we have no idea whatsoever of the context of this collected set of series of images. Moreover, images do not exactly speak for themselves. The fact that a person is crying on camera does not provide us with the reasons. In sum: we get a construct of images that literally build an image-story accompanied by supporting image-words and phrases. The latter are carefully crafted to give the images their full effect by means of a variety of well-documented techniques.[35] There is no speaker or listener in the sense of a face-to-face conversation, but only a spokesperson and a construct aimed at a statistically determined "typical" listener.

As noted in the Introduction, if we track what a newscast covers in relation to human life and the world over a period of time, we quickly realize that it portrays the most fragmented, discontinuous, and incoherent view of it all. Events come out of nowhere, are covered for one or more days for usually no more than one or two minutes, and disappear to be never heard from again. We have no idea of the context that led up to them and the long-term trends they contribute to. The reason we have not stopped watching these newscasts is that this reality, with its simple complexity to which image-stories can be added or subtracted to fill the newscast, is held together because each fragment is explained as part of what resembles a kind of mythology. Its commonplaces roughly correspond to our intuitions of how our society has symbolically grounded our life in the world, so that we see them as the sameness we are familiar with.[36] Once again, children who eventually grow up and watch these newscasts will have to learn to deal with the double discontinuity in order to make any sense of it.

We can multiply these kinds of examples almost without limits. Everything we see on the screens of our computers, whether we are playing a game, doing our work logged into an enterprise system, surfing the web, or anything else, is a construct with an architecture of reality. The same holds true for the layout of any store and the positioning of the merchandise on the shelves. It is a construct accomplished by a myriad of techniques.[37] Our involvement in many daily-life activities as encounters with these constructs is the result of a long preparation of growing up to deal with the double discontinuity in our lives.

It is important to recall that the dialectically enfolded architecture of any symbolic language is capable of sustaining our lives in an open-ended relativity. The fact that we need to impose something absolute on this relativity will force endless distortions in a language – a reflection of the alienation caused by symbolically anchoring our relativity and finitude in myths. However, even with these distortions, the organizations of the brain-minds of toddlers who playfully interact with an open-ended relativity can act as a metalanguage. This happens in a manner that is so effective that virtually all children can learn a language and, via this language, a culture, even though it is probably the most complex skill we all acquire.

Two Streams of Experiences

Ever since the introduction of an economic order, followed by a technical order based on an architecture of non-life, babies and children growing up have faced a stream of experiences with these constructs. In our present anti-societies, these developments have reached the point that toddlers are confronted by two streams of experiences, which at first they will be unable to differentiate. For the purpose of analysis, the two streams may be distinguished by the architecture of their constituent experiences. In the stream that corresponds to the kinds of experiences babies and children had in pre-industrial traditional societies, both the foreground and the background have the architecture of a dialectically enfolded human life in the world. It is the stream that sustains and enhances their symbolic capacities in becoming members of a symbolic species. The second stream of experiences is mostly confined to the first generation of industrial societies building an economic order and the so-called industrially advanced societies participating in a global technical order. The constituent experiences are characterized by a foreground with an architecture of reality and a background divided between two architectures. Consequently, these experiences reflect the double transition children must learn to make in order to be able to fully participate in their society. These experiences are highly desymbolized, since they correspond to a participation in or a being on the receiving end of the constructs of a global technical order. The influence of this stream of experiences is the undermining of the development of the symbolic capacities of children. The strategic implications are far-reaching. If the first stream of experiences can dominate the second stream until toddlers and young children have sufficiently mastered the learning of a language and the accompanying highly desymbolized culture to the point that the second stream of desymbolizing experiences can no longer decisively undermine this development, they can gradually learn to differentiate the two streams and proceed to learn to treat them differently. However, if the desymbolizing influence of the second stream on the first prevents this from happening, a range of new mental difficulties can arise that would appear to correspond rather well to some learning disabilities and the treatment of language as images.[38] It may also lead to a negative impact on the child's development of the metaconscious knowledge of a social and cultural self. Possibly one of the most troubling developments is the spatial orientation of our contemporary cultures, which is affecting the ability of some children to symbolize their lives relative to time. As a result, periods of time become spatial time-objects. These are inflexible and solid in a way that cannot be disturbed without creating crises, as if something of their lives has been broken.

We will refer to experiences that sustain symbolic development and those that undermine it, according to the architecture of their foregrounds. There are thus "real" experiences associated with encounters with the technical order, and "true" experiences – or simply experiences associated with encounters with what remains of a symbolic culture in the form of a kind of personal "space" shared with significant others. These experiences are true only if we disregard the alienation that comes from symbolically grounding human life in the technical order, which is why the designation of true is either omitted or put in quotes.

Generally speaking, when we direct our attention to a domain of a discipline or the content on a screen, or in any other way limit our symbolic ability to relate everything to everything else in our life, the metaconscious differentiation, integration, and symbolization are no longer open-ended. They become fundamentally restricted to a domain, micro-world, or construct that is separated from experience and culture by having been built up with the architecture of reality.[39] In such cases, the architecture of the focus of attention is the diametrical opposite of that of an open-ended relativity, which has been closed by discipline-based approaches and everything built up with them to contribute to the technical order. This order is further closed by symbolically grounding it in our contemporary secular myths, which separate it from our personal and private "space" and devalue it as purely subjective and thus of no consequence to what really matters.[40]

A simple example can illustrate the difference. When a student listens to a lecture on physics, his attention may be distracted by someone he would very much like to meet. At this point, his focus of attention has returned to the daily-life world, with the result that the foreground and background of the resulting experience belong to the same world. When his attention goes back to the lecture, the foreground will now represent a portion of the domain of physics, thereby excluding everything from the world except for its physical phenomena. Consequently, this real experience has a discontinuity between its foreground and its background. Moreover, the foreground is made up of a further distinction between whatever portion of the domain of physics is the focus of attention and whatever part of it remains in the background. The role language can play in these very different experiences will differ significantly. The language used in association with any constructs will heavily rely on image-words that are differentiated from all the other image-words whose joint scope is limited to the domain(s) of the discipline(s) or to a fragment of the reality a society constructs through the widespread use of techniques. The language associated with our personal space is desymbolized by its limited association with this space, which limits its scope, with the result that these language-words are also desymbolized. In addition, there are the issues related to an interpenetration of the two kinds of words. We will return to all these issues in the following chapters, but we will now return our attention to language acquisition by babies and toddlers.

Language Acquisition in Anti-Societies with Three Frames of Reference

Throughout this and my previous works, I have insisted that our relationships with our surroundings are reciprocal in character because of our being a symbolic species. Consequently, when the first generation of industrializing societies symbolically grounded their ways of life by means of the secular myths of capital, progress, work, and happiness, one way in which the resulting alienation manifested itself was in a reinterpretation of "human nature" in terms of human work. It had become the most important resource in the service of the economic order of that time.[41] The people of those days began to see themselves as *homo economicus*. The sameness created by these societies was projected onto all of human history, with capitalism and communism being the two dominant forms of their reinterpretation of history.

The next generation of so-called industrially advanced societies, characterized by their participation in an increasingly global technical order, symbolically grounded this order in the secular myths of technique, the state-nation(s), science, and history. Once again, "human nature" was reinterpreted in terms of what was most decisive in our alienation and reification by this order, namely, the ability to deal with information. We began to see ourselves as *homo informaticus*. It thus became intellectually and scientifically essential to reinterpret all human functions and activities in terms of the all-important component of information processing. This summed up the significance of everything relative to ways of life that expanded and advanced the technical order as their highest priority out of the fear of being "left behind."

Dismissing these sociological and historical correlations as evidence of pessimism or reactionary backwardness is to make a comprehensive secular religious judgment that attempts to avoid the equivalent of a sacred taboo from the past. Until the emergence of the above two forms of "human nature" as a result of what we have referred to as "technique changing people," what was regarded as human nature had always been very different. This was equally true for Western civilization when it gave birth to the economic and technical orders. It is somewhat embarrassing how uncritical we have become of reducing a great deal of human life to a predominance of information processing. I have shown that when people talk face to face, a dialectical tension is required between their being different enough to be able to enrich each other's lives with something new and their not being so different that they might as well live in different worlds.[42] A dialectical tension is thus required between individual differences and similarities to allow a meaningful communication to take place. In contrast, the exchange of information as defined by information theory has no such constraints, to the point that computers have no difficulty "communicating" the same string of digits indefinitely. It may well take decades before our enslavement to this new technical order can be weakened by resymbolizing it in terms of its meaning and value relative to human life and society.[43] Surely, there is a great deal more to human life than information processing, important as it may be; and this reductionism must not be intellectually and scientifically legitimated by our enslavement to the technical order.

We have become spectators to one generation of techniques after another, creating a mesmerizing spectacle of efficiency and performance. We have also mostly approved of the state-nation's sacrificing almost everything that could be turned into a resource in order to feed our new scientific, technical, and economic fertility, as well as its efforts to have us devote ourselves to it with all our hearts through a secular political religion. Thinking of ourselves as *homo informaticus* amounts to a highly destructive way of thinking about ourselves – as reified non-beings made up from dead information processors and dead information flows, all wrapped in a living skin. Doing so completely disregards our physical, social, and cultural embodiment, which, despite considerable desymbolization, continues (for now) to enable us to symbolically ground the technical order and thus prevent the cultural chaos that would bring a secular hell on earth. What we refuse to face is that an alienated life can at least be liberated, but a reified life requires a resuscitation from non-life. Is this how we will finally overcome our relativity and finitude – by destroying what is alive in us and our children? Will this be the ultimate triumph of the order of non-sense? Are we trapping ourselves into what amounts to a kind of collective autism? Perhaps I am getting carried away with my feelings, but as a relatively successful engineer, I am entirely bewildered by our confusion between life and death.

I know very well that these questions do not have answers. Hopefully, they may motivate us to attempt to recover an understanding of what remains alive in our children, and what may undermine it. In other words: in terms of the two streams of experiences to which we expose them, we will first focus on the stream that allows them to live and become members of a symbolic species. We will then attempt to face our responsibility of dealing with what may undermine their development, in the knowledge that we are not inevitably destined for a death by information. The battle for the human spirit will rage on unless we take a stand in all of this. Our primary line of defence is to respond the way humanity always has: we need to symbolize what is happening to individual and collective human life in order to distance ourselves from our primary life-milieu and to gain a margin of freedom that will open up new possibilities by transcending the sameness created through myths.

Let us return to the toddlers who learned to recognize squirrels in the park. They may first have been taken there as babies in their "snugglies." Their early experiences may initially have been differentiated from the previously discussed experiences of babies who had not as yet developed an awareness of their physical selves. Their experiences in the park would have stood out by the very different tactile sensations and a great many moving visual blurs and sounds. Assuming that these babies had already learned to follow movement and sound, their ability to turn their heads, focus their eyes, and bring certain sounds into the foreground had gradually improved in the less complex setting of their homes. Their experiences in the park would have provided further opportunities for these developments.

Making sense of what was going on in the park may have begun by differentiating moving things from static ones. They may thus have learned something about birds relative to the sky, and something about the sky relative to birds. Similarly, they may have learned something about birds relative to trees, and vice versa. The same developments may have occurred in relation to squirrels and trees, squirrels relative to the dogs that chased them, the dogs relative to their more slowly moving owners, and so on. The babies' involvement in all these developments would remain relatively passive until they began to differentiate their bodies from the world, and vice versa. The growth of the corresponding metaconscious knowledge would provide these babies with an intuitive sense of being a physical self relative to their surroundings. Eventually, this awareness would become externalized in their behaviour by, for example, showing a particular interest in squirrels. The child's experiences of being in the park would eventually lead to several metaconscious clusters of experiences that would jointly begin to constitute a metalanguage in the context of which the word "squirrel" could become a vocal sign.

Some time later, when these babies had grown into toddlers capable of moving around the park by themselves, both their metalanguage and the usefulness of vocal signs would grow, especially in encounters with their parents and others. The beginnings of a symbolic development would follow. For example, a toddler may interrupt her cautious approach to a dog when she hears her mother call to her saying "Look! A squirrel!" In response, the toddler will look at her mother to see where she is pointing and shift her attention in that direction, given that her interest in squirrels is greater than in dogs. In the same vein, to motivate the toddler to walk all the way to the park, her mother may remind her of the squirrels she will soon see there. At this point in her development, the use of vocal signs is being symbolically "stretched" because it cannot take on a meaning in the context of the immediate situation at hand. The same is true when the toddler gets tired and hungry following a lot of exploring in the park, and her mother responds to her fussing by promising to go home and give her something to eat.

Gradually, toddlers, who have learned a significant number of vocal signs and short phrases, will differentiate these not only from the immediate situation at hand but also from all other sounds as having a unique significance in their interactions with others. The vocal signs thus become directly differentiated relative to the situations in the context of which they took on their meanings, as well as being directly differentiated from each other, since the toddlers do not confuse them. This latter differentiation begins to build a new form of metaconscious knowledge that will gradually provide them with intuitions about a language. Such intuitions will become externalized in their behaviour as they exhibit a growing interest in what everything is called, thus boosting the importance of verbal communication in their lives.

An important question remains unanswered. How are toddlers going to learn words and phrases that have no relationship whatsoever to the immediate situation at hand? Up to this point, vocal signs and phrases have taken on their meanings relative to the contexts of the child's experiences of the situation at hand. However, when words or phrases refer to something beyond their immediate context, it is no longer possible to make sense of them in this way. First, these words and phrases must become an embryonic language foreground, with the remainder of the experience of the immediate situation forming the background. Second, all experiences with an embryonic language foreground must be interpreted directly relative to each other and indirectly relative to all the other experiences without a language foreground. Initially, the words and phrases that reach beyond the immediate context at hand take on an indexical character by pointing to the memories of earlier experiences in the context of which they would make sense. Eventually, these words and phrases become symbolic when they take on new meanings by being directly differentiated from one another, that is, in the context of a language. At this point, experiences of spoken language and children's corresponding memories can be made sense of even when they reach beyond the life of a child. All the experiences with a language foreground lived by a child have become directly differentiated from each other, and jointly from all the other clusters of experiences without a language foreground.

The new experiences and their associated memories permit children to reach into the symbolic universe of a traditional society, or into its desymbolized remnants enveloped by a technical order in our anti-societies. This capability results from the discontinuity between the language foregrounds and the backgrounds of the new experiences made possible by means of symbolization, language, and culture (desymbolized or not). Early in the emergence of these kinds of experiences, they represent a tension between the meaning of what is being said by others and the meaning of the immediate situation at hand. The conflict between these meanings can be sorted out only by focusing one's attention on what is being verbally communicated and thus relegating everything else to the background. The result is the emergence of a secondary frame of reference. The experiences with a language foreground continue to be differentiated from all the others lived by means of a primary frame of reference acting as a metalanguage for the acquisition of vocal signs. The new experiences lived by a focus of attention on language, thus resulting in a discontinuity between their language foregrounds and their backgrounds, will increasingly differ from those experiences lived by means of the primary frame of reference. The meaning of what is being spoken can thus take on a meaning relative to everything a child has learned about language but not to the immediate situation at hand. As a result, the secondary frame of reference remains grafted in the first but also transcends it. The meanings of the language foregrounds derive directly from being differentiated from each other, and only secondarily from all the other experiences of a child's life.

The emergence of a symbolic language manifests itself in two ways. First, the new level of differentiation appears to be concentrated in a different portion of the organization of the brain-mind, which becomes increasingly associated with language. Second, it gives rise to entirely new kinds of memories. These are the semantic memories, distinguished from the episodic memories – those resulting from what was directly experienced. For example, teenagers learn about corporations as legal fictions by having this explained to them, and such explanations can be remembered as semantic memories. Their episodic memories of a corporation would be the experiences of visiting a plant or an office, meeting its employees, or using its products. The distinction between semantic memories and episodic memories is confirmed in people who, because of brain injuries, can sometimes lose one but not the other.[44] This also confirms that the secondary frame of reference is distinct from the primary one. The former could not have developed had their experiences with language foregrounds not become directly differentiated from each other, and jointly differentiated from the experiences that give rise to episodic memories. It is important to emphasize that the development of the organization of the brain-mind up to this point is able to act as a metalanguage onto which any human symbolic language can be grafted. Likely for this reason, almost all children can learn a language and thus acquire a culture, even though these are possibly the most complex skills they acquire during their lifetime.

These developments have far-reaching implications that may be illustrated by historically unusual situations. For example, children who are homeschooled by parents who live on the land in an isolated location may acquire many semantic memories resulting from having been told how other children live in faraway urban centres. When they eventually visit a city and join their lives to those of the people who live there, their lack of episodic memories may create misunderstandings, for example, when they misinterpret their emotions in relation to another person or situation. They may know about these possibilities based on their semantic memories, but they have never *lived* them. Hence, they lack the metaconscious knowledge that is essential for living their lives with others.

In our contemporary anti-societies, language acquisition by toddlers and children has become a great deal more complex because they are surrounded by adults who constantly divert their attention away from them toward their screen-based devices, whenever these appear to make sounds. To these children, these devices may appear as portals into another world that clearly has priority over the world they share with these adults. Their curiosity will thus be aroused, and they may want to get their hands on one of these devices to look for themselves into this other world. When they finally get a chance, they quickly discover that they are encountering a world that is different from and yet related to the one of which they have learned to make some sense. The experiences they derive of this new world and the developments in their brain-minds that result from the metaconscious symbolization, differentiation, and integration of these experiences are as fundamental as those that resulted from their acquiring a language as a gateway into a culture.

All the elements of the experiences associated with toddlers and children making sense of screen-based devices must eventually be interpreted as jointly belonging to one world, whereas all their other experiences must be made sense of as belonging to the world in which they live (which includes their "inner" world). In order to accomplish this, children must learn to focus on anything that appears on a screen and the associated sounds coming from the speakers as a foreground against the primary background of everything else on the screen. Everything else they experience about the world in which they are embodied as they are making sense of this screen-based device must jointly be interpreted as a secondary background. Learning to focus their attention in this way involves learning to impose a discontinuity between the foreground and its primary background (which are separated from experience and culture), and the secondary background that is embedded in experience and culture. The former results from children learning to make sense of a constructed reality, while the latter would have contributed to their making sense of a lived symbolic universe if they were growing up in a traditional society. This lived symbolic universe has now become a personal world shared with intimate others. They will gradually learn that this personal world is embedded in and enveloped by the constructed reality accessed through the portals of screen-based devices. The imposition of this discontinuity is integral to sorting out two different worlds to which they must learn to relate with very different levels of embodiment, participation, and reciprocal engagement and commitment.

There is a clear parallel in this development with those of children learning to make sense of experiences that can only be dealt with as having a language foreground that is discontinuous from the background derived from the situation in which the verbal communication takes place. However, there is the important difference that in these situations, the language foreground and background usually derive from the same world, while in the case of screen-based devices, this is more complex. The experiences derived from conversations thus have a language foreground and background that have the same architecture, while experiences from screen-based devices have a foreground and primary background with an architecture that is different from that of the secondary background.

The development of the skills required to make sense of screen-based devices results in and is sustained by metaconscious modifications to the organizations of the brain-minds of children. Whereas their secondary frames of reference are well suited to the symbolic universes of traditional societies, they cannot take them very far in learning to make sense of experiences of a constructed reality accessed through screen-based devices. The situation is similar to and also different from the fact that the primary frame of reference is unable to make sense of experiences with a language foreground traditionally associated with a symbolic universe. In the same way, the secondary frame of reference cannot make adequate sense of the experiences of the world accessed through screen-based devices, because such experiences will have to be directly differentiated from each other as belonging to one world, and jointly differentiated from the semantic and episodic experiences as belonging to another world. All three metaconscious frames of reference are developed with symbolized experiences that have incompatible structures because they symbolically mirror the different worlds from which they have been derived. The tertiary frame of reference will emerge out of the secondary one (into which it will remain grafted) as children learn to focus their attention on screen-based devices and make sense of them. In other words, the tertiary frame of reference operates with the secondary one working in the background. In the same vein, the secondary frame of reference operates with the primary one working in the background.

Together, the three frames of reference involve a reciprocal relativity in which an experience is metaconsciously made sense of relative to a frame of reference, while this frame of reference derives its sense from the experiences of the same category of which the child has made sense previously in his or her life. Hence the circularity of this interpretation. Moreover, the three levels of this reciprocal relativity remain grafted into one another, and thus must be jointly defended against cultural chaos. This is achieved through the integral development of ever deeper metaconscious knowledge that eventually gives birth to the myths of their society. This will be explained in subsequent chapters.

The tertiary frame of reference is thus the gateway into a constructed reality. The secondary frame of reference is the gateway into the world of a culture. The primary frame of reference was the gateway into the world of immediate experience prior to acquisition of language. However, the development of the tertiary frame of reference goes far beyond learning to make sense of screen-based devices.

Further developments in the tertiary frame of reference will occur when teenagers enter the domains of the disciplines they learn in high school. Since such domains exist nowhere and must be imagined, teenagers cannot make any sense of them other than by a similar development that I have described as the emergence of the secondary frame of reference. They must learn to make sense of these new experiences by focusing their attention on what they encounter in these domains, with some of the "surroundings" becoming the primary background and relegating everything else they experience in the classroom to the secondary background. It becomes possible to make sense of these domains, which exist nowhere in the world, only by interpreting them relative to their own corresponding experiences. Teenagers' resulting experiences will be directly differentiated from each other, and jointly, indirectly differentiated from their experiences with the language foreground lived by the secondary frame of reference as well as those lived by the primary frame of reference. These experiences thus represent another discontinuity between them and the symbolic universe of a traditional society or its highly desymbolized remnants within the technical order of an anti-society. This discontinuity is rooted in the one between the foregrounds associated with the domains of disciplines and the backgrounds of life in the classroom because of their diametrically opposite architectures. These foregrounds cannot be interpreted relative to anything in the world of the teenagers and thus must be separated from the highly desymbolized experiences of their lives in this world. This development has been referred to as the separation of knowing and doing from experience and culture.

What has mostly escaped the awareness of parents, and (especially) educators, is that the secular religious attitudes that are being transmitted during all these developments are not found in the places governed by their traditional counterparts, but in the interrelating of the three frames of reference and how they are to be used in the conduct of our lives in our contemporary anti-societies.

We thus live with three complementary frames of reference. The primary is associated with our episodic memories. The secondary and tertiary are associated with our semantic memories, which they divide into those memories derived from the experiences of our lives that remain embedded in a highly desymbolized language and culture, and those experiences that are separated from this language and culture because their foregrounds interpret reality or our involvement in our discipline-based knowing, doing, and organizing. The experiences lived by the secondary frame of reference have foregrounds with a diametrically opposite architecture to the architecture of the foregrounds of the experiences lived by the tertiary frame of reference, with the result that the metaconscious knowledge built up within each cluster of experiences inherits this discontinuity. There are many manifestations of this discontinuity in our parallel modes of knowing and doing, exemplified by the well-established distinction between school physics and intuitive physics. This interpretation of growing up in contemporary anti-societies is very far removed from those of cognitivism. The former is an attempt at resymbolizing our situation, while the latter attempts to force individual and collective human life into the architecture of non-life by means of discipline-based approaches. This latter attempt results in human activities being remade in the image of computers, whether they are slow, fast, or neural net-based.

In our contemporary anti-societies (in the sense that their characteristics are generally the diametrical opposites of their traditional precursors), events of the following kind may interfere with the kinds of developments described above. Suppose that the toddler who had been somewhat baffled by his parents' use of cellphones and screen-based devices had moved into a new house with his family. He had explored the entire house with a great deal of excitement. This fascination led to his learning some new vocal signs such as doors, windows, walls, and stairs. One day his mother turned on the new television, and it showed some kids playing in an urban setting. The toddler got very excited and, while repeatedly saying "window," rushed to the screen and began to look through it from different angles as if it were a window; noting that something was somehow different, he also looked behind the screen. He quickly learned that, unlike a regular window, this window did not reveal anything more of the world behind it as a result of changing his vantage point. Moreover, he was baffled by being able to look at the back of it. His mother came to the rescue by telling him that this was not a window but a screen. Initially, the toddler could not make much sense of this. The same kind of situation was repeated when, one evening, his father had to catch up on his work, having lost a lot of time as a result of the move. As he opened his laptop on the dining room table, the toddler rushed over, looked at it carefully, and again asked: "Window?" It is these kinds of experiences that may further reinforce his intuitive feelings about another world to which he has no access. He knows very well that when he looks out a regular window and notices the neighbour coming with her daughter, his mother only has to open the door and his friend will step in, and they can begin to play together. However, the kids he saw on the television screen could not enter into his world.

The symbolic developments described thus far place a person at the centre of the following interactions. A person evolves through a growing self-awareness. This self-awareness expresses itself in a diversity of relationships that are woven into the fabric of a life. The objects of these relationships were previously drawn from undifferentiated stimuli that were no more than experiential noise, of which the person could make no sense and with which he thus could not establish a relationship. The objects of the relationships of a person's life jointly constitute a cultural envelopment that becomes the person's "world" and eventually the symbolic universe of the community relative to which his or her life is lived. In due course, the unknown will become symbolized as more of what is known and lived in order to sustain the person's life and allow it to proceed with complete confidence despite its relativity. Finally, all these relationships, the objects of these relationships, and the "world" (or symbolic universe) are internalized, causing the person's life to adapt and evolve. There is a constant unfolding of the individual's potential in relation to an unfolding of the relativity of the world that becomes eventually transformed into the sameness by means of myths. There is thus a correspondence between the level of differentiation of the one relative to the other, and vice versa. The unfolding of a person's life is particularly marked by the presence or absence of an awareness of being a physical self, a social self, and a cultural self. It cannot be adequately emphasized that the social embodiment of children is entirely centred in their physical embodiment. Similarly, their cultural embodiment is centred in their social embodiment and, via it, in their physical embodiment. Even though, in our contemporary anti-societies, there is a great deal of decentring, any fantasies of eternally living on the net, or preserving a person's life in terms of information, are symptoms of our enslavement – an enslavement that takes it as self-evident that we are *homo informaticus*. These thoughts merely serve to make our enslavement liveable.

The above developments have become a great deal more complex in our contemporary anti-societies. As noted, toddlers and children now encounter two streams of experiences which they may be incapable of differentiating: the experiences with a language foreground and those with an "engineered" construct as foreground. Initially, stories and image-stories as well as words and image-words may be differentiated as being different forms of the same thing. Some indications as to how these two streams of experiences become gradually differentiated from each other would require careful studies, but there is no question that by the time young people enter high school, this has been accomplished. Without it, they would be unable to learn disciplines such as physics and chemistry. They have developed a tertiary frame of reference that allows the technically mediated experiences to be directly differentiated from each other, and jointly differentiated from the culturally mediated experiences with a language foreground. The metaconscious knowledge built up with the former stream will be separated from experience and culture, while the metaconscious knowledge built up with the latter stream will be embedded in them. Similarly, the semantic memories will be split into those that are embedded in experience and culture and those that are separated from them.

These distinctions may provide us with a clue as to how young people learn to differentiate the two streams of experiences. The two streams differ greatly in terms of the level of embodiment (physical, social, and cultural), participation, commitment, and alienation.[45] The research of Sherry Turkle has shown that extroverted children tend toward the more socially and culturally embodied experiences that also require a fuller participation and commitment.[46] When we meet another person and then begin to like one another a great deal, we embark on a relationship that deepens as we share more and more of ourselves and our lives. Entrusting another person with so much of ourselves implies a confidence that it will not be abused because the other person is as committed to us as we are to him or her. In other words, these kinds of relationships can be ambiguous, risky, but also immensely beautiful. It is for these very reasons that other persons may look for safer relationships which, in our anti-societies, increasingly involve a construct of one kind or another. The "world" of such constructs is a great deal simpler: things are either black or white, there is little ambiguity, and it is possible to fully win or lose. In other words, reality also has a lot to offer in the eyes of other people. This has been made a great deal more complex because our extensive use of screen-based devices tends to strengthen some aspects of our personalities and weaken others. These processes can become self-reinforcing; and when they do, it appears to be mostly in favour of reality.[47] As noted, this work seeks to understand what is happening to human life relative to the conflict between technique and culture that we have created through our secular myths. We live with the one as if it had no limits and with the other as capable only of sustaining our personal and subjective needs. If we were able to truly symbolize technique as well as culture in terms of their meaning and value for individual and collective human life, we would turn to each one for what it truly is and what it could accomplish. Relying on them within each of their limits would then lead to a complementary and harmonious relationship. Nothing of the kind is now possible; and this will become increasingly evident as we turn to specific aspects of growing up in an anti-society. As has been the case throughout human history, we are trapped in our present situation because of our ultimate commitments to our secular myths. However, our present participation in building a global technical order and our secular religious allegiance to the state-nation means that if indeed our ways of life turn out to be neither liveable nor sustainable, there will not be any alternative in the form of another civilization. Humanity is increasingly committed to our universal approaches to knowing, doing, and organizing and, in the process, desymbolizing all symbolic and cultural alternatives.

Before we turn to anti-societies in the next chapter, one of their most important features has gradually become apparent. Its development has been a major factor in transforming traditional societies into their diametrical opposites. Prior to the great cultural divide, language had been the basis for establishing, maintaining, and evolving all the relationships between the members of any society. Paradoxically, a language backed by a culture was possibly the most fragile way of doing so. What speakers said always implied more than what they were aware of because their entire lives and cultures were always working in the background. Consequently, it needed to be interpreted by the listeners. Such interpretations also implied more than what the listeners were aware of because these interpretations were also arrived at by their own lives and cultures working in the background. A symbolic distance was thus created between speakers and listeners, which allowed for a significant margin of freedom for all parties.

In contrast, most of the relationships between the members of anti-societies are mediated by images backed by image-words and image-stories, or myth-information and myth-stories, which completely alter the involvement of both speakers and listeners, in terms of embodiment, participation, and commitment. In a great many instances, these images are not backed by speakers, but by the results of the application of public relations techniques presented by spokespersons on behalf of corporations, governments, or other institutions. This technical mediation greatly delimits both their freedom and that of their listeners because of the power of the techniques being used. Alternatively, when these relationships between the members of anti-societies involve the intermediary of social media, the same kinds of issues occur to a somewhat lesser extent because of self-branding aimed at projecting a certain image of a person and a life similar to those projected by public relations techniques, advertising, and so on. Once again, the independence of both speakers and listeners is undermined. It is difficult for us to appreciate how language and culture acted as a kind of "social lubricant" prior to the great cultural divide. We fail to appreciate the extent and depth to which the power of technique has become internalized into most relationships between the members of anti-societies. It has completely altered the fabric of social relationships, in such a way that is very difficult to sustain the lives of individual members. It does so by decisively weakening the dialectical tension between individual diversity and a shared cultural unity as a consequence of desymbolization. This lack of social and cultural support had to be compensated for by integration propaganda – with significant negative consequences for the well-being and mental health of us all. As a result, technique has diminished not only the capacity of the biosphere to sustain human societies but also the ability of societies to sustain their members.

This change in collective human life is further rooted in the way the members of anti-societies use their senses. Before the great cultural divide, the visual dimension of experience was interpreted by means of the secondary frame of reference. This interpretation included the elimination of the "edges" around people's fields of vision, for example, because such edges could have no meaning or value for their lives. People knew very well that all they had to do was to move their heads to see more of what was all around them, but the secondary frame of reference helped to establish a deeper metaconscious knowledge that implied that this "more" of the visual world remained within the culture: the visual dimension of the symbolic universe in which people lived.

In contrast, the members of anti-societies primarily live their lives relative to what appears on screens and what is contained in the domains of disciplines – in which they either participate professionally or on the receiving end of which they find themselves in their daily lives. Their deepest metaconscious knowledge informs them that this is their "real" life, and that their personal life matters a great deal less to their society, even though it may be subjectively more rewarding and interesting. It is now the tertiary frame of reference that integrates both what is seen and what is perceived via language and culture, but does so in favour of the former at the expense of the latter. Although the tertiary frame of reference continues to interpret out the "edges" around the field of vision, what lies beyond it has now become the reality in which the person lives by means of a way of life that is integral to the constructed global technical order. This reversal in the relationships between our two most important senses, those of seeing and hearing, permeates all experiences, relationships, and the fabrics they weave in individual and collective human life in anti-societies. In addition, what is interpreted by means of a tertiary frame of reference involves a highly desymbolized life and a self that, by operating this "radar," becomes minimally embodied, restricted in participation, and limited in commitment. It is an entirely different relationship from what was typical in traditional societies, in which people used the organizations of their brain-minds as complete mental maps of their lives, into which were enfolded their sociocultural selves as the map readers.

3 Living with a Dual Relativity beyond Cultural Embodiment

A General Interpretation of Our Dual Relativity

We have thus far developed a general interpretation of the relativity of human life and the world. In the first two chapters, we have essentially focused on the "internal" relativity of babies and toddlers and how their growing awareness of it begins to open up a greater access to the "external" relativity of human life and the world. In their own playful explorations they encounter this relativity as open-ended, while in their contacts with significant others they encounter the sameness of this relativity resulting from their community's symbolical grounding of it. Before language acquisition, their making sense of these relativities completely intermingles the two streams of resulting experiences because of their shared dialectically enfolded architecture. Beginning with the acquisition of vocal signs, but especially with their transitioning into language, most people will interact with small children on the basis of their (highly desymbolized) primary and secondary frame of reference. When, in some situations, these grown-ups rely mostly on their tertiary frame of reference with screen-based devices and the reality these devices display, it is highly unlikely that toddlers will be able to make much sense of this. Consequently, their earliest development resembles that of growing up in a traditional society with a symbolic universe, except for the high levels of desymbolization of the language and culture being used.

At this point in our analysis, it becomes essential to recognize that, via the intermediary of the significant others in the lives of these toddlers, the influence of an anti-society will be very different from that of a traditional society. In an anti-society, the level of dialectical enfolding of adults into this society has been greatly weakened through desymbolization. To be sure, there still is no separate society "out there." These people remain both individual and society, as individually unique expressions of the latter. Nevertheless, following the great cultural divide, the introduction of the technological and economic orders into the cultural orders of the industrializing societies split the general relativity of human life and society into what we have referred to as its culture-based connectedness and its technology-based connectedness, each with diametrically opposite architectures. The limitations of culture-based approaches in these new orders gave rise to alternative rational technological and economic approaches, which still later were replaced by discipline-based technical ones. When the latter spread into all areas of a way of life, the cultural orders were replaced by a constructed and universal technical order. This development has had far-reaching implications for the growing up of children and teenagers. They will increasingly contribute to the adaptation and evolution of their society's cultural relativity of human life and the world as the significance of their participation grows. We will now turn to the vantage point of society to examine how children become individually unique expressions of the anti-societies in which they now grow up.

The members of each new generation born into an anti-society no longer grow up to participate in the historical evolution of a cultural order (with a dialectically enfolded architecture) embedded in a natural order (with an enfolded architecture). Instead, they grow up to participate in a *constructed* technical order, one with an architecture that exhibits the characteristics of reality: non-contradiction, separability, closed definitions, openness to measurement (and thus to quantification and mathematical representation), and having a simple complexity. Table 1 shows some of the fundamental differences between the living cultural order of a traditional society and the constructed technical order of an anti-society.

What we habitually refer to as the science, technology, economy, social structure, political institutions, legal framework, morality, religion, and artistic expressions of a traditional society were lived as dimensions of a dialectically enfolded culture that mediated all relationships engaged in by its members and integrated them into their individual lives as well as into their collective way of life.

Industrialization, urbanization, rationalization, and secularization helped to create a cultural divide beyond which all such dimensions of mediation of a traditional culture were unfolded from a dialectically

Table 1

Cultural dimensions of mediation or technical constructs	Civilization based on culture	Civilization based on technique
Science	Embedded in experience and culture, local, culture specific	Separated from experience and culture, non-local, universal
Technology		
– knowledge	Embedded in experience and culture, local	Separated from experience and culture, universal
– organization	Non-systematic	A system
– regulation	By means of cultural values	By means of non-cultural performance ratios
Economy	Embedded in and inseparable from a way of life and regulated by its culture	Initially a separate and distinct economy ruled over by the universal institution of the Market and based on the economic approach to life involving the maximization of utility or profit. The remainder of society ruled over by the cultural approach. Later, the economic approach mutates into the technical approach and rules over all of society
Social structure	Tradition-based societies	Mass societies with weak culture supplemented by integration propaganda
Political structure	A small and distinct political sphere because most of human life is self-regulating by means of a culture	Everything is political. The state is involved in almost every area of human life
Legal institutions	Laws are a strong expression of, and an extension of, the metaconscious values of a community. Common law tradition	Laws have an organizational character reflecting the technical orientation of society
Morality	Group morality anchored in the cultural unity of a society	Individualistic, statistical morality; what most people do is normal, and what is normal is normative

(*Continued*)

Table 1 (Continued)

Cultural dimensions of mediation or technical constructs	Civilization based on culture	Civilization based on technique
Dominant personality type	Tradition-directed	Other-directed, reality-directed
Religious institutions	Traditional religions	Secular political religions
Art and literature	Symbolizing the cultural unity and changes within it; what is the most profound in a civilization	The arts depicting non-sense, non-cultural
Relations with the biosphere	Generally sustainable	Unsustainable
Overall characteristics of a way of life	Culture-based connectedness dominates technology-based connectedness	Technique-based connectedness dominates, permeates, and envelops culture-based connectedness, universal

enfolded individual and collective human life, to become relatively distinct and separate constructs. For example, the introduction of the technical division of labour unfolded human work from a dialectically enfolded life, thereafter to be thought of and dealt with as a construct built up from a sequence of distinct and separate production steps that produced a final output.[1] This unfolding of work from people's lives necessitated the previously noted triple abstraction that conceptually, theoretically, and practically excluded whatever remained of their lives.

When people assigned to the different production steps laboured in this manner, everything that had been externalized did not magically disappear, however. The tensions between what had been taken into account as technically divided labour and the remainder of their lives turned into a major source of physical and mental illness. Moreover, a greater labour productivity increased the local throughput of matter, energy, and their composites in the technology-based connectedness of society at the expense of its culture-based connectedness, which needed to be sacrificed to varying degrees.[2] The introduction of the technical division of labour was rapidly followed by mechanization and eventually automation and computerization. These developments spread through an industrializing society, resulting in a technological order characterized by the domination of the technology-based connectedness over the culture-based connectedness, while in the remainder of society the reverse remained the case. It thus became increasingly difficult to evolve this technology-based connectedness by means of a traditional culture. A new economic approach emerged that unfolded the economic dimension of mediation from the dialectically enfolded cultural order. The result was a construct built up from markets that each independently organized one good or service and did so relative to the narrowest possible context of the supply and demand of that one good or service. Such a piecemeal economic organization required the unfolding of all economic activities both from each other and from the way a culture related everything to everything else by means of meanings and values. This evolving over time was now guided in a different way.

The technological and economic dimensions of cultural mediation were thus turned into constructs built up with human activities that were similarly unfolded from people's lives, to the degree that they resembled the actions of an abstract *homo economicus*: a utility-maximizing algorithm in the image of a wage earner, and a profit-maximizing algorithm in the image of an entrepreneur. Their actions were mediated by the market mechanism of supply and demand one good or service at a time. All this is very far removed from the possibility of such activities being genuine individually unique and culturally typical expressions of individual and collective human lives working in the background. As for technically divided labour, this unfolding of economic activities was not total, with the result that these activities also became a source of many tensions between the emerging economic construct and the cultural order of a society. The tensions multiplied over time, to a degree that transformed these economic constructs from wealth-producing into wealth-extracting entities. The costs of all these tensions grew out of control, and eventually overtook gross wealth production.[3]

In addition to the well-known market externalities and their synergistic aspects cumulating into Market forces, there were also what may be called "internalities": the results of a limited unfolding producing internal tensions within an economic construct and the lives of those participating in it. Wherever the new Market displaced a traditional culture in organizing an economy, it contributed to the desymbolization and eventual collapse of the traditional culture and its values.[4] There was a significant incompatibility between the market prices of goods and services, on the one hand, and the meanings and values of these goods and services for individual and collective human life as established by the culture-based symbolization of everything relative to everything else, on the other. Nevertheless, people increasingly had no choice but to govern their behaviour according to Market prices – at the expense of individual and collective human life.

It thus became impossible to live a human life relative to an economic construct other than by commoditizing and reifying that life. It had to be divided between those aspects that were the "human resources" for the technological and economic constructs and those that could be excluded from them, because the architecture of these constructs could not accommodate a dialectically enfolded individual and collective human life working in the background of all technological and economic activities. People could no longer live their lives relative to the technological and economic constructs of the industrializing societies, other than by means of a technical mediation requiring the commoditization of technological and economic activities. A person could participate only as a hand, a brain, or an intermediary coordinating or management function, to "produce" the aspects corresponding to the necessary "resources" for these constructs. This situation contrasts sharply with that of people participating in the technological and economic dimensions of cultural mediation, mostly by means of metaconscious knowledge embedded in experience and culture and thus with their lives working in the background as members of a symbolic species.

Traditional activities were grounded in the relativity of people's lives and their world through traditional myths; and this made it impossible to participate in traditional technological and economic activities other than as dimensions of human lives enslaved by these myths. In contrast, living human lives relative to technological and economic constructs superimposed commoditization and reification on the alienation brought by secular myths. The market externalities and the costs they create are but symptoms of the deeper problem of turning life into non-life by imposing a change in architecture on it.

As the technological and economic constructs grew within the cultural orders of the industrializing societies, the above tensions multiplied in scope and depth. The radical differences between the architectures of these constructs and those of their enveloping cultural orders were confirmed by the rapidity with which the symbolic cultural approaches reached their limits, which placed substantial constraints on the early development of these economic constructs. This is hardly surprising, because symbolization, language, and culture had evolved to make sense of and interact with a living world. Further technological and economic development required that the relevant human knowing and doing be separated from experience and culture. This was done by successively transforming human knowing and doing, first into an economic approach, then a rational approach, and finally, a discipline-based approach that no longer made any reference to sense.[5] During the Second World War, but especially following it, discipline-based approaches spread into all areas of human life without any regard for their limits. When nearly all aspects of a way of life became organized and reorganized by means of discipline-based approaches without any reference to sense, the results took the form of a technical order of non-sense – the diametrical opposite of a cultural order of sense. Since a society's technical order was no longer built relative to a time, place, and culture, it took on a universal character, one that permitted all such orders to be seamlessly integrated into a global order. People needed to mediate their relationships with the technical order by the domains of disciplines, and this became the cause and effect of all cultural dimensions of mediation becoming constructs acting as building blocks of a global technical order.

The development of this global technical order made it essential to treat all life as non-life in ever greater detail and precision. It is no exaggeration to sum up the achievements of our current global civilization as the creation of a mega-project by which humanity organizes all life in the image of non-life by relying on technique rather than on culture, although the latter had always permitted human life to be lived relative to local conditions. Nearly all the accomplishments and failures of this mega-project derive directly from the limits of technique relative to culture, and the limits of culture relative to technique.

Such a dialectically enfolded individual and collective human life in relation to a constructed order of non-life needed to be made liveable by means of secular myths in order to ground that life. Without these myths and the secular religious attitudes they sustain, the lives of children and teenagers would have become unbearable, because they would have progressively discovered that their lives were being turned into expressions of non-life. It would have become unliveable to watch themselves being turned into bundles of commoditized aspects in tension with the remainder of their lives in order to become the necessary "resources" for the global technical order. By acquiring our myths and secular religious attitudes, their lives became oblivious to all this except for a variety of symptoms; but these were interpreted and trivialized in a way that posed no obstacles to future technical solutions. Individually and collectively, each new generation is thus unaware of the war we have unleashed on ourselves, including a loss of what it is to be human as a symbolic species. Because we live with our discipline-based knowing, doing, and organizing as if they had no limits, the difficulties and crises we face are deemed well within our abilities to overcome. In this way, our contemporary anti-societies have prevented human life from being plunged into cultural chaos – by means of secular myths generated by our highly desymbolized cultures that delegate all knowing, doing, and organizing in the public sphere to discipline-based approaches under the authority of institutions accommodated to this purpose. Each new generation is therefore unaware of humanity's having put everything on the line for a limitless future made possible by equally limitless discipline-based approaches.

It would be very difficult to overstate the critically important role played by secular religious attitudes in the way children and teenagers learn to live their lives relative to constructs and to a global technical order of non-sense. Ever since the first phase of industrialization, the growing politicization of everything and the battles between the political Right and Left formed a perfect veil that obscured what was happening to human life and society. It was not the deep tensions that were placing individual and collective human life on a rack, but a simple dispute over how the new technological and economic future was to be organized and how the benefits were to be shared. The redirections of human attention away from what was truly happening were endless. The introduction of the technical division of labour and everything that followed in the forms of mechanization, automation, and computerization would not alienate factory workers, provided that the means of production were collectively owned. The accumulation of capital necessary to make the huge investments to build the infrastructures of industrializing societies led to a capitalist system that was opposed to socialism and communism, as if the governments of these political systems could make the capital they required appear by magic out of nowhere. On the political Right, the struggles between management and labour were only over power and the way the benefits were to be shared. Going to the root of many problems by joint ventures, such as the creation of healthier workplaces and different ways of organizing the work, rarely came to mind; and when they did, they were usually dismissed as being economically impractical. The distraction of public attention away from our alienation – and eventual commoditization and reification – continues, and will doubtless outlive us all. The Left and the Right can blame each other, thus collaborating in ensuring that we do not face what is truly happening to all of us.

Is this any different from the past, when agricultural societies engaged in bloody rituals to convince their gods that rain was urgently needed for their agricultural endeavours? Rituals equally prevented those societies from tackling the issue of drought head-on. The present extreme political polarization in the United States, and its upsurge in Canada, are accompanied by active support from Christian social conservative movements that legitimate all this in the name of God. Religion continues to be an extremely successful "opium for the masses," allowing them to avoid dealing with what we are doing to ourselves and the planet. All this politicization by means of secular religious attitudes – now increasingly reinforced by traditional religious Christian attitudes – of what is fundamentally non-political reflects how we (supposedly secular people), like those who went before us, continue to be enslaved to myths. There is nothing new under the sun, other than the unprecedented power of our global technical order over everything it touches.

Each new generation growing up in our contemporary anti-societies must learn to deal with a dual relativity: that of a constructed reality and that of the desymbolized and weakened dialectically enfolded activities of human life and the world related to people's personal lives lived with significant others. The former has been integrated into a global technical order, while the latter continues to be lived by people relative to their time, place, and a highly desymbolized personal sense of space on the margins of society. People continue to encounter constructed situations as well as greatly weakened dialectically enfolded ones, which gives rise to their previously noted two streams of distinct and separate experiences with diametrically opposite architectures. These architectures correspond to those of a technical order and to those of a personal space shared with significant others as a subjective sphere, giving rise to parallel modes of knowing and doing that coexist in all our lives. The previously noted example is typical: after having taken a few physics courses, people will respond to daily-life situations configured as physics problems by giving a "school physics" answer by means of the tertiary frame of reference. However, if these situations are described in daily-life terms, an "intuitive physics" answer is given with the use of the secondary frame of reference. These answers may be entirely contradictory. The brain-minds of the people giving such answers make either technical significance or cultural sense of a situation depending on what is appropriate in terms of their dual frame of interpretation (one embedded in experience and culture, and the other separated from them). It is as if their brain-minds have learned to function as a kind of bifocal mental lens, of which one part is built up from scientific and technical discipline-based education and the other part by symbolizing daily-life experiences by means of a desymbolized culture.

This dual mediation is learned by acquiring the previously discussed tertiary frame of reference that essentially differentiates our experiences, giving rise to semantic memories forming two constellations of clusters of such experiences. One of these contains the clusters of experiences interpreted by means of the secondary frame of reference that remains embedded in experience and culture, while the other contains the clusters of experiences interpreted by the tertiary frame of reference dealing with what is separated from experience and culture, and thus is characterized by the dual architecture with a foreground and a primary background separated from a secondary background. Consequently, the organization of a person's brain-mind will act as one part of the bifocal mental lens when an experience is differentiated as being radically different from all those that have formed the one constellation, and thus as being different from, but not unlike, one or more experiences belonging to the other constellation. It functions as the other part of the bifocal mental lens when the reverse is the case. The part of the bifocal mental lens associated with the secondary frame of reference deals with what can be made intelligent through language and culture as an extension of what is aurally and temporally perceived. The other part of the bifocal lens extends our spatial perception of human life and the world by dealing with whatever cannot be seen and must be imagined via the intermediary of a scientific or technical discipline with a domain having the same architecture as the world we see.

Children growing up with the above kind of dual relativity of human life and the world, corresponding to their anti-society's participation in a universal technical order and in a local (desymbolized) cultural life, develop very differently from children growing up in traditional societies. Their entirely new forms of socialization have been in the making for some two centuries, but they have gone largely unnoticed because we live as if we have simply been adding all kinds of new discoveries and possibilities to our ways of life while subtracting nothing essential. Doing so is commonplace because what we have summarily referred to as "technique changing people" through the vast influence of the technical order on human life, society, and the biosphere cannot be directly experienced and is blocked by our failure to symbolize this technical order in terms of the meaning and value it has for our lives. Ever since humanity began to construct an economic order followed by a technical order, everything related to these efforts amounted to a preparation for an almost universal use of computers.[6] It is quite remarkable that the original estimates of their usefulness were exceedingly limited, and even their use in large organizations initially led to only modest productivity increases – until it was recognized that the entire corporation also had to be reconstructed in the image of the computer.[7] The computer and the accompanying information technology have little scope for application when the relativity of human life and society has a dialectically enfolded architecture, but it has almost unlimited applicability when much of that human life and society is dealt with by means of the tertiary frame of reference, which is ideally suited to the architecture of reality. Since this preparation for the computer and information revolution can thus be interpreted as having been in the making for nearly two centuries, their bursting on the scene in the late twentieth century and their rapid spread into every aspect of society through the World Wide Web should have come as no surprise. Along with this development came the rapid expansion of the duality of the relativity of human life and society to an unprecedented level. It is increasingly permeating the socialization of toddlers and children, a situation for which as a society we are entirely unprepared, especially our educational system.

Living and Constructed Entities

In order to further clarify the implications of all these developments, the differences between a constructed entity and a dialectically unfolding and enfolding entity cannot be overemphasized in our technical age. We do well to constantly remind ourselves that everything is relative. What something *is* must be understood in relation to its "internal" relativity and its "external" relativity, but never in terms of itself. Consequently, anything comes into existence either by being constructed or by contributing to its temporal unfolding and enfolding evolution and adaptation. For example, a mechanical construct is built up from separate and distinct parts that in their turn were constructed through separate and distinct processes that may have depended on others, and so on, all the way back to the society-biosphere boundary. It is across this boundary that we borrow all the required matter and energy that we are unable to either create or destroy. Hence, our constructed entities are a rearranging of what is already present.

It may well be objected that we have created a great many new things. However, do we ever truly create something? Or do we, in more or less ingenious ways, simply rearrange the existing relativity to "create" something new? From a cultural perspective, do our ideas and inventions not build on the relativity that came before us, and will future generations not build on the relativity we have passed on to them? Can anything ever be considered as entirely new *in itself* to the point that it requires no reference to anything from which it came and relative to which it exists? For example: Does technology not advance to a considerable extent because the relativity within which it makes sense has changed, thereby changing the effects of technology, and this in turn drawing our attention to the point of possibly convincing us that a better fit ought to be possible by reinventing it? This component of technological evolution is now increasingly dwarfed by another, which emerged when discipline-based approaches gained the upper hand, making many changes attributable to the flows of technical information within the global system. A great many things are now undertaken because it has become technically feasible to construct them as opposed to responding to the changing relativity of human life. It has resulted in technique's taking on a considerable autonomy with regard to this relativity. Moreover, this autonomy could arise only because of our being alienated by the system of technique as a consequence of technique's changing us into its faithful servants, thanks to our secular myths. In other words, technical advances have become related to the constructed relativity of the system of technique and the global technical order it adapts and evolves.

In contrast, living entities do not come into existence in such a piecemeal fashion. Each one is a whole from the very beginning, which unfolds itself by progressive differentiation that in most cases involves stem cells becoming unique and particular expressions of this whole. This unfolding is enabled by the DNA enfolded in each cell, which allows the whole to unfold relative to each cell, and each cell to unfold relative to the whole. The integrality of the whole can thus be maintained on a level that is entirely unachievable in a constructed entity. The internal relativity is organized by a multiplicity of stem cells expressing the whole as a tissue, an organ, and larger "subsystems" that enfold them to carry out functions of the whole that would otherwise be impossible.

This internal relativity is lived in relation to a larger external relativity, of which it constitutes an integral whole.

E.O. Wilson has succinctly expressed our inability to "construct" an ecosystem as if all its constituent organisms were its "parts":

> In the forest patch live legions of species: perhaps 300 birds, 500 butterflies, 200 ants, 50,000 beetles, 1,000 trees, 5,000 fungi, tens of thousands of bacteria and so on down a long roster of major groups. Each species occupies a precise niche, demanding a certain place, and exact microclimate, particular nutrients and temperature and humidity cycles with specified timing to trigger phases of the life cycle. Many, perhaps most, of the species are locked in symbioses with other species; they cannot survive and reproduce unless arrayed with their partners in the correct idiosyncratic configurations.
>
> Even if the biologist pulled off the taxonomic equivalent of the Manhattan Project, sorting and preserving cultures of all the species, they could not then put the community back together again. It would be like unscrambling an egg with a pair of spoons. The biology of the microorganisms needed to reanimate the soil would be mostly unknown. The pollinators of most of the flowers and the correct timing of their appearance could only be guessed. The "assembly rules," the sequence in which species must be allowed to colonize in order to coexist indefinitely, would remain in the realm of theory.[8]

In other words, the only way of "constructing" this ecosystem would be to go through the equivalent of the evolutionary process all over again, during which the potential of each organism unfolds relative to the ecosystem, and as the ecosystem develops relative to each organism. As the DNA expresses something of the internal relativity of each whole, it also expresses something of the relativity of the ecosystem by belonging to its DNA-pool.

The same kind of argument can be made regarding the growth of embryos into newborn babies. Their internal relativity is an expression of their DNA made possible by the matter and energy supplied through the umbilical cord in the womb of the mother. We have attempted to briefly sketch how the social relativity of babies emerges from their physical relativity, and how their cultural relativity emerges from the social relativity. We now turn to this cultural embodiment from the perspective of the relativity of the society in which each new generation grows up. We have already hinted at the relativity of a traditional society finding its expression in a culture symbolically grounded in myths. The relativity of so-called anti-societies is symbolically grounded in secular myths, but their architecture has been transformed from a dialectically enfolded kind into that of a constructed reality. From a historical perspective, the potential of a traditional society thus unfolded itself relative to its "cultural DNA," and that of an anti-society unfolded itself relative to its technical order and reality. An individual member is a biological whole made up of cells that share a unique DNA, as well as a cultural whole as an individually unique expression of the cultural unity of the society or anti-society. Furthermore, because of the internal and external relativity of its members, their potential is launched toward something that is never entirely fixed or closed. There can be no fate or destiny other than what we invent and impose on ourselves. The potential remains open-ended because the sameness that marks every historical period of the journey of a community comes to an end when its relativity becomes unliveable as it is pushed out by another relativity, provided that the society does not disintegrate or collapse from a lack of a new emerging cultural unity. When ways of life became constructed by means of discipline-based approaches, the character of their relativity changed from a dialectically enfolded one to a dual relativity. This dual relativity adapted itself and developed in very different ways from the relativity of traditional societies. How do these and other changes affect the cultural embodiment of the generations growing up in anti-societies?

This interpretation of the general relativity of human life may be summed up as our lives not being life itself. We do not have life within us any more than we are able to create the matter and energy on which our lives depend. We are not life itself. We can only transmit life through sexual reproduction, but we cannot *have* life any more than the matter and energy that we borrow to sustain it. It is integral to our being finite and relative human beings.

Beyond the question as to whether or not our lives can one day be captured by millions of lines of computer code lies a much more important question. Can we reverse-engineer such a construction in order to extract from it the equivalent of its DNA and its cultural complement? Will we be able to express its potential and then launch it toward something that is open-ended? How would we next incorporate this potential into what amounts to a mega-algorithm? If we could be successful, we may be able to launch this potential in an imitation of life itself. However, our ability to construct anything of the kind would be light-years away from an ability to transform the architecture of such an algorithm into an enfolded and dialectically enfolded architecture that characterizes our lives. Doing so would almost certainly involve a radical shift away from our dominant spatial perspective to a more temporal one. For example, the above perspective on an ecosystem is more temporal than it is spatial, and this is equally true for our lives.

The Emergence of Cultural Mediation in a General Relativity

As Arnold Toynbee and others have shown, civilizations are born, grow up, mature, stagnate, decline, and collapse.[9] This is hardly surprising, since they are made up of a cluster of interacting societies with similar cultural unities, and these societies in turn are made up from people. As a result, civilizations may be interpreted as living entities that cannot be constructed. The members of each society individually and collectively mediate all their relationships by means of a culture as a shared way of making sense of and living in the world.[10] Each generation is born into the relativity adapted and evolved by preceding generations through the culture that also symbolically grounds it. This cultural mediation has a dialectically enfolded architecture that unfolds its potential through progressive differentiation and occasional re-enfolding to facilitate and sustain the historical journey of a society. Generally speaking, this cultural mediation became differentiated into a variety of dimensions in response to particular obstacles, necessities, and opportunities that presented themselves.[11] Jointly, these dimensions of cultural mediation express what is commonly referred to as "the spirit of an age," which mutates into the spirit of another age during the transition from one historical epoch to another. As a result, there was no possibility of a universal technology, since, as archaeology demonstrates, each technology and its artifacts bore the stamp of a specific time, place, and culture. The technology was thus one dimension of the cultural mediation of a society. The same was true for science and the economy.

Gilbert Simondon has suggested that the earliest human groups culturally mediated all their relationships by means of a primitive magic.[12] He then showed how this became untenable when early human groups had to deal with what we have attributed to the relative and open-ended character of human life. The result was that, over time, the original magical cultural mediation gradually became differentiated into several dimensions. What each dimension expresses of this cultural mediation implies something of how it is differentiated from all the others and how they are from it, as well as how it is differentiated from the overall cultural mediation, and vice versa.

We have previously discussed how our symbolic species had no choice but to confront the relativity of human life and the world by symbolically grounding in myths the relativity of what a group or society knew and lived. That their members did so was evident from their behaviour, which implied that the entities corresponding to these myths were no longer relative but absolute. The gods they thus created as a community could then be obliged by serving them through a morality and a religion. This absolutization was the equivalent of symbolizing the unknown as more of what a group or society knew and lived, and thus as something implied in every experience, which made anything radically "other" unknowable and unliveable. It is metaconscious, and thus the opposite of a conscious ontology.

In the same vein, we have also noted that any society's confrontation with the relativity of human life and the world had to be made more liveable by the introduction of artificial juridical models for a variety of relationships that stabilized, and made human life as it was known and lived a little more predictable and reliable. Consequently, this primitive magical mediation began to be differentiated into its religious, moral, juridical, and magical dimensions; and their universality appears to indicate that no group or society has ever escaped confronting the necessities resulting from the relativity of human life and the world.

Simondon suggests that the original cultural mediation by primitive magic also needed to be further differentiated because different religious and technical approaches became necessary.[13] These were increasingly differentiated from the cultural mediation, although they continued to enfold something of the magical wholeness out of which they emerged. Even our contemporary technique retains an aspect of magic, according to Richard Stivers.[14] It has also regained a secular religious character.[15] Moreover, throughout human history, traditional technologies that were culturally mediated and the objects they produced always bore the stamp of their time, place, and culture.

According to Simondon, the aesthetic expressions of human life in the world were also increasingly differentiated from the original magical mediation, while continuing to enfold something of its wholeness.[16] The resulting works of art sought to reconstruct the original magical universe into a limited aesthetic one.[17] Sacred and profane art did so in different ways. The former encompassed attitudes of ritual and mysticism, making them comparable to the acts of priests in seeking to recover some of the lost wholeness. Profane art expressed itself in aesthetic works that were situated between what people lived as being true in their lives and their world and what was good in response to the moral necessities imposed on their lives. This form of art thus mediated between knowing and willing.[18]

Eventually, a political dimension of cultural mediation was differentiated from the religious one.[19] The latter emerged by acting on intuitions based on the deepest metaconscious knowledge as to what was so meaningful and valuable to human life and the world that its withdrawal would open the door to cultural chaos, and thus to the impossibility of human life as it was known and lived. Such entities not only made human life and the world what they were during a time, place, and culture but also created and maintained that life and world. As these metaconsciously created gods, they could not ever err. Consequently, religious approaches had to be separated from political ones, which were often misguided or mistaken.

It is for these reasons that every culture had to create a religious dimension of mediation to appease the gods it had created for itself. For example, in a totemic symbolic universe these powers had to be reached in order to ensure the regularity of all natural processes that were at the whim of the spirits, but on which human life depended. However, the institutionalization of a religious authority had to be distanced from a political authority, for the above reasons. Nevertheless, the legitimacy of the political institutions invariably derived from a religious authority.

From Cultural to Technical Mediation

Possibly the most important aspect of Simondon's interpretation of how the cultural mediation of all our relationships became increasingly differentiated is due to its need to respond to the necessities imposed by the relativity of human life and the world. Such developments led to further necessities of creating, maintaining, and adapting a way of life that matched the aspirations of a community to the moral and religious obligations it had imposed on itself. In turn, all this had to be matched to what the group or society was able to accomplish based on what it knew and did about everything in its surroundings through symbolization. Just prior to the beginning of industrialization, the traditional cultures of West European societies had created ways of life that mediated all their relationships by means of scientific, technological, economic, social, political, legal, moral, religious, and aesthetic dimensions of mediation.[20] During the past two hundred years, this cultural mediation was turned into a technical mediation utilizing discipline-based approaches, complemented by a highly desymbolized cultural mediation restricted to people's personal lives involving significant others. We have referred to prior analyses of how the introduction of the technical division of labour and subsequent mechanization and industrialization introduced an economic order which included sectors dominated by a technological order. This economic order was split off from the remainder of society, with its increasingly desymbolized cultural order. This transformation implied a reversal in the priority of the culture-based connectedness of human life and society over the technology-based connectedness, in favour of the latter.

When discipline-based approaches to knowing, doing, and organizing spread throughout all areas of human life, the cultural dimensions of mediation were transformed into relatively distinct constructs of technical mediation. It is as though our contemporary ways of life were constructed out of building blocks that we live with as if they were relatively separate and distinct from one another. These include: a universal science, technology, and economy backed by the resources of a social structure, political framework, legal organization, statistical morality, secular political religions, and entirely new forms of art. The structure of our contemporary universities amply confirms this. Each former dimension of cultural mediation has thus become comprehensible by and amenable to discipline-based approaches that are universal in character, separated from experience and culture, and autonomous from each other. We will illustrate this with the emergence of an art of non-sense – art that is separated from experience and culture and contributes to the global technical order with its own unique desymbolized elements.

Although the constructs of the technical order are dominated by techniques of all kinds, they do retain aspects of a past cultural and historical character. We must not imagine that our anti-societies participating in the building of a global technical order have turned themselves into mega-machines.[21] Doing so would be the equivalent of arguing that, during the eighteenth and nineteenth centuries, markets comprehensively regulated the distribution of goods and services without any market externalities. We know very well that each and every market transaction generates externalities, and that these jointly aggregate into Market forces, which in their turn jointly undermine the effect of the only positive Market force – the so-called great invisible hand, which was supposed to create the best possible world for most people. Similarly, the use of technical approaches generates technical externalities. If these are not taken into account, they accumulate disorders in the relativity of human life and the world that jointly undermine it.

We have already furnished several examples. The technical division of labour created both a spectacular increase in labour productivity and a shocking disordering of human life that manifested itself as physical and mental illness, nervous fatigue, anxiety, alcoholism, drug abuse, family breakdown, and more. Anything but this complete picture shows up as myopia due to the servitude of our secular myths. In the same vein, a megamachine society represents the local manifestation of the global technical order and disregards that it is indissociably linked to the disorder it creates in everything living. When discipline-based approaches are used beyond their limits, they remain very successful because they so simplify everything, by externalizing everything having to do with life, that the improvement of efficiency and performance becomes a relatively straightforward task. However, what these technical approaches theoretically externalize does not magically disappear during their application. These theoretical externalities are turned into technical externalities that create disordering effects of all kinds. These are accumulated in the relativity of human life and the world to produce forces of technique that are the equivalent of the Market forces of the eighteenth and nineteenth centuries. They all act as powerful selective pressures on human history.[22]

The Economy, Art, and the Order of Non-Sense

During the first phase of industrialization, art began to explore the growing sphere of non-sense within human life and society. Like the economy, it transformed itself from a dimension of cultural mediation into a construct of non-sense – another sphere with a relativity whose architecture made no reference to sense.[23] Initially, the art of the industrializing societies of the eighteenth and nineteenth centuries enfolded something of what was happening to the cultural mediation as a whole and thus to the relativity of human life and society. At the same time, the cultural unity of the industrializing societies expressed itself through the dimensions of cultural mediation. This influence tended to constrain the acceptable as well as the unacceptable analogies between these dimensions. What was happening in one sphere of human life thus could, or could not, be used to make sense of what was happening in another sphere. For example, Siegfried Giedion showed that works of art could express something of the deep metaconscious knowledge of what was happening to human life and society as a consequence of mechanization. The artists expressed their intuitions of it in a seismograph-like manner.[24] This also helps to explain the remarkable correlation between the developments in science and in art, as noted by many observers.[25] For example, it was not necessary for Picasso to be reading the latest publications in physics, or for the physicists to be looking at Picasso's paintings, for this correlation to be possible. However, Jacques Ellul argued that the correlation between the developments in literature, music, and the arts, on the one hand, and those in technology, on the other hand, was far more significant.[26] The lives of artists were much more deeply influenced by what we have summarily referred to as "technology changing people" than by "science changing people."

In any case, the emerging new cultural unities that were turning the cultural orientations of the pre-industrial societies inside out confronted artists with an impossible dilemma. The growing technological and economic orders were being constructed at the expense of the cultural orders. The result was that the latter could no longer be counted on as a source for permissible analogies for making sense of what was happening to human life. Simply put: both the subject itself and the cultural means for making sense of everything were disappearing. This was expressed in a variety of ways: paintings were becoming abstract and human subjects were being painted in terms of geometry or movement. It also affected the perception of nature, which had previously become differentiated from humanity, and vice versa.[27] In sum: the world of sense could no longer be the guide for aesthetic sensibilities. Other aesthetic expressions had to be invented. Through a meticulous analysis, Jacques Ellul showed that this dilemma greatly constrained the aesthetic approaches, which eventually constituted a sphere of nonsense in anticipation of the technical order of non-sense.[28] It reflected the growing dominance of the architecture of the relativity of non-living entities over the architecture of the relativity of living entities, which is extraordinarily difficult to express aesthetically. Nevertheless, painting, architecture, music, and literature have provided a particularly clear seismographic interpretation of what was happening to our symbolic species, because the architecture of the relativity used in their forms of expression first matched the architecture of the technological and economic orders of the first generation of industrial societies and then that of the technical order of the so-called industrially advanced societies of the second half of the twentieth century. It thus transformed the symbolic character of art, realigning it with the two successive secular cultural unities that had metaconsciously acted as "blueprints" for relating these orders to the increasingly desymbolized cultural orders and the correspondingly diminished human subject. Any remaining symbolic art was thus relegated to people's personal lives.

The aesthetic expressions perfectly manifested how first the technological and economic orders and then the technical order had little use for human subjects, their lives, and their cultures. These had essentially been transformed into the necessary resources for these orders, thereby perfectly reflecting the alienation and reification of human life. In order to be turned into a resource, any required aspect of human life was first reified though the application of various techniques to make it more efficient and thus compatible with these new orders; this was accomplished one aspect at a time. People attempted to escape in a variety of ways. For example, today the most permissible form for "living" our lives is to do so on the Internet by means of the Facebooks of this world. Everything is reduced to images and image-words that can be engineered into fascinating "pages" of our lives to illustrate their rich and exciting character. Any significant embodiment, participation, and commitment in such "lives" lived on the Web is impossible. The image-lives exhibit an enslavement that goes deeper than what can be said, often accompanied by a well-hidden anxiety about the highly desymbolized lives "behind" these images. I recognize that in a so-called mass society this may well give us the best possible chance at encountering others with whom we may have a great deal in common, and who may therefore be most able to enrich our lives as we may enrich theirs. All this might reduce our solitude were it not for the vast difference between the technical mediation of the Internet and the cultural mediation that symbolizes an embodied, participating, and committed subject. The latter mediation made a human subject possible, although it may have been enslaved to the traditional gods of the past. However, the technical mediation cannot admit the subject, but only reified fragments of what is left of our being a symbolic species.

We are left with little choice but to turn all the events of our lives into "photo-ops" for the pages of our "life" on the Web. We must do everything possible to obtain sefies that are always different and exciting. For example, the selfie-stick can now be replaced by a hand-launched drone capable of recognizing our face, allowing it to zoom around us to obtain the most wonderful images. These can then be enhanced by the Photoshops of this world. Image-words can then be added to further enrich these images, a task for which language-words related to a dialectically enfolded life are ill suited. Our disembodied experiences thus become increasingly technically mediated, to the point that culture-based participation and commitment can play only trivial roles. This is but one example of how humanity is becoming a spectator to the technical order it is building for its supposed benefits. We stand at a distance and admire all the spectacular successes this technical order accomplishes, but we appear to be untroubled by the way it profoundly disorders the human subject and life itself. Does this herald the end of our history as a symbolic species? Is this the beginning of a mutation so profound that the implications remain difficult to foresee, let alone understand?[29] The aesthetic expressions of the second half of the twentieth century anticipated many of these developments.

We must not get bogged down in judging one another as to how much we are or are not involved in the frontier of these developments. We are all being swept along by a kind of desymbolizing tidal wave that began by undermining our cultures. As noted, the mass media became indispensable in diffusing a technical complement to an increasingly desymbolized culture. It took the form of a "soft" social integration by means of a bath of images that more or less told people what, in a traditional society, they would have acquired metaconsciously though the way the organizations of their brain-minds symbolically mapped their tradition-based way of life. The mass media, now complemented by the new social media, are replacing our symbolized experiences with a reality that is externally and technically constructed for us.[30] We ourselves are no longer symbolizing our life-milieu and system of technique. As a result, we are no longer able to symbolically distance ourselves from our life-milieu the way groups and societies did in relation to the natural life-milieu and the life-milieu of society. By surrendering this essential task for a symbolic species to the technical mediation of the media, we appear to have paralysed a margin of freedom relative to our life-milieu and have thus blocked the path toward a more liveable and sustainable future. The only way of out of this situation is to recognize that we have ceased to be human subjects, that what we know or do is of little consequence, and that our values are our private affair. It is the direct result of the way we have symbolically grounded the relativity of our lives by excluding it from the technical order through our secular myths. This is exactly what the art of non-sense is telling us.

According to Jacques Ellul, in the industrializing West European societies, the first major transformations in literature, music, architecture, arts, and painting occurred between 1845 and 1885.[31] The new technological, industrial, and economic changes were beginning to affect people's lives and societies. As a consequence of "technology changing people," artists began to express their intuitions regarding their deepest metaconscious knowledge. The technical division of labour had begun the transformation of individual and collective human life by removing work from the cultural order, reorganizing it in the image of the machine, and then inserting it into a technological order. It separated the hand from the brain on the factory floors and the brain from the hand in the offices. Because work was no longer a daily-life activity as part of a human life, it could be based on endless repetition instead of evolution and adaptation. It atrophied the brains of the workers and the hands of the people in the offices. They in effect "died" to their cultural order to become reified as functions of a large organization in which technically mediated relationships all but replaced the culturally mediated ones. As the output of local production grew, it necessarily had to decouple itself from serving local needs that had been expressed through a local culture-based connectedness. It had to be mediated by markets for goods and services that were driven by increasingly non-local needs. Consequently, to fit this technological order into the cultural order of society, an intermediary economic order was required. In the technological and economic orders, people could no longer participate as human subjects, while in the cultural order they now had to participate as subjects that had been banished from the most important activity of their lives. People thus became spectators and banished subjects. The cultural order had to be kept at bay to a corresponding level; and this was not a simple task for the members of a symbolic species. All traditions were shattered because past experience was of no value to the building of the new orders with their radically different architectures of relativity.

Jacques Ellul showed that between 1890 and 1940, following the discovery of the limitations of the cultural approach based on symbolization, experience, and tradition, art became influenced by the new rationalistic orientation that was beginning to characterize individual and collective human life.[32] It necessitated an ever-greater reliance on rational approaches, and later on technical approaches, as human knowing and doing became reorganized by means of disciplines. This development completed the break with the past, as a technical order now integrated and transformed the previous technological and economic orders. This technical order also spread beyond industry and the economy into the state, which was becoming the principal locus for organizing society as its highly self-regulating cultural mediation became desymbolized. The architecture of the relativity of this technical order could not furnish the permissible analogies for artistic activities the way daily-life activities were lived in the past. The new emerging world had to be seen in new ways, and the eye was much more suited to expressing this than was human language.

During this time, a search began for a new habitat for humanity. People did not look to the traditions of city building that had been embedded in experience and culture. Instead, they invented completely abstract models that had never been built or lived in before. These included the Garden City and the "radiant city" with its towers in the park.[33] These kinds of visions were an expression of how much production mattered. The Garden City divided human life into four categories of functions: working, shopping, relaxing, and domestic living. It thus paralleled the technical division of labour now applied to all of life. Building radiant cities with towers in the park was as close as building construction could come to mass production, giving priority to the flows of goods and services, and establishing a new authority over human life.[34] This period was also a time when the technical approach produced techniques in all dimensions of cultural mediation, thereby transforming them into functions, structures, and subsystems of a growing technical order. Individual and collective human life became less and less dependent on culture-based adaptation and evolution, and instead was surrendered to repetition that could be specified in terms of rules, the "one best way," algorithms, or programs. It thus prepared society for the computer and information revolution that would burst on the scene a few decades later.

Following the Second World War, human life was increasingly lived in relation to the life-milieu and system of technique. Jacques Ellul showed that at this point, art came into the grip of abstract forces and became a kind of "field of forces."[35] It was as if the artist had become driven by technique, using technical media and equipment as an agent of the technical system. The artist was not conscious of this change, because the relationships with the primary life-milieu and system of technique were no longer symbolically mediated.[36] At this point art had detached itself from culture, which had become permeated by technique.[37] Art now took its place within the system of technique as a force of social integration. Art no longer had a magical quality. It had become the expression of means and was less and less concerned with content. Art itself became a universe of means from which the ends had disappeared, as occurred everywhere else in the technical order. Artists thus became engaged in an endless search for new techniques, and the consumers of this art quickly tired of yesterday's techniques. These works no longer manifested human intentions but were only expressions of the reign of technique. Jacques Ellul meticulously examined it as "the empire of non-sense."[38] This analysis made the extraordinary diversity of artistic expression more intelligible than any other that I am aware of. It is far too complex to be briefly summarized; therefore I will limit myself to a few remarks. This art must be understood in relation to the life-milieu of technique, just as the very earliest art had to be understood in relation to the natural life-milieu, and the art that succeeded it had to be interpreted in relation to the life-milieu of society. The art relative to the first life-milieu had a magical-religious character, and was integral to dominating this life-milieu by means of symbolization.[39] With the formation of societies, a new primary life-milieu was created that interposed itself between the group and the life-milieu of nature. The organization and governing of these societies became more important than surviving in nature. During this time, art became a reflection of the society that produced it. Its magical-religious character was now related to that society, and its cultural mediation was now directed toward this society. It gave this art a more political character.[40]

In today's life-milieu of technique, art is no longer magical-religious, symbolic, or political. It takes on a more functional character in relation to *artifice*, which is no longer symbolized.[41] It derives its character from a life-milieu in which forces play on one another. Art now deals with a technical order that makes no reference to and excludes the sense derived from symbolizing human life and the world. The architecture of the relativity of this technical order, based on the principle of non-contradiction, excludes everything symbolic associated with the incommensurate architecture of the relativity of human life and the world. All aesthetic expression must now be based on the permissible architecture of the new relativity that is technically constructed, and must reject the architecture of the symbolized relativity of human life and the world, including the art associated with it. Art can also express the disorder that the new technical order brings out in human life, society, and the biosphere.

Since the "distance" between the actual artistic works and the corresponding deepest metaconscious knowledge associated with personal experiences symbolized by the secondary frame of reference is very great, we can expect a range of interpretations. These collectively express how the relativity of human life is being constrained by a different deep metaconscious knowledge, the one associated with experiences made sense of by the tertiary frame of reference and the secular myths that prioritize it over the "subjective" and metaconscious knowledge associated with what remains of the secondary frame of reference.

For example, the secular sacred of technique has an opposite pole: the sacred transgression of sex.[42] It is as if people have an intuition of how deeply our lives have been reified by technique, and that they could act on this by retreating into their physical selves in order to recover a measure of freedom through sexual expression. When this happens, it takes on an overriding concern with performance and intensity, thus becoming assimilated back into the technical order as its sacred transgression. In this manner, our most intimate relationships have become affected. For decades, however, this was interpreted as a sign of freedom celebrated through rituals of free love and a great deal of sexual experimentation to increase pleasure and performance. We may thus expect in art also an imitation of technique, an escape from it, or a sacred transgression of it.[43]

We can also expect explanations of works of art to take the form of a sacred transgression of the state-nation through politically revolutionary messages, especially in Europe.[44] It may be expected, therefore, that the art produced in the life-milieu of technique will express how the relativity of human life has become constrained, by expressing the relationship between the technical order and our being a highly desymbolized species with a secular political religion.

In one way or another, we have all learned to make liveable the nonsense of our technical order. If we participate as scientific or technical specialists, we go about our work without feeling much of a need to symbolize what we are doing in order to achieve an intellectual grip on what our work means for human life and society. When someone points this out, we may cynically say that we leave this to the philosophers and theologians and that, in any case, the scientific and technical difficulties we struggle with are challenging enough by themselves.

Even in our homes, we are more or less convinced that we do all this out of our love for our partners and children and for our common future. We are not bothered that the daily lives lived with our significant others remain opposed to the very conditions essential for the use of discipline-based approaches to our knowing and doing. Even our highly desymbolized daily-life activities involve a diversity of categories of phenomena that all make non-trivial contributions to our lives, with the result that discipline-based approaches are neither scientifically acceptable nor professionally responsible as to their effects. Nevertheless, we are surrounded by situations that have been carefully structured by means of discipline-based approaches whose rational character distorts the way everything is related to everything else in our lives.

Let us imagine ourselves in the daily lives of scientific and technical specialists. We, like everyone else, are on the receiving end of countless techniques that organize and reorganize everything around us. Our deepest metaconscious knowledge is extremely successful at eliminating the kinds of analogies that, as members of a symbolic species, we could be expected to employ to make sense of and get on with our lives. It shows the extent of our desymbolization. Any interpretation that hinges on our being a human subject able to make sense of and live in the world by means of a culture has all but disappeared from our participation in the technical order. Our lives are deemed to be so subjective and trivial that they merit no serious consideration. We cannot know anything other than through scientific disciplines. We cannot do anything other than through technical disciplines. Nevertheless, in our personal lives and our private spaces, we somehow cope with all the disordering that accompanies our collective technical ordering. We ought to carefully reflect on what we mean when we talk about freedom and democracy; and we will perhaps come to the conclusion that these are no more than secular religious expressions that correspond to little in our lives. It is perhaps the ultimate confirmation of the rule of non-sense. Like the artists, we have all learned to live with it, but we continue to observe how difficult this is for our teenagers, as their rising levels of anxiety, depression, and suicide indicate. What is the sense of living a life in an anti-society dedicated to the building of a global technical order of non-sense?

Making Sense of Non-Sense

What the artists in the industrializing and so-called industrially advanced societies have intuited and expressed in their work is indicative of what each and every human being professing to be interested in freedom ought to be concerned about. Is it still possible to make sense of our being human subjects continuing to live lives when we are constantly on the receiving end of all manner of techniques that reify every aspect of all of this? Since we are unable to explain or make sense of anything in itself, what kinds of analogies will we be able to use when our present secular myths rule out our doing so by comparing ourselves to a symbolic species making sense of and living in the world by means of symbolization, experience, and culture? It is all very well for traditional or socially conservative people to believe that we can strive for living as we did before; but the most elementary psychological and sociological analyses of what this involves will reveal endless contradictions of which they are not aware. It is impossible to conserve a past, or to think it can be recovered in our time, when this past was symbolically grounded in cultural unities very different from our own; and this difference becomes one of the sources of these contradictions. All we are able to do is to face our present situation as best we can understand it just short of plunging ourselves into cultural chaos.

If we take another look at what has been happening during the last two hundred years of our successively building a technological, economic, and technical order within the cultural order of our societies, it becomes apparent that the above analysis of what happened to the aesthetic dimension of cultural mediation is exactly what happened to all its other dimensions. Let us go back to the nineteenth century to uncover and reinterpret the source of our mistaken conceptions. We will limit ourselves to considering the two dominant interpretations. The one proposed by Karl Marx, which gave rise to communism, interpreted the separation of the economy from the cultural orders of the industrializing societies through the secular myths of his time. These myths symbolically grounded what was happening on a "sameness" that Marx applied to all of human history. Consequently, apart from the primitive commune, all the stages of human history were characterized by a unique economic base comprising the forces and relations of production, on which a superstructure of everything else was erected.[45] There was thus no more to explain, However, with a great deal of hindsight and with the disappearance of the secular myths of that time, it appears that Karl Marx had explained nothing more than the capitalist economic organization of his day. He cannot and will not help us understand the kind of economic organization that emerged in the twentieth century, as some of his successors have convincingly demonstrated.[46]

As far as the liberal or conservative explanations of capitalism are concerned, both put the economy at the centre of human life as the most important dimension of cultural mediation. The economy had to generate the wealth to increase and accumulate capital, without which nothing significant could happen. This is just another interpretation of the sameness created by the first generation of secular myths. The entire political spectrum remains enslaved to these myths, which exist today only as a mythology, since they have been displaced by a new generation of secular myths. In these democratic traditions, John Kenneth Galbraith, perhaps more than any other economist, saw that the growing reliance of the economy on highly specialized scientific and technical knowledge changed almost everything; but his analysis deliberately excluded the effects of discipline-based approaches beyond the economy and the state.[47]

With a great deal of hindsight, it has become a little clearer what was happening. The "sameness" of the cultures of the industrializing societies made it impossible to recognize that the economy was being "spun off" from the cultural orders of societies by taking on a diametrically opposite architecture that eventually became universal. With growing rationalization being applied to science, the military, the state, and education, as it had been in industry and the economy, the bond between whatever area of human life was being rationalized and the cultural order of the society was being weakened, setting it on a course that would turn out to be not very different from the one the economy had already taken. When discipline-based approaches transformed all human knowing, doing, and organizing following the Second World War, each and every dimension of cultural mediation spun off from the cultural order to become a construct or building block of a technical order by sharing the same universal architecture of the relativity of human life and the world. The cultural order was thus emptied of everything related to the collective life of a society, causing it to collapse and remain as highly desymbolized fragments only in people's personal lives and "spaces." The results were highly constructed societies in terms of a global technical order corresponding to a vast disorder in human life, society, and the biosphere, which was made sense of as if the cultures of the participating societies had not been decisively desymbolized. In other words, societies became complex and divided in a mechanistic fashion. The enfolded and dialectically enfolded character of human life, society, and the biosphere was taken for granted as if (from a cultural perspective) it was "business as usual." The constructs that became the constituents of a global technical order were thus *unfolded* from everything living. The other constructs became organized to efficiently "mine" the required resources for this global order from human lives, societies, and the biosphere.[48]

Most of the constructs of our contemporary anti-societies restrict people's participation to one or more aspects that are essential for their functioning. Everything else related to their persons and their lives is externalized in the process. It is a generalization of the kind of situation we have examined in relation to work, which requires a person to behave as a part of a collective brain, hand, or coordinating management function. Everything else has to be suppressed to the greatest extent possible in order not to disturb the technical functions of the construct. When this happens in most areas of a person's life, the organization of the brain-mind can no longer work in the background in its entirety, and begins to function as a plurality of "radars" each associated with its participation in a different construct.[49] The metaconscious social self is thus decentred into a plurality of social selves, each associated with a technical role carried out in relation to a particular construct. The corresponding experiences are differentiated from each other and form constellations of similar clusters of experiences that jointly make up a person's life.

As the dimensions of cultural mediation were gradually turned into constructs, both society and the lives of its members became divided. The other-directed person, using the organization of the brain-mind as a radar, now became a bundle of technical functions performed in various constructs, each associated with a particular "area" of the person's organization of the brain-mind acting as a specialized radar tuned to a particular technical role.[50] This change must not be interpreted as giving rise to a situation somewhat analogous to a technical multi-dimensional schizophrenia. There may well be tendencies in this direction that, in the extreme, can lead to new forms of mental illness. However, this situation is avoided in the lives of most people by the new secular myths. The kinds of intuitions to which these myths give rise express something of how the relativity of human life has taken on a different architecture, and that people's daily-life experiences now correspond to this. Hence, things are normal and exactly what should be expected even though we may find it hard to live with. In any case, attempting to make sense of our lives as if they continued to have the relativity of sense and culture has been banished to people's personal lives and private spaces.

People's reactions to all this will vary. Some may behave in a way that seeks to inject a little more sense into our collective life ordered by non-sense. Others may appear to be entirely comfortable with the situation, interpreting it as the dawn of a new age with a new secular human awareness. Many others are somewhere in between these tendencies. The situation may have to be deduced from very careful studies of the new kinds of mental disorders that appear to be occurring. We are still far from understanding what is happening to our lives and societies.

In the course of our daily lives, we move from participating in one construct to another; and each time we participate in terms of one or more aspects by suppressing a great deal of our person and our life. For example: The store where we shop for our groceries is a highly technicized milieu whose significance can be explained in terms of efficiency and performance. This often translates into our leaving with more than we thought we needed. Shocking as this may appear, we enter into a very similar situation when we are hospitalized. We are immediately classified in terms of one or two aspects of our health and treated accordingly, with the greatest possible efficiency and performance. When we attended school, we were constantly evaluated in terms of our academic abilities and streamed accordingly. The same was true when we tried out for a sports team or in the way we were coached once we joined one. Whenever we carry out a search on the Internet, every move is being tracked and classified in terms of those aspects that may interest vendors, who may then pay in order to create the appropriate advertising to frame future searches. We are being enclosed into what has been referred to as "filter bubbles" constituted of the kinds of news we prefer, the kinds of things we like to read, the products that interest us, the services we may need and like, possible "friends" that we may enjoy, and much more.[51] When we visit our bank, the advisor may know more about our "financial life" than we are aware of, and we are dealt with according to these aspects. When we obtain a quote for insurance of one kind or another, we are increasingly placing ourselves in the hands of what we have previously discussed as a "weapon of math destruction," which creates our profile by mining various data, some of which will almost certainly be associated with our postal code.[52] When we visit a funeral home to make arrangements for a relative or friend whom we have just lost, everything (including the casket-viewing room) has been carefully arranged in accordance with those aspects that we are likely to respond to, for the purpose of expanding the scope of the services beyond what we had intended and deemed essential. We can multiply these examples almost indefinitely. With the growing prevalence of enterprise systems acting as weapons of math destruction, the level of sophistication of these technical constructs is significantly enhanced each and every year.[53]

We can expect a significant influence on individual and collective human life as a result of the transformations of the dimensions of cultural mediation, into either a construct of the global technical order or a construct that transforms aspects of human life, society, and the biosphere into its required resources. Each time we participate in these constructs, we do so on their terms, and we end up doing things that go well beyond what we intended, desired, and expected. One of the most important aspects of this transformation is what we have referred to as the separation of human knowing and doing from experience and culture in each and every dimension of cultural mediation. It has transformed much of the social dimension of cultural mediation into a mass educational construct that has fundamentally changed the way the members of each new generation are allocated to their places, first in the social and then in the technical division of labour.

Insight into this transformation may be gained by asking the following question: From a historical perspective, why did the traditional preindustrial societies and civilizations almost always restrict social mobility to the greatest extent possible, and legitimated doing so in moral and religious terms? Once again, the universal character of this approach to allocating the members of each new generation to the social division of labour of their ways of life hints at an equally universal necessity for doing so. When human knowing and doing is embedded in experience and culture, this knowing and doing depend almost entirely on acquiring the associated metaconscious knowledge – knowledge that cannot be transmitted directly and explicitly. Hence, by having children follow in the footsteps of their parents, their upbringing would include a kind of informal apprenticeship into what their parents contributed to the way of life of their society. This could then be followed up by a formal apprenticeship or internship; and by the time they reached adulthood they would have acquired much of what was essential to make their contribution to evolving and adapting the social division of labour of their society. In other words, this strategy ensured the best possible success at transmitting a way of life from one generation to the next, with minimal losses. Moreover, its strategic importance for the survival of a society was such that, without exception, it was defended in terms of the highest moral and religious values and beliefs.

When the knowing and doing involved in every dimension of cultural mediation were separated from experience and culture to be reorganized by means of disciplines, the above strategy became a disaster. Discipline-based knowing and doing are separated from experience and culture and thus can be acquired only by classroom- and textbook-based learning. The universality of these approaches further weakened their connections with any particular dimension of cultural mediation. Moreover, the number of people now requiring a formal education in discipline-based approaches was far greater than the number of people in the upper strata of society who traditionally were the only ones having access to higher education. New necessities were thus imposed, and these now had to be overcome through entirely new approaches for assigning the members of each new generation to a way of life. A mass education construct had to be created to which all children had compulsory access. Once they had entered the construct, their abilities to deal with discipline-based knowing and doing were constantly assessed and reassessed in order to stream them appropriately. When they exited this construct, they had become extraordinarily unequal in terms of the positions in the technical division of labour accessible to them. It was as if the educational construct had put them onto different "social elevators" that provided them access to different social strata. As long as young people's talents for contributing to discipline-based approaches are exceptional, the social position of their parents ought not to interfere. There is little doubt that when these educational constructs became increasingly dominant following the Second World War, a greater social equality and resistance to any form of discrimination became a technical necessity for participation in the building of the global technical order. At the same time, this higher level of social mobility enormously weakened the social fabric of society and contributed substantially to the emergence of what are referred to as mass societies.[54]

We have suggested that all these developments transformed the political dimension of cultural mediation into a political illusion.[55] However, this illusion still required important resources, especially our public opinions and our secular religious support. These were essential for legitimating the state-nation and everything associated with it. It is important to emphasize that our public opinions are the diametrical opposite of our personal opinions. The difference stems from the extent of our embodiment (physical, social, and cultural), participation, commitment, and alienation. A personal opinion is arrived at by exploring an issue that, for whatever reason, matters to our life. Hence, we read as much as we can about it. If possible, we talk to people who were directly involved in it, or to people who know such people, and so on. We invest a great deal of thought in attempting to make sense of it all, and finally arrive at an opinion. In contrast, as the literature shows, public opinions are the diametrical opposite. When people are approached by a television crew about some issue on a street corner or in a subway station, for example, they always have an opinion; and if we are honest with ourselves, we are not quite sure where these come from. Much of what we have opinions about does not have the backing of the kinds of efforts required to arrive at a personal opinion. It is a feature of human life in a mass society where the media provide us with the "reality" of our lives and our world, including the commonplaces, stereotypes, and mythologies by which we can make some sense of the barrage of image-stories accompanied by image-words.[56] Søren Kierkegaard was probably the first to foresee and warn us about the rise of public opinion. Hubert Dreyfus has interpreted Kierkegaard's insights about our use of the Internet.[57] From the perspective we are developing in this work, the Internet presents us with a reality that is vastly greater in scope than the one created by the traditional and the new media. Nevertheless, there is no change in the architecture of this reality. It remains entirely detached from our lives, which continue to be somewhat dialectically enfolded even in our anti-societies with their highly desymbolized cultures. It induces a wide range of public opinions in all of us that, as all the studies show, is like the wind that comes and goes – and we have no idea where it came from or where it is going. These public opinions turn over at a great rate, but our adopting or discarding them makes no difference whatsoever to our lives.

If we compare this situation to one in which we have spent many years of higher education to gain a deeper insight into some area of human life, we can understand what Kierkegaard was concerned about. For example, I have taught engineering, sociology, and environmental studies in order to make some sense of our lives and our world and to develop more socially viable and environmentally sustainable practices. If I constantly changed my opinions during the academic year, I would soon become the laughingstock of my students, and no client would solicit my advice knowing that I would abandon it the next month or the next year. When it comes to our public opinions, we have become physically, socially, and culturally disembodied spectators of a reality that makes no sense; or, it could not have any sense without the commonplaces, stereotypes, and mythologies that embody something of our secular myths by which we avoid cultural chaos. Adopting or changing any of my public opinions commits me to absolutely nothing. I can tell other people that I've simply changed my mind; and in any case no one is interested in which public opinions I held last month or last year. It changes absolutely nothing; and when the public opinions in society change, I can adopt new ones because I am committed to nothing. Hence, all our discussions about the news do not make us more informed citizens. They do not strengthen democracy, and they are of no value whatsoever when we as a community have to make a genuine political decision between options that each have significant and far-reaching consequences, but of which the details are largely unknown. The public created by shared public opinion is not a community in which we all have a lot at stake when such political decisions are made. On the contrary, the public is a crowd of spectators of a reality that would drive them toward nihilism were it not for the fact that the architecture of this reality has been legitimated as a reflection of our secular myths. We would not dream of putting our public opinions into practice in our lives, nor would we back those opinions with our word and our lives. There is a complete disconnect between our public opinions and the way we live our lives.

As a society, things are even worse. Our beloved democracies have been decisively weakened by the rise of public opinion. It has presented our governments with a terrible dilemma. On the one hand, they cannot govern against public opinion and maintain their legitimacy for very long. On the other hand, it is equally impossible to govern on the basis of public opinion. Imagine a government tackling an important issue that will require years of study, many months of consultation, and then many more years for an adequate implementation of the best strategy for dealing with it. From the perspective of public opinion, the time span of this government action would appear to be an eternity, during which public opinion would fluctuate in all directions relative to this issue and how it was being handled. The only practical way out of this dilemma is for governments to manage public opinion by means of political advertising. Together with the corporate complement of managing public opinion, it creates a vast extension of what John Kenneth Galbraith has referred to as the revised sequence,[58] which undermines the economic and political sovereignty of the members of contemporary so-called democracies. When all this is understood in conjunction with our lack of guidance from our highly desymbolized cultures, the influence of the technical complement of social integration propaganda first examined by Jacques Ellul takes on its full significance.[59] It reveals our continued alienation and reification by our secular myths, a situation that would be unliveable without a secular political religion.[60] Our governments, trapped in the dilemma posed by public opinion, want our allegiance – and especially in times of crisis, our love, our devotion, and possibly our lives. These are the times when many of us will be told that we must either love our country or leave it. There simply is no other way to justify and legitimate our dependence on the state following the desymbolization of our lives and cultures. We cannot act as solitary intellectual Robinson Crusoes and invent our own culture – a task to which everyone in the past contributed on a daily basis. We thus have to surrender ourselves to highly abstract currents and forces; and this is possible only by dedicating our loyalty and love to them.

Can we make any sense of the technical non-sense of the technical order and the state-nation? The simple answer is that we can make some sense of our lives and our world by resymbolizing them. However, doing so involves questioning everything to which we have given our ultimate commitment by symbolically grounding our lives. In other words, any attempt at resymbolization involves cracking open the door to cultural chaos. And who or what will give us the existential courage to undertake it? If we desire to be truly serious about this business, we must close all false escape hatches of the kind we see all around us. In North America, evangelical and fundamentalist Christianity is probably the most disturbing example. For them, there is one "spiritual" life and another "worldly" life. All the talking and judging that come from the one life are not backed by the other life, and vice versa. It is an extreme example of something that afflicts us all.

4 Mathematics as the Non-Language of Science and Technique

Mathematical Foundations and Truths

In order to put the role mathematics plays in our contemporary anti-societies in a historical perspective, we do well to remind ourselves of the great cultural divide. The emergence of our global civilization and its complete dependence on discipline-based approaches, at the expense of the cultures of its constituent societies, has created a new role for mathematics: the non-language by which the domains of all disciplines are interpreted, organized, and developed. For the first time in human history, mathematics has become the dominant non-linguistic alternative to human languages and cultures in the ways of life of all societies. It marked a fundamental transition in the way humanity lives as a symbolic species – a way increasingly transformed under the pressures of desymbolization and extensively sustained by this non-language math. The desymbolization of human languages and cultures has now progressed to the point that many mathematicians no longer appear to be aware of mathematics being a non-language. It is opposite to all highly desymbolized human languages because of its diametrically opposite architecture.

We must keep this new situation in mind as we begin our analysis with an interpretation of the conventional relationship between mathematics and society prior to the great cultural divide and to a short period following it that marks the transition toward our present situation. This conventional relationship applies to the period of human history when there was a minimal level of desymbolization of human languages and cultures. Even then, the diametrically opposite architecture of mathematics compared to the architectures of languages and cultures appears not to have been appreciated in most studies. One of the primary consequences is that the influence of "mathematics changing people," which became a dominant component of "technique changing people," continues to be thought of in terms of its most superficial aspects.

Mathematics is commonly defined as the outcome or product of a scientific approach to the study of numbers, quantities, and space. Definitions of this kind are highly misleading in relation to the general relativity of human life and the world. To put it simply: the meanings of the terms mathematics, numbers, quantities, and space are relative to the vocabulary of a language and, via it, to the culture of a society. This language and culture are acquired by growing up in it, which includes becoming physically embodied relative to an unknown, socially embodied relative to the lives of others, and culturally embodied relative to symbolically grounding in myths the general relativity of this life in the world. In traditional societies, this was accomplished through a secondary frame of reference, which is required for the acquisition of the language and culture of any society. In contemporary societies, this growing up becomes quickly influenced by the acquisition of a tertiary frame of reference, which is required to access all the constructs that help to build the reality of a global technical order that has been almost completely separated from experience and culture.

In other words, the study of mathematics must take place in a context of a particular society's having made sense of and living with the general relativity of human life and the world by establishing either a cultural order or a technical order. We then need to discover how mathematics, numbers, quantities, and space are embedded in one or in both of these orders relative to everything else. We must also be mindful that the discipline-based approach to human knowing and doing is a relatively recent development; and we must avoid projecting the idea of mathematics as a discipline on most of its history. Until recently, mathematics was generally unique relative to the culture of a society.

Since this work is focused on understanding human life and the world of today, we will limit ourselves to a few highlights of how Western civilization developed mathematics as a unique activity relative to all the other activities that jointly constituted the ways of life of its constituent societies. From this perspective, Euclid sought to extend the cultural order of his time, place, and culture by building a geometry of what was widely regarded as true; and for a very long time no one came up with any reasons for doubting it. It would have been almost impossible to imagine how space could be anything else than what was described by

Euclid. The properties of this space were built up with the principle of consistency or non-contradiction, thus establishing a universal knowledge of it. In other words, Euclid attempted to extend or refine some self-evident aspects of what anyone could see, as if it was an extension of the general relativity of the cultural order – even though his geometry had a diametrically opposite architecture. Euclidean geometry did not describe the space in which his fellow Greeks lived their daily lives, as a consequence of having grown up in that society and thus having acquired a symbolic metaconscious knowledge of space. As we will see, he essentially described the space of the order of the stars, which he brought down to earth.

In most cultures, this metaconscious knowledge of space was externalized in their making of paintings, for example. In medieval paintings, things were commonly not depicted in relation to their spatial positions, but according to their meaning and importance for the painting. Generally speaking, things continued to be painted according to their meaning and value for human life, and were positioned in space to reflect this.[1] It must be remembered that it was not yet possible for the painter to behave as a relatively detached observer of the world "out there."[2] People still saw themselves as God's creatures integral to his Creation and the land. It was not until this dialectically enfolded whole became differentiated into a humanity and a separate and distinct nature (later depicted as a clockwork) that people could begin to think of themselves as detached observers. They could engage in a new activity, that of tourism – a kind of staring at this marvellous separate and distinct nature.[3]

We have become so accustomed to time, space, and human life as separated from experience and culture that we do well to attempt to imagine the general relativity of time-space-body(including the brain-mind)-culture-society as it was lived in earlier societies. Suppose that a few of us became stranded on an isolated island and that we had agreed on a way of marking time because we could no longer rely on our time-keeping devices, those having been lost or destroyed. We would likely agree that noon would be taken to coincide with the sun having reached its highest point before beginning its descent, and would mark this as the midpoint of the day. Such a time would be relative to our position in space as well as to our bodies making the necessary observations. For example, if we had perfect instrumentation, it would be possible to demonstrate that when we moved ten metres in any direction, the time would have shifted slightly, even though this cannot be detected by symbolizing the associated experiences. Nevertheless, this time and space would be dialectically enfolded into our lives and culture. They would thus become indivisible, inseparable, impossible to define in a closed definition, impossible to measure in an absolute manner, and impossible to represent as part of a simple complexity. The same argument can be made to show that this time would also be relative to the natural cycle of the seasons and our position on the planet. Consequently, the metaconscious knowledge of space of a traditional culture did not correspond to the space described by Euclid's geometry. The former is dialectically enfolded in the general relativity of a cultural order, while the latter constitutes a domain built up with the principle of consistency. Sooner or later, the inconsistencies between them were bound to be discovered.

Historically speaking, this happened when non-Euclidean geometries were discovered, which raised the question as to which geometry was "real." It triggered a crisis in what was supposed to be the most reliable human knowledge, namely, mathematics. The consistency of non-Euclidean spaces could be guaranteed if Euclidean space was consistent. This could only be ensured by the consistency of arithmetic, which in turn was based on set theory founded on a number of axioms that were believed to be ultimately founded on common sense – regarded by some as a kind of intuitive logic. However, Bertrand Russell showed that this involved irresolvable contradictions.[4] The result was that mathematics was believed to be founded on consistency, whereas Euclidean geometry had simply been considered to be true. The decisive blow to any possibility of arithmetic's having a foundation came from Gödel's incompleteness theorem.[5] It showed that any consistent formal system containing an elementary arithmetic would not be able to prove its own consistency. Following this discovery, the search for the foundations of mathematics was abandoned. All this would have come as no surprise if we had accepted the general relativity of human life and the world, including the necessity to symbolically ground it in myths. Culture is non-logical and logic is non-cultural. It does not mean that culture is illogical and that logic is nonsensical. What it does mean is that logic does not belong to a cultural order and lies outside of it as a separate and distinct domain built up with the principle of non-contradiction, which gives that domain an architecture that is diametrically opposite to the architecture of any cultural order. As Bertrand Russell showed, the domain of logic had to be completely separated from experience and culture.[6]

It should be noted that for the Greeks, mathematics meant geometry, given their cultural orientation, which gave priority to seeing over listening as the access to the world by means of language and culture. In other words, the Greeks had the exact opposite orientation to the Jews; and this was partly responsible for the difficulties of integrating the so-called three perfections on which Western civilization was supposed to be based. This priority implied a corresponding one between anything that extended the architecture of human life and the world apprehended through seeing over the architecture of human life and the world acquired by listening to language to become a member of a symbolic species.

What we are beginning to discover is that our society has produced a certain kind of mathematics; and this unique mathematics has contributed to the kind of society in which we live. As people change mathematics to bring it to bear on a variety of activities, it simultaneously changes people. This reciprocal interaction became historically significant following the Middle Ages. Three primary contributing factors stand out. The first was the demands placed on mathematics during the Renaissance by the navigation of the seas in order to discover and exploit the world by commerce and trade, and by the creation of the necessary financial infrastructure. These demands would soon be greatly expanded by industrialization and the creation of an economic order within the cultural order of societies. The second factor was the influence of the dominant religious tradition on West European societies following the collapse of Christendom that marked the end of the Middle Ages. The third factor, closely linked to the second, was a renewed influence of Greek philosophy, which held the universe to be rational, accessible through universal knowledge, and built up with the principle of non-contradiction. These last two factors reoriented the cultures of West European societies away from symbolization, language, and culture as the way to understand human life and the world, and toward seeing and all its extensions for apprehending human life and the universe. It also shaped Western philosophy. As noted in the Introduction, the dynamic character of the history of Western civilization is sometimes attributed to the contradictions between the "three perfections" relative to which its societies sought to make sense of human life and the world. The Christian tradition quickly distanced itself from its Jewish roots, which were based on the importance of listening to the Word of God. It increasingly sought to understand itself relative to, and by means of, Greek philosophy in a Platonic or Aristotelean manner.

Culturally stirring Roman law into the mixture complicated things even further. Toward the end of the Middle Ages, it became apparent to many that an entirely new start had to be made; and these efforts helped to bring this historical epoch to an end.

One aspect of this transformation relates to the possibility of the existence of "natural laws." Such laws were unthinkable for Plato and Aristotle.[7] Under the influence of Christianity, God was reinterpreted as a kind of lawgiver who thus brought order to his creation. This interpretation created serious theological difficulties because such laws would limit God's sovereignty. These were set aside by the appropriate theological metaphysics suggesting that natural laws could be regarded as evidence of God's ongoing involvement in his Creation as an expression of his will for that Creation to continue. These new images of God helped to pave the way for the mathematization of the order of creation.[8] During the late Middle Ages there was a growing conviction that the key to God's design of nature was its mathematical laws. Galileo thought that science was not to seek physical explanations of nature but mathematical descriptions of experimental data.[9] Newton went further when he posed the mathematical law of gravity and the attraction of planets, without anyone having any idea at that time of its physical basis. Many people felt that this posed no problem because the explanation would follow later. Hence, by describing experimental results in mathematical terms that may have reflected little more than an intuitive sense of what was physically happening, scientists were discovering the mathematical regularities of creation. It has been argued that in the fifteenth and sixteenth centuries, there was a mathematical revolution in Western Europe that was possibly more fundamental than the scientific revolution.

Centuries later, the concept of an underlying order describable in terms of mathematical laws was extended to the living world. The contribution of Charles Darwin may be interpreted as the discovery of the law of the development of organic nature.[10] It gave rise to many ideological extensions into the social world by extreme interpretations such as sociobiology and social Darwinism. In the same vein, Friedrich Engels, speaking at the funeral of his friend Karl Marx, pointed out that one of Marx's contributions had been the discovery of the law of the development of human history with its five stages, each built on a unique economic base.[11] This extension into the living world of the concept of a law manifesting an underlying order, most likely expressible in mathematical terms, had far-reaching consequences. It implied a new metaphysics of human life and the world.

These kinds of developments led to a technical orientation in the cultures of West European societies.[12] They fundamentally contributed to the conditions that created a fertile soil for the introduction of the technical division of labour, mechanization, and industrialization.[13] However, it took a crisis in the putting-out system of textile making to trigger industrialization and build an economic order within the cultural order of English society. The architecture of this economic order matched and reinforced the technical orientation in every industrializing West European society. It soon became apparent that the approach for making sense of and dealing with the world based on symbolization, language, and culture encountered significant limitations when dealing with these economic orders.[14] It led to the separation of human knowing and doing from experience and culture and their reorganization by means of disciplines that technically mediated all relationships with the world. As a result, the secondary frame of reference continued to culturally mediate the relationships in people's personal lives, but a tertiary frame of reference technically mediated the relationships in people's professional lives.[15]

Without the emergence of the second generation of secular myths, these developments might have given rise to a profound existential crisis about what kind of order sustained individual and collective human life. According to the secondary frame of reference, it would be what little remained of the highly desymbolized cultural order in people's personal lives. According to the tertiary frame of reference, it would be the global technical order. The second generation of secular myths transformed this latter order of reality into the true and primary order, making the other purely subjective and personal.[16] In other words, it was as if our symbolic species was beginning to acquire a kind of cultural bifocal lens, with one part constituting the secondary frame of reference and the other the tertiary frame of reference.[17]

The Emergence of a Secular Religious Daily-Life World

With hindsight, the process of industrialization can be interpreted as one of desymbolization unfolding the cultural orders of all participating societies.[18] It began by unfolding the economic dimension of mediation from all the other dimensions. With the separation of human knowing and doing from experience and culture and the use of discipline-based technical approaches to adapt and advance every area of the ways of life of participating societies, every other dimension of cultural mediation was spun off from what was left of its cultural order to become a construct adapted and evolved largely on its own terms. Consequently, a reciprocal interaction began to transform people's lives and their societies, which has been summed up: "as people change technique, technique simultaneously changes people."[19] As a result, the dialectically enfolded character of individual and collective human life was unfolded by desymbolization to a historically unprecedented level. The new emerging order of reality, built by all the constructs spun off from what had been an already highly desymbolized cultural order, thus became plausible even though it had the diametrically opposite architecture to the cultural orders of all societies that preceded our present global civilization. In turn, these developments were reinforced by almost every discipline that had succeeded in describing the contents of its domains by the non-language of mathematics. This success further reinforced the perception of an underlying mathematical order.

However, this order did not resemble in any way the one proposed by Galileo, which we noted above. There are two reasons for this. First, there has never been a "science of the sciences" capable of integrating the findings of all disciplines into an overall scientific interpretation of human life and the world. Hence, we no longer possess the kind of understanding that science might have produced if all our attempts at unifying it through interdisciplinary, cross-disciplinary, and multidisciplinary approaches had succeeded. Consequently, we have no physical or sociocultural understanding of what this order might be. Second, this does not in the least deter us in advancing our discipline-based approaches in science or technique. As noted, all of them seek to impose a mathematical order on every category of phenomena they study or use in their applications. The only consistency between the findings of all these disciplines is found in the consistency of the mathematics they all employ. Our civilization thus moves forward in the confidence that all of life and this entire universe can be adequately understood through mathematics. In practical terms, this means that the tertiary frame of reference mediates all these efforts in a neutral fashion and does not filter out anything essential. That this is the case is highly implausible, given our structural economic, social, and environmental crises. Every living system is being weakened in some essential fashion.[20] One way of interpreting this situation is to acknowledge that the consistency of the mathematics underlying all discipline-based approaches accomplishes the equivalent of filtering out the enfolded and dialectically enfolded character of all life in general, and human life in particular. It amounts to the most fundamental re-engineering (in the literal sense of this word) of ourselves as a symbolic species relying on symbolization, language, and culture to make sense of and live in the world.

It is the second generation of secular myths that prevents us from living with our most powerful creations without any secular religious attitudes. Otherwise, we would employ our cultural bifocal lens in a much more appropriate manner. We would rely on symbolic approaches to go beyond the limits of discipline-based ones, and vice versa. Doing so is entirely possible, as I have shown through the development of preventive approaches in relation to engineering – which is to technique what physics is to science. Such approaches employ design exemplars on the level of symbolization, language, and culture, and analytical exemplars on the level of discipline-based approaches. Since the latter cannot be integrated to produce anything equivalent to the former, our symbolic approaches, which make the greatest use of context possible, would be used first, followed by discipline-based approaches for analysis and refinement.[21] Hence, this is not a question of science or technology bashing, which would be the secular equivalent of a taboo on touching anything sacred.

In other words, our unprecedented reliance on mathematics in most disciplines is related to our global civilization shifting from a primary reliance on symbolization and culture to discipline-based approaches and technique following the Second World War. Much of the success and power of technique depends on its use of mathematics; and in turn, much of the applicability of mathematics depends largely on technique. To be sure, this is a circular argument associated with all relationships of reciprocal interdependence without any clear causality; and it is ultimately symbolically grounded in the second generation of our secular myths. It and it alone gives priority to our knowing and doing. We treat everything living, which has an enfolded and dialectically enfolded architecture, as if it can be understood in terms of the diametrically opposite architecture that characterizes mathematics, machines, information machines, and everything non-living. We are thus enslaved to our collectively looking for what is living among what is dead.

This approach toward life is made self-evident and true in our daily lives lived through the secondary frame of reference, in part because life has become regarded as taking place in a time and space that is separated from experience and culture. Moreover, society itself is now regarded as built up from relatively distinct constructs that make up its science, technology, economy, social structure, political democracy, legal system, mass education system, the military-industrial complex, statistical morality, and a public secular political religion (complemented by a private traditional religion). What little remains of the highly desymbolized cultural orders has been reduced to cultural-aesthetic expressions. Our daily lives thus appear to confirm our being in the grip of technique. This correlation was to be expected because, once again, as people change technique, technique changes people through its vast influence on human life, society, and the biosphere.[22] We will briefly highlight some key aspects that have affected our perceptions of daily life, going back to the beginning of industrialization as the emergence of our present global civilization.

As noted, industrialization, and everything that came with it, profoundly affected the lived general relativity of human life and the world mediated through the secondary frame of reference. Humanity had already become differentiated from nature. With the latter being increasingly interpreted as a gigantic mechanism by means of our first generation of mechanistic worldviews, it and all the other mechanisms assumed to be produced in abundance by industrialization could only be imagined and dealt with as existing in their own three-dimensional Cartesian space. The metaconscious concept of space derived from the secondary frame of reference thus began to undergo fundamental changes. This became particularly evident when people began to make railroad journeys, which further changed people's time-space-society relativity, since travellers were now embodied in the landscape via the train. It led to what has been referred to as the panoramic perception of space,[23] which is comparable to what Georg Simmel[24] called urban perception. Industrialization also necessitated the standardization of time to permit the coordination of people's lives in terms of their participation in the economic order of their society in general, and in industrial production in particular. Within the economic orders of the first generation of industrial societies, human activities were embedded in time and time was embedded in money, with the result that time became money. Money as capital became the new secular sacred that symbolically grounded the new economic orders. Because human activities require time, and because time was money, any activities that did not earn money were a waste of time; and this in turn became expressed in an entirely new morality.[25] Hence, in the new economic orders, time-space-society became embedded very differently as each one had now become separated from experience and culture.

Elsewhere,[26] we have shown that industrialization may be interpreted as the progressive unfolding of the dialectical enfoldedness of what remained of the cultural orders following their separation from the growing economic orders. As we have already noted, once human knowing and doing became reorganized by means of disciplines, this unfolding resulted in all the cultural dimensions of mediation being spun off from what remained of the cultural orders into relatively distinct constructs. All this constitutes the background against which the socialization of children needs to be understood. It also forms the essential context against which their learning of mathematics, science, and (later) discipline-based professional skills needs to be examined. Everything in human life must now be interpreted in terms of its significance for the secondary frame of reference (language and culture), the tertiary frame of reference (discipline-based approaches and technique), and their interplay. It is also in this context that we must understand any definition of mathematics, including its terms.

There is a complex interplay between our making sense of and living in the world by means of the equivalent of the two parts of our "cultural bifocal lens." We have noted that there is a coexistence of what has been referred to as intuitive physics (arrived via the secondary frame of reference) and school physics (arrived at via the tertiary frame of reference). There are many other such parallel forms of knowing and doing.[27] Moreover, contemporary physics is now drawing mathematical images of the universe. The gap between these and our intuitive understanding of the physical world is now so great that experience and culture can no longer make sense of it. Will the end of science be a collection of mathematical formulae?[28] Will these and other issues lead to the end of science?[29] However, when it comes to a conflict between what we are able to understand through our highly desymbolized cultural mediation based on the secondary frame of reference and what we can comprehend through a technical mediation by means of the tertiary frame of reference, our secular myths have symbolically grounded the latter at the expense of the former.

What we have attempted to establish is a relationship between the internal consistency of mathematics and its external consistency or sense relative to human life and society. Mathematics is developed, applied, and transmitted from one generation to the next by means of a tertiary frame of reference, which became directly differentiated from the existing secondary frame of reference onto which it was grafted and in relation to which it developed within every person's life. Consequently, all activities that jointly constitute what we refer to as mathematics are carried out relative to a domain that is separated from experience and culture. As a result, culture is non-mathematical and mathematics is non-cultural. What remains of our highly desymbolized cultures is dealt with via the secondary frame of reference while mathematics is dealt with via the tertiary frame of reference. Any discussions regarding the foundations of mathematics cannot be restricted to mathematics itself but must refer to its highly desymbolized cultural context. Shocking and unacceptable as this will undoubtedly appear, contemporary mathematics is thus ultimately symbolically grounded in our second generation of secular myths. We will now turn to the exploration of the external consistency of mathematics.

Science and Mathematics

In our explorations of mathematics, it is essential not to lose track of the general relativity of human life and the world's symbolical grounding in a second generation of secular myths. We must therefore be very sceptical of anything that would appear to have no limits, of anything that exhibits a great deal of autonomy from outside influences, and of anything that in any other way is incompatible with the general relativity of human life and the world. We will begin with the obvious example we have probably all encountered in our high school days: the detached and objective observer studying something in a value-free manner by means of the scientific method and recording the results in an equally neutral fashion. The general relativity of human life and the world would immediately suggest the presence of secular religious attitudes toward science, resulting in scientism as well as a possible ideology for legitimizing the autonomy of the scientific discipline of this observer under all circumstances. It is not deemed necessary to involve a psychologist, sociologist, cultural anthropologist, economist, political scientist, lawyer, philosopher, theologian, linguist, historian, or a member of any other discipline to verify whether what the person really does is actually as neutral and detached from this relativity as we claim. We simply carry on as if none of these disciplines can contribute anything that is intellectually essential for understanding any scientific activity. Such a view is so asocial and ahistorical that the actual scientific activity has ceased to be a human activity. It is apparently not carried out by an embodied human being whose participation involves him or her with a level of commitment of some kind. The activity can thus be understood in itself, making it autonomous from the relativity of human life and the world. Any other interpretation would question the autonomy of both the activity and the discipline and thus weaken the legitimation of our discipline-based scientific and technical division of labour. By now it will be clear to my reader that this roughly corresponds to the interpretation referred to as logical positivism. It could not withstand any social or historical analysis, with the result that eventually scientific activities were restored to being human activities of a unique character. It was Thomas S. Kuhn[30] and his discovery of the remarkable earlier work by Ludwik Fleck[31] that opened up scientific activities to sociological and historical analysis. At first, the grip of logical positivism and all its variations was so strong that it gave rise to a great deal of confusion. Possibly the most decisive one centred on Kuhn's most central concept, namely, that of a paradigm – to a degree that he abandoned it in the second edition of his work. In the process, I believe something essential was lost that I will seek to recover in the following paragraphs.

The lengthy intellectual development required to reach the frontier of a particular scientific specialty is usually acknowledged by the awarding of a doctorate. It may be interpreted as a unique process of secondary socialization into a "world" constituted by a domain of a discipline, followed by a growing specialization that narrows the intellectual focus on one of its subdomains. The corresponding developments in the brain-minds of the persons undergoing this secondary socialization would be grafted onto those resulting from the primary socialization centred on the acquisition of the secondary frame of reference necessary for learning a language and a culture. Because the relevant experiences have a constructed foreground corresponding to what belongs to the domain of the discipline, a tertiary frame of reference will have to be developed to interpret them. As noted, this development would have its origin in making sense of experiences that were technically mediated through screen-based devices. The person's experiences related to the learning of arithmetic and geometry in elementary school, as well as mathematics and science in high school, would have led to the formation of unique clusters of experiences associated with the semantic memory that were clearly differentiated from all the other clusters of the technically mediated experiences of reality. Without these three developments within the brain-mind, parallel forms of knowledge such as intuitive and school physics could not have occurred. Our behaviour rarely externalizes any confusion regarding the three differentiated constellations of clusters of experiences. These correspond to: culturally mediated experiences involving a secondary frame of reference, technically mediated experiences of reality involving a tertiary frame of reference, and technically mediated experiences related to the domains of disciplines involving a further development of the tertiary frame of reference.

It is this third development that roughly corresponds to what Kuhn originally designated by his concept of a paradigm. The acquisition of a paradigm within the organizations of the brain-minds of scientific specialists would be externalized in their activities and behaviour patterns in a variety of ways that would be enormously diverse, although not as diverse as the externalization of their desymbolized cultures. It would include the articulation of explicit concepts, theories, and knowledge as well as a great deal of metaconscious knowledge (separated from experience and culture) such as implicit scientific "domain-values," beliefs, intuitions regarding the plausibility of analogies, and so on.

To arrange all this as a disciplinary matrix, as Kuhn did in his second edition, loses something very essential. It is precisely because all the elements of such a matrix are not separate and distinct entities in the organization of the brain-minds of scientific specialists that they are able to do their work. In other words, the difference between a paradigm and a disciplinary matrix is that the former clearly implicitly acknowledges its embodiment in the brain-minds of scientists, while the latter has eliminated this embodiment in favour of a "list" of its contents, as it were.

The theoretical implications are far-reaching. The concept of a paradigm maintained the possibility of recognizing how it was embedded in and dependent on the earlier developments of the brain-minds of scientists, including those of primary socialization. It left scientific activities as fully embodied human activities with a highly unique participation through a portion of the tertiary frame of reference and a commitment backed by a metaconscious knowledge associated with the domain of a discipline and thus separated from all other metaconscious knowledge related to the lives of these scientists. This helps us to understand that logical positivism was indeed a kind of ideal type of a human activity that socially and historically was impossible to live and achieve. The gap between them is full of significance for our understanding of scientific activities.

Consequently, what culture is to a society, a paradigm is to a "community" of scientific specialists. From a sociological perspective, this "community" has been more appropriately referred to as a kind of invisible college, since its members are faculty or researchers at different institutions. Both the concept of a culture and that of a paradigm have their roots in our being a symbolic species. We have no direct access to the unknown from which flow all discoveries. We must mediate our relationships with this unknown, either by means of a symbolic culture or by a subsequent development grafted onto it by means of a tertiary frame of reference that interprets this unknown as reality or as the domain of a scientific or technical specialty. We have no direct access to this unknown, as logical positivism and much of Western philosophy have implied.

Disciplines, Games, and the General Relativity of Human Life

The Greek concept of universal knowledge built up with the principle of non-contradiction could obviously not be entered into directly by Greek babies and children and consequently was embedded into Greek culture, much as games are embedded in all cultures. Games constitute a separate and distinct micro-world that admits only those events that conform to its rules and rejects all others as inadmissible. In other words, the events of a game cannot be in any way affected by anything other than its rules during the time that people play it. Almost any culture has made room for people to leave its symbolic universe and enter into the domain of whatever game is played. In other words, game-based activities are differentiated from all other human activities as involving the entry into a micro-world restricted to events that follow the rules of the game, with the result that the "world" of the game is separated from experience and culture. What games have in common despite their extraordinary diversity is this necessary entry into the micro-world of the game.

To push the analogy a little further: if a culture is a way of making sense and living in what is ultimately an unknown, then any games that are played by the members of a culture involve their stepping out of all of this to amuse themselves by creatively playing by its rules, which have no relationship whatsoever to this unknown. In contrast, the creation of Greek universal knowledge by means of the principle of non-contradiction was unlike any game, in the sense that it was supposed to improve on what could be achieved by means of Greek culture. In this way, the domain of a scientific discipline is built up with the principle of consistency to provide its practitioners with what they believe is a better and more accurate understanding of those aspects of the unknown associated with the category of phenomena studied by the discipline. It seeks to displace its cultural alternative on the basis of its scientific exactness. This exactness relative to a domain is entirely different from its being true relative to a cultural order. Since we have no direct access to the unknown, what constitutes a more true or more accurate insight cannot be established relative to it. All we are able to do is to compare various paradigms that have succeeded one another in the course of the history of a particular discipline.

For example, in Western civilization, physics has had an Aristotelean paradigm succeeded by a Newtonian one, in turn followed by an Einsteinian one; and today we have a paradigm of several fragments that no one has been able to mathematically unify.[32] The Aristotelean paradigm sought to systematize people's experiences of the physical behaviour of their world. In the Newtonian and Einsteinian cases, a discipline-based approach mediated all relationships with the unknown through a tertiary frame of reference differentiated from a secondary frame of reference. This continues to be the case today, except that there is a level of fragmentation not encountered before. As Kuhn has correctly pointed out, the result is that today we know a great deal more about the behaviour of physical phenomena, but we cannot scientifically conclude that this means that we know these phenomena any better.[33] This latter evaluation would require a direct access to the unknown, which is impossible for us as a symbolic species.

It is thus impossible to claim that a scientific fact is true relative to the unknown. The only conclusion that we can reach regarding a fact is that it is exact relative to the theories that explicitly express the ruling paradigm and that it is consistent with that paradigm as a whole, and vice versa. A fact can be true only if we can guarantee that no more paradigm changes will take place in the future and that, at the end of time, some authority will declare the unknown as it has become known through that discipline to be compatible with all of the unknown. In other words, a scientific fact can be held to be true only by means of a metaphysics. Within the domain of a scientific discipline, the only claim that we can make is that all its entries are consistent with one another, which includes the theoretical, experimental, and procedural ones. Science cannot be related to truth other than through a metaphysics.

Culture by means of symbolization attempts to relate everything to everything else relative to human life in the world. It thus deals with what is relatively true and not true for that life within a cultural order.

Humanity mistook it for absolute truth because that order was symbolically grounded in traditional myths. Today, the discipline-based approaches of science establish what is real, which we confuse with what is true because of the manner in which our ways of life blur discipline-based approaches with cultural approaches jointly grounded in secular myths. Scientists have confused truth with reality, and many Christian theologians confuse reality with truth.[34]

When we are attempting to discover the foundations of science, it is essential to distinguish the unknown that we have come to know and live within by means of a secondary frame of reference, the unknown that we have come to know by means of scientific discipline-based approaches, and the unknown that we have come to know through technical discipline-based approaches. All of these must jointly be distinguished from the unknown that lies beyond, with which we can interact in new ways that will lead to new discoveries and possibilities. Since the latter is not directly accessible to us, nothing can be said about the foundations of science relative to this unknown. We are limited to making comparisons between what we know through cultural mediation and what we know through discipline-based mediation. Any claims regarding a foundation of science are nothing but metaphysics, given the relativity of human life and the world. All we are able to do is to impose theories on the unknown in order to establish new relationships with it by developing the experimental implications of the theory, and vice versa. Hence, the theory guides the development of the relationships made possible by implementing the theory in experimental ways; and the findings can in turn guide the elaboration of the theory, or if they are anomalous with respect to it, necessitate a paradigm change to reorganize what has become known in a different way. It can then be expressed mathematically by means of a different theory.

Hence, it seems to me that Kuhn's original use of the concept of a paradigm would not have been in the least confusing had it been interpreted as something equivalent to, but also very different from, a culture. Both concepts refer to ways of making sense of and acting on the world, by symbolic or technical mediation. The postscript of the second edition of Kuhn's work begins to move in this direction. Scientific education is essentially examined as a unique secondary socialization process into this "culture" of a discipline or paradigm.[35] However, I believe it would have to be complemented by the kinds of analysis made by Ludwik Fleck, to whom Kuhn acknowledges his deep indebtedness.[36]

We will return to this, but for now it is important to use these remarks for clarifying our terminology.

Science may be interpreted as a system whose constituent disciplines' interactions with the unknown are guided by their domains. Each discipline is autonomous from all the others, with the result that they do not jointly build a comprehensive understanding of human life and the world. Each discipline's domain is internally consistent as expressed through mathematics to the greatest extent possible. The external consistency of each domain relative to all the other scientific disciplines is mostly ensured through the consistency of the mathematics on which they depend. The external consistency of a discipline with the unknown involves a reciprocal interaction with it in which the theory guides the experimentation, and the experimentation guides the refinement and elaboration of the theory until an anomaly is encountered. The external consistency of each domain is thus limited to the range of relationships this domain can mediate with the unknown; and it is this set of relationships that limits its external consistency with the unknown. Possibly the most important flows of information between autonomous disciplines involve changes in the use of mathematics, which has implications for its consistency with other disciplines. Therefore, science cannot establish the truth of anything but only confirms its exactness or consistency relative to one or more domains of its constituent disciplines. Anything else is metaphysics but not science. Consequently, science is non-cultural, and culture is non-scientific.

The specialists who participate in the building of this discipline-based science by virtue of their lengthy educational training can have only a partial knowledge of the scientific frontier with the unknown, limited to a subdomain of a discipline or several subdomains of "neighbouring" disciplines. Consequently, no scientific specialist is able to assess the significance of particular advances on this frontier for all of science. They only know what may be regarded as small pieces of a gigantic mathematical puzzle in which all the disciplines participate. They cannot turn to mathematicians to advise them on the scientific significance of any advance because they are as specialized as themselves, unable to assess the significance of mathematical contributions for mathematics as a whole. There is thus a strong technical character that may even dominate the scientific character of their work. Perhaps this is the meaning of the concept of a techno-science.

This technical character of science manifests itself most of all because its intellectual division of labour is incapable of scientifically dealing with relationships with the unknown that appear to have an enfolded or dialectically enfolded architecture, or both. Discipline-based science deals with the unknown as if it had the architecture that appears to be limited to non-living entities and relationships as well as everything built up with them. For our current situation, this means that science cannot conceptualize any relationships with the unknown that, at least in principle, cannot be represented by mathematics. Whatever is qualitative must thus be reduced to something that can be expressed quantitatively, and thus in the image of non-life. In this respect, discipline-based science is unique but nevertheless a part of the system of technique made up of all discipline-based approaches.[37]

It also follows that the frontier of science is not the boundary between what we have come to know scientifically and the unknown that lies beyond it. The actual existence of such a boundary is scientifically untenable, but it constitutes a metaphysics that we continue to rely on. Under the interpretation of logical positivism, science was pushing back this boundary, thereby exponentially expanding what was known. It was assumed that this would progressively diminish our reliance on religion and would lead us to a new secular paradise free from religious persecution and violence. It was a wonderful consolation, but it was utterly mistaken in its interpretation of both science and religion.

It follows from what we have noted above that our concept of the unknown is what cannot yet be detected by our relationships with it that are guided, but also limited, by the domain of a discipline. When an anomaly is discovered relative to the domain of a discipline, for example, its constituents and relationships must be reinterpreted and reorganized by means of a new theory. The reorganized domain then makes possible a new set of relationships with the unknown that will partially overlap with the previous set of relationships that were organized in a theoretically different way. There was thus something of the unknown embedded in this previous set of relationships, which did not become apparent until the anomaly necessitated their rearrangement on a different theoretical basis. It can be argued that something of the unknown was embedded in, or enfolded in, what was previously known; and the same kind of thing can be reasonably be expected following the discovery of a new anomaly in the future. Hence, the so-called exponential growth of knowledge is indissociably linked to, and permeated by, an exponential growth of scientific ignorance. This permeation of the known by the unknown affects everything that we scientifically know, which is hidden from our awareness by the way we symbolically ground all our scientific knowledge in the second generation of secular myths. This grounding is done via the tertiary frame of reference grafted onto the secondary frame of reference, in turn grafted onto our embodiment in the world. Much of the philosophy of science has, for a very long time, abstracted scientific activities from human life in the confidence that nothing essential would be lost. I will therefore insist on scientific activities being unique among all other human activities in the way they are embodied (physically, socially, and culturally) in our lives, in the way we are restricted in our participation in them by means of a tertiary frame of reference, and the way this limits our commitment to science because we cannot back it with our lives but only with our secular myths.

We have argued that discipline-based approaches are limited to those situations in which the corresponding category of phenomena dominates all the others, to the point that they can be neglected.[38] Georges Devereux has further shown that the physical, social, and cultural embodiment of a scientific observer can play a significant role where there is a conflict between this embodiment and what is being observed.[39] In such cases, the common reaction of the observer is to metaconciously symbolize, differentiate, and integrate what is observed into his or her brain-mind in a way that, in subsequent behaviour, reduces the level of anxiety. Even where such tensions do not occur, some disciplines have recognized that the observer affects what is being observed. Furthermore, the difficulties of reproducing the domain of a discipline to the extent required for an experiment are in some instances so great that the question must be asked as to what is being studied: the effects of the experiment on the unknown, or the unknown itself (if it could be assumed that it was unaffected by the experiment).

It is ironic that in the model discipline of physics, there probably have been more questions raised regarding the scientific validity of discipline-based approaches than in any other discipline. Some have challenged the assumption of divisibility and separability and thus the absence of enfolding.[40] Others have questioned whether it will ever be possible to know what scientists commonly refer to as reality.[41] Still others have raised the possibility of the limits of science and thus the possibility of reaching a point beyond which we cannot go.[42] It is more than a little perplexing that our having surrendered all human knowing to discipline-based approaches, as if they had no limits, has given rise to intuitions about the complexity of the way everything appears to be related to, and evolves in relation to, everything else, which appear to raise serious questions about their scientific validity. It may well be that, jointly, our discipline-based approaches are uncovering the most fundamental anomaly possible: the architecture of the unknown may well be incompatible with the one implied in our discipline-based science. Autonomous disciplines are possible only if the architecture of the unknown has a simple complexity. If it turns out not to be the case, our discipline-based science will have been just another metaphysics. We ought to have been much more sceptical about autonomous disciplines that have no limits because of the relativity of human life and the world. They are surely a secular religious distortion of human knowing that has its symbolic grounding in the second generation of secular myths. We can begin to address this metaphysics by withdrawing our discipline-based approaches in science and technique from the examination of anything living. Since this would include all of human life and our biosphere, it would not be unreasonable to argue that much of what we urgently need to know if we are to move toward a more viable and sustainable way of life in the world has been sacrificed on the altar of our secular myths. It is time we stopped serving our secular equivalents of the traditional gods.

Mathematics as a Discipline

We can now begin to apply these findings to the discipline of mathematics in the recognition that it is unique, but a discipline nevertheless. Mathematics would be indistinguishable from scientific disciplines if we accepted that the advances made by mathematicians are the discoveries of things that are already inscribed in the unknown. It is our contemporary version of behaving as if human life and the universe were written in the language of mathematics. However, when mathematicians are challenged on whether they truly believe this, they argue in favour of a more formalistic interpretation. It is as if they were playing the equivalent of a game in the domain of mathematics according to its rules. Hence, it has been said that mathematicians behave as Platonists during the week and as formalists on Sundays.[43]

In other words, the discipline of mathematics is unlike all the scientific disciplines in that it is not seeking to establish relationships with the unknown as it exists "out there." What the others do resembles the development of new entities and relationships in the domain of mathematics by creatively applying the principle of consistency.

When published mathematical papers are examined, it becomes immediately apparent that they are written as if the activities that produced them can be summed up as a lengthy formalism. There may be several pages of definitions, a number of lemmas, a theorem (often requiring a lengthy statement), and the proof of the theorem. The proof will generally involve the applications of a number of specified lemmas to a number of specified definitions – a lengthy mechanical procedure in which the mathematician as a human being behaves in a manner that may be compared to a kind of logic machine. Nevertheless, the understanding of such articles requires an extensive education, leading to the practice of mathematics in a highly specialized area. This education involves the acquisition of sets of motivations, accepted arguments, and exemplars, ways of thought and modes of reasoning shared by fellow practitioners. This background is so important that, if one day all the practitioners of a specialty happened to die, it could well be possible that no other mathematician could make sense of the proofs described in these published articles.[44]

When mathematicians demonstrate proofs on the blackboards in their classrooms, they do not carry them out step by step, since this would take forever. What they are essentially demonstrating is that it is possible to undertake such a proof. What happens in these articles and in these classrooms is a presentation of an argument that ought to convince someone who knows the subject.[45] As students advance toward the frontier of mathematics, the arguments with which they are presented become increasingly incomprehensible to anyone who has not received a similar education.

In the same vein, the reviewers of mathematical papers do not reproduce all the steps required to prove a theorem. They essentially assess the plausibility of the argument that a proof could be completed as described in the paper. No one, without having developed the appropriate tertiary frame of reference acquired through a lengthy education and extensive practice, would be able to decide whether or not the argument presented in a paper can or cannot be translated into a complete proof. The acquisition of this tertiary frame of reference involving countless "experiences," first of the domain of mathematics in general, and then of a subdomain in particular, will have developed the required conscious and metaconscious knowledge to reach the level of expertise that permits these people to make the kinds of judgments that reviewers and readers of mathematical papers are required to make. It confirms the presence of what Kuhn originally referred to as a paradigm, which we have reinterpreted as the portion of the organization of the brain-mind built up with a tertiary frame of reference, the functioning of which may be compared to one part of a "cultural bifocal lens." It confirms the work of Michael Polanyi and his insistence on the essential role of tacit knowledge.[46] It also corresponds to the findings of Hubert and Stuart Dreyfus in their five-stage model of human skill acquisition. The last two stages of this model fundamentally depend on what, in this work, is referred to as metaconscious knowledge, which is equivalent to their term, intuitive knowledge.[47]

In the context of our global civilization having entrusted all human knowing to discipline-based approaches, the significance of mathematics is that it provides a very powerful reasoning tool capable of guiding experimental investigations and using the results for refining this reasoning and the mathematical models it is able to construct in both science and technique. It has been a spectacularly effective way of imposing a simple architecture on human life and the world, resulting in both an enormous increase in the efficiency and power of our undertakings and the fact that their theoretical and practical externalities distort and undermine everything living. In conformity with the second generation of our secular myths, mathematics has proven itself to be critically important for our global civilization. We have become so mesmerized by the success of our mathematics that we are generally content with going no further than observing how well it works. However, this success is entirely due to our having surrendered all human knowing and doing to discipline-based approaches. The replacement of culture by technique in our ways of life is unthinkable without our mathematics, and in turn the success of this mathematics is unthinkable without technique. This reciprocity is symbolically grounded in the second generation of our secular myths.

We are now in a position to deal with the relationship between mathematics and truth. No entities or relationships within the domain of mathematics can be held to be true for human life and society. What is true or false belongs to the realm of sense, now diminished to what is left of our highly desymbolized cultures. What is true is reserved for any element of human experience belonging to a cultural order that is not experienced as being incommensurate with it. Its status of being true does not extend beyond this cultural order, since it is symbolically grounded in myths. Hence, the status of being truthful is relative; it may extend to other cultural orders only if they share its symbolic grounding and thus belong to the same civilization. Mathematics is non-cultural, and culture is non-mathematical. Culture and mathematics belong to different spheres of individual and collective human life, which is further confirmed by their incommensurate architectures. Consequently, nothing within the domain of mathematics can be true. All that we can say regarding one of its constituents or relationships is that it is exact or consistent relative to all the other constituents and relationships within the domain. The domain of mathematics as a whole, insofar as it can be known by the members of our global civilization, may be resymbolized as a kind of game whose events cannot be affected by anything that is not consistent with it.

Mathematics, Languages, and Games

Since no foundations of mathematics could be found within the world in which we live, it may be concluded that there is no point of contact between the domain of mathematics and this universe. If this is the case, it raises some rather troubling issues for our being a symbolic species fundamentally dependent on language and culture. Most of these issues have largely been made invisible by assertions that mathematics is a language. For example, in an honest and self-critical examination of mathematics by two practising mathematicians, the one issue that is never raised is the possibility that mathematics is not a language at all.[48] This appears to be inconceivable. Similarly, the leading researchers associated with the first generation of general artificial intelligence regarded computers as symbol-manipulating machines.[49] Furthermore, computer languages are not distinguished from natural languages, as if they were their artificial counterparts. There is assumed to be a mathematical "discourse" involving mathematical objects and their more complex structures, "verbs" such as equal signs, lesser-than or greater-than signs, and so on, mathematical "adjectives" such as restricters or qualifiers specifying subgroups, and so on. It is apparently possible to take any written or spoken text and formalize it in mathematics by means of a mathematical language; and this language is the language of formal set theory. There is also a dialectic mathematics distinguished from algorithmic mathematics, the existence of a mathematical object and its significance, and a great deal more that would suggest that the vocabulary of the language of mathematics partly overlaps the vocabularies of "natural languages."

However, if mathematics is non-cultural and culture is non-mathematical, the difficulties of "translating" from a culture into mathematics, or from mathematics into a culture, may not be self-evident at all. In the domain of mathematics there are no speakers of a language, infinities abound, and independent variables are common. Within the general relativity of human life and the world, the closest we can come to infinity is the mistaken idea that regards the Jewish and Christian concepts of eternity as a kind of infinite amount of time and thinks of eternal life as being equally infinite. Eternity means no such thing in these traditions: it refers to what is outside or beyond the general relativity of the time-space-society-biosphere-universe continuum because it was not created. In the same vein, an independent variable within this general relativity of human life and the world would correspond to the traditional gods: omnipotent, without limits, and independent from everything else. There appear to be some very difficult, if not insurmountable, issues of translation here.

Possibly the most insurmountable of all such translation issues derives from the diametrically opposite architectures of the general relativity of human life and the world expressed through a culture but symbolically grounded in myths, and the simple complexity of the domain of mathematics based on the principle of consistency. As noted in the Introduction, despite vast research funding that has attracted some of the most gifted academics in the world, attempts to build the first generation of a general artificial intelligence by dividing, separating, and then mathematically representing human life as a construction of micro-worlds, frames, scripts, or narratives remain without any success whatsoever. The general relativity of human life and the world presents us with an indivisibility and an inseparability that would have stopped all mathematization of it, were it not for our distorting this general relativity by means of our current secular myths.

If we are to avoid the traps of metaphysics, it is best to remind ourselves how members of a symbolic species become pure or applied mathematicians. Hopefully, we can all agree that they are born like everyone else, and then physically, socially, and culturally embody themselves in the world by developing a secondary frame of reference that manifests itself in their acquiring a language and a culture. The development of this secondary frame of reference is undermined from a very early age onward by the coexistence of two streams of experiences. In the first, the foregrounds are arrived at through symbolizations that respect the general relativity of human life and the world until the metaconscious knowledge begins to go deeper and deeper in anticipation of people's acquiring myths. In the other stream, the foregrounds of the experiences are constructed as a simple complexity instead of a dialectically enfolded one, with the result that this stream is essential for building the tertiary frame of reference provided that it remains sufficiently marginal until the secondary frame of reference has firmly established itself. In other words: in a traditional society, the first stream of experiences would have been of the cultural order of the community in which people grew up. In our contemporary anti-societies, children's experiences with a constructed foreground force them to compromise their secondary frame of reference and then graft a tertiary frame of reference onto it to begin building the reality that has displaced the cultural order. For some, this may lead to significant learning disabilities. Following a great deal of higher education in mathematics, this tertiary frame of reference will differentiate this reality from the growing domain of mathematics. Both this reality and the domain of mathematics are separated from experience and culture. It is exactly the same development that occurs when people pursue a higher education in scientific or technical disciplines to acquire a "paradigm" within the organization of their brain-minds.

All these developments fundamentally depend on the use of language as well as the use of images. We have noted that no person can become a member of our symbolic species without listening to a language, acquiring it, and beginning to enter into the cultural "world" of the significant others in their lives. Even when the language is highly desymbolized along with the culture, there is no other way of becoming members of our symbolic species that we are aware of. When babies and toddlers are not exposed to language for whatever reason, they build up the organization of their brain-minds by extending their visual experiences. They will eventually look like grown human beings, but they will not behave like them. Even though mathematics extends the architecture of the world implicit in our visually apprehending it, the development of a tertiary frame of reference cannot occur without the prior development of a secondary frame of reference into which it must be grafted because it fundamentally depends on it. Of course, when mathematicians are engaged in the kinds of activities that belong to mathematics, they can enter the domain of mathematics only by their mathematical imaginations and thus with minimal embodiment, participation, and commitment, but these limitations play a role that cannot be neglected. Such mathematical activities take their place in human lives in accordance with the developments associated with primary and secondary socialization referred to above.

As a result, we can expect that whatever symbolic language mathematicians have acquired will "bleed" into the development of the tertiary frame of reference and, via it, into their "experiences" of mathematics. In a society like ours, where many members engage in mathematical activities, we may expect that whatever language has "bled" into the domain of mathematics will be significantly affected by this very different context, and that something of this will "bleed" back into society through primary and secondary education, mathematicians talking about their work on the media, and so on. In other words, it should come as no surprise that a portion of the vocabulary of the symbolic language that mathematicians grew up with overlaps with or fundamentally affects their professional vocabulary. Nevertheless, within the domain of mathematics, the significance of any object or relationship is not dialectically enfolded into those of all the others, and vice versa. Whatever professional vocabulary is used by mathematicians is partly or wholly unfolded from the dialectically enfolded symbolic language; and this will significantly affect the meaning of many words when it "bleeds" back into human life and society.

These influences of a symbolic language on the domain of mathematics and the effects of the domain of mathematics on a language and culture will thus weaken the autonomy of the discipline of mathematics. My argument converges with that of Didier Nordon, who argues that pure mathematics does not exist.[50] He shows that there is a close relationship between mathematics and the kind of society we live in. He quotes J. Berg, stating that the incorporation of the laws of dialectics into our logical mathematical system would result in a catastrophe for mathematics and for science in general, and would mean the end of human civilization.[51] However, the laws of dialectics referred to are almost certainly those implied in the work of Karl Marx, which gave rise to a reigning ideology in East Germany of that time.[52] It was shown to be completely untenable in a reanalysis of his work in the context of the twentieth century, undertaken by some Marxists.[53] Even this analysis has some serious limitations.[54] I would therefore rephrase the above observation as follows: If we recognized the enfolded character of life on the biological level and the dialectically enfolded character of human life on the cultural level, mathematics would be transformed from an autonomous discipline into a powerful and useful tool that, like any other tool, has its limitations. The recognition of these limitations could begin to transform our civilization into one that would be more socially viable and environmentally sustainable.

Our current metaphysics of mathematics is helping to block our access to a more desirable future. Unfortunately, this is exactly what all myths (traditional or secular) have always accomplished. They represent entities that we live with as if they have no limits, with the result that what we expect from them cannot possibly come to fruition; and in the meantime, we have not done what we could to deal with our situation. It is hardly surprising that the Jewish and Christian traditions have always considered the creation of these kinds of unlimited entities as the greatest threat to any society. In the past, mathematics and science were generally unique to a culture and associated with one of its dimensions of cultural mediation. Hence, they were limited and did not pose the threat they do today.

We have given far too much credit to what philosophers have concluded about these matters. What is important is not their philosophical interpretations in themselves, but why these ideas were or were not getting the attention of a significant number of the members of their society. Every age has its "crazy" people, but what drives a society to either ignore them or listen to them? Similarly, in politics, it is not the ever-present extremists that matter but the spirit of an age that can bring a society to the point where it begins to listen and to support them.

If we extend these ideas to specialists in mathematics and science, we will have to acknowledge that we have bought into the notion that discipline-based approaches are objective. This illusion is reinforced by the fact that the practitioners of a mathematical or scientific specialty all appear to more or less agree with each other, with the result that what they are doing must be taken seriously. It is also implicit in the way the publications in mathematics and science are written, the way these subjects are taught, and the way students are evaluated in these disciplines. All this goes hand in hand with the notion that these disciplines are free from any metaphysics and ideology, and because of their neutrality and objectivity cannot have a significant effect on human life and society.

Nordon's work has a unique French perspective on mathematics as a kind of product designed for a particular kind of consumption.[55] Every mathematics article is essentially written for and is going to be read by a handful of specialists who work in the same narrow area. These articles reply to questions posed by these practitioners and thus have little or nothing to do with the general questions that in the past drove such publications. It gives them a highly technical quality in conformity with the technical spirit of our age, as described by Jacques Ellul.[56] The system of technique increasingly becomes the life-milieu in which we live; and because this life-milieu is not in any significant way affected by influences from the outside, it can be described in terms of its own internal "laws."[57] As such, mathematics becomes the non-language of technique.

Few administrators within the university care about anything other than how many articles the professor has published; and the content of these articles is immaterial except to a handful of other specialists around the globe. Few people, if any, are concerned with what the broader development of mathematics implies for human life and society. Most are concerned only with the small and highly specialized improvements made in a very narrow subdomain of mathematics. Mathematicians know this very well, but they hope that one day a modest contribution of theirs will somehow lead to a breakthrough. All this is part of a civilization that is entirely lost on the frontiers of mathematics, science, and technique in terms of what their significance is for human life on this planet. Few, if any, exercise any responsibility in what they do, with the result that the system of technique is able to function according to its own internal "laws." In this context it is very difficult to believe that mathematics is able to perform the task that is even remotely close to that of a symbolic language.

Moreover, all disciplined-based approaches contribute enormously to the desymbolization of language and culture, which are now torn between a secondary frame of reference and what little remains of a cultural order; a tertiary frame of reference and the buildup of a reality with a simple complexity; and a tertiary frame of reference and the building of the domain of a discipline through a lengthy educational process complemented by the practice of activities separated from experience and culture. There is thus a general desymbolization of language and culture and a proliferation of image-words and image-language that have been unfolded from a dialectically enfolded vocabulary and language. Consequently, even pure mathematics contributes to "technique changing people" through its influence on already highly desymbolized languages and cultures.

For more than two centuries, there have been three converging developments. First, as a result of "people changing technology" and "people changing technique," there has been the equivalent of "rewriting the book" of human life and society in an architecture that matches that of mathematics by the building of first an economic order and then a technical order within the cultural order of industrializing societies. Second, as a consequence of "technology changing people" and "technique changing people," languages and cultures have been desymbolized, thus immersing people into a reality that has replaced the cultural orders of these societies. Third, mostly through the replacement of culture-based approaches by discipline-based approaches, the architecture built by mathematicians has become universal. Mathematicians from all over the world produce articles of exactly the same form regardless of their desymbolized language or culture. This universality adds to the self-evident objectivity of the discipline of mathematics, even though it has no foundations.[58] The convergence of these three developments was ensured by their being symbolically grounded in two successive generations of secular myths.

Mathematics and Time

The universal order of mathematics built up with the principle of consistency has a unique relationship with time, which also converged with the above developments. It excludes time as we live it. For example, a mathematical object cannot be consistent at one time and become non-consistent at another time. Logic cannot evolve in time, but psycho-logic of the brain-mind does evolve in time. If mathematical objects and logic are outside of time, as it were, the domain of mathematics is unfolded from time and thus from everything that has time enfolded into it. It is for this reason that infinity presents relatively few problems in the domain of mathematics.[59] It further confirms that mathematics extends the architecture of human life and the world apprehended through seeing at the expense of the architecture apprehended by acquiring a language and culture. The universality of the order of mathematics is thus also established relative to time, which is separated from experience and culture and replaced by a mathematical time that is linear or cyclical and is not enfolded into evolution and adaptation. The order is thus asocial and ahistorical in character. Consequently, the application of this mathematical order to anything living, whether it is enfolded, dialectically enfolded, or both, amounts to intellectually "squeezing" the life out of it and practically reifying it. It ought to make all life and thus the planet tremble with dread, were it not for the successive generations of secular myths that prevent us from understanding what we are doing to ourselves and all life. It further reinforces a civilization that gives preference to seeing over listening, and to images over the word.[60]

Nordon questions whether or not there is a possible correlation between the way mathematics and music deal with time, given that many mathematicians are also excellent musicians.[61] However, he recognizes that a distinction must be made between the way mathematicians deal with time, their concept of time as it is externalized in their practice of mathematics, the concept of time that mathematicians defend following their critical reflection on the subject, and time from the outside, as it were. To this we can add the metaconscious knowledge they have of time as members of a society. Clearly, all these "bleed" into one another, making it next to impossible to clearly distinguish them without a great deal of careful study.

In any case, it is noteworthy that the history of mathematics tends to be rather asocial and ahistorical, frequently taking the form of a series of questions and answers with few, if any, surprises.[62] The context is stripped away to leave only the "order" of mathematics without any of the disorder of historical events. Once again, we must ask ourselves: If mathematics is incapable of taking lived time into account and is obliged to desymbolize it into reified time, is our civilization, by its reliance on discipline-based approaches instead of culture-based ones, condemning itself to focusing on what is non-essential for human life and all life in the biosphere? Nordon suggests that much of what our civilization is all about is related to the central place it accords to mathematics "in pursuit of a kind of knowledge whose efficiency makes the earth and its people tremble."[63] All this is matched by practitioners participating in this order of mathematics by means of a mathematical imagination that requires only a minimal embodiment (physical, social, and cultural) – a minimal participation that is mostly centred in the portion of their brain-minds acting as their "mathematical paradigm" as a result of the tertiary frame of reference, and is backed by a life enslaved to secular myths.

Mathematics and Daily Life

We need to examine the way mathematics aids our civilization in recreating the architecture of human life and society in its own image, that of mathematics. How can it be applied to human life and society, which, despite high levels of desymbolization, continue to exhibit an architecture of a dialectically enfolded character, be it greatly weakened as a result? We have now arrived at a point where this question can be answered. Since the architecture underlying all mathematics has a simple complexity, human life and society must be translated into that simple complexity; and this can be accomplished by treating it as a kind of game. From a historical perspective, this became entirely plausible as a consequence of the first generation of industrial societies symbolically grounding their ways of life in the first secular sacred, that of capital. In other words, life had been alienated into a kind of money game. All of daily life could now be interpreted in terms of "money points": some activities earning money points and others costing money points. The members of each new generation begin at the starting line of this money game according to the money points their parents had bestowed on them. Once they grew up, they would be playing this game on their own by collecting as many money points as possible, since almost everything was denominated in these terms. Becoming affluent or even wealthy was a condition of having enough money points to procure whatever satisfied their needs, wishes, and desires. They would reach the end point of this game at their deaths; and what they had accomplished was frequently reflected in the money points they could afford to spend on their funerals.

Between the starting point and the end point of this money game, almost every daily-life activity involved a direct or indirect interaction with a large organization or institution, particularly in our contemporary anti-societies. These activities became part of the money game. For example, a credit rating might be interpreted as our daily-life activities being scored in terms of credit points. A good education, a steady job, a relatively permanent address, and paying bills on time might earn points, while encounters with the law or failing to repay a loan would lead to the subtraction of points. School activities would earn points toward an IQ, GPA, or SAT score, and so on. Applications for an insurance or travel policy would be decided on by rewarding certain activities with points and others by the loss of points; with so-called big data the possibilities are becoming almost endless.

As these large organizations and institutions are able to make their services a great deal more efficient by creating the kinds of systems that adjudicate these money games, we are just beginning to wake up to our defencelessness against what will turn out to be "weapons of math destruction."[64] Their collateral damage will be enormous. We will lose our ability to understand why our dealings with some organizations and institutions appear to work well, while others turn into a disaster. We are creating a society in which we are being rewarded or punished by means of a statistical morality hidden in assumptions, algorithms, and policies that are almost entirely hidden from us. For example, we may suddenly be denied home and fire insurance after making a third claim for a bicycle stolen from our garage. A "three strikes and you're out" policy would almost certainly have fewer people make such a claim, but we will not discover this rule, by which we were playing without knowing it, until it is too late. The denial of insurance could trigger a chain of events leading to the bank recalling the mortgage on our house, which could put us out on the street. These kinds of strikes by our "weapons of math destruction" are going to become increasingly more common, thanks to "big data." Imagine what may happen when information about our DNA is extrapolated by employers, health insurance companies, banks, and landlords in order to place us in a particular category and deal with us accordingly. What about the weapons of math destruction employed by the police to determine when and where a crime is likely to be committed? We would be entirely defenceless against the collateral damage of these weapons. It would also further aggravate the growing inequality gap in North America, which is already well beyond what any civilized society ought to tolerate.[65]

These developments must be understood in the context of a society that is becoming increasingly more informatized and computerized, one in which the social sciences have largely turned themselves into disciplines that make the greatest possible use of mathematics. It matters little whether a sociologist or political scientist has a qualitative understanding of what is happening to human life and society. What does matter is the development of the potential of big data. This potential must be mined by means of survey methodology that classifies all of us in terms of categories that can make our large organizations and institutions more efficient, powerful, and profitable in their dealings with us. No general qualitative understanding is thus required; and specialists in these and many other fields are turned into a kind of mathematical technicians who can apply their "tools" to corresponding problems. Operations research, with its many branches and game theories, provides some obvious examples. In other words, expertise in a highly specialized and narrow area of mathematics is sufficient to allow people to make all kinds of assumptions and decide on procedures amenable to mathematical treatment with little or no understanding of what is really happening and what will be the consequences of their actions. Superimposed on all of this is an illusion of democracy, while this technical rationality is shutting us into the prison of technique.

The money game of our daily-life activities is matched by another money game – the one played by large organizations and institutions. The rules of this game match those of technique: divide, separate, and classify daily-life activities according to those aspects that determine their efficiency. Everything is thus measured in itself and unfolded from everything else. Only efficiency and power matter. Thus far, our human rights codes, legal protection, and democratic influence lag far behind the latest developments in the way we are literally re-engineering all our daily-life activities and relationships through the organizations and institutions of our anti-societies. The game will literally destroy the lives of some people, impair the lives of many others, and reify the lives of all of us. However, with our eyes on the stock market, we will most likely continue to treat all this as inevitable collateral damage. It must be remembered that, even if initially these new systems are well designed, they cannot evolve and adapt other than by means of patches on the software. Such interventions rarely improve the coherence of these systems.

We are back to how we deal with time, and how our lives and societies are dialectically enfolded into it. Of course, the evolution and adaptation of our lives will be tracked by the big data that all our systems increasingly depend on. Nevertheless, there is something about the inability of mathematics to deal with lived time that ought to raise concerns. The domain of mathematics is not a "world" that evolves and adapts as does the dialectically enfolded world of life. This is evident from their diametrically opposite architectures. The more we make aspects of our collective life asocial and ahistorical, the more some of this will "bleed out" into the applications of mathematics to human life and society in a self-reinforcing cycle, supported by symbolically grounding all of this in our second generation of secular myths. It is ironic that we entirely failed to "put human beings into machines" in an attempt to create what was supposed to be a general artificial intelligence, but we are now rushing into a surrender of human life to information machines on a scale that is unprecedented. Everything will become "intelligent," even our cities. But it involves a complete and total change in the meaning of what it is to be intelligent. We do well to remember an earlier pattern of events. When every cultural dimension of mediation was unfolded from a cultural order to become a construct in a global technical order, our relationship with time was fundamentally altered. The separation of knowing and doing from experience and culture also included the separation of knowing and doing from time. All this is not limited to mathematics but extends to each and every discipline.

In other words, the growing use of mathematics as a result of big data and its exploitation by every organization and institution is another big step forward in unfolding human life and society by re-engineering them in the image of the architecture of mathematics and thus further reifying life toward death. We do well to remember how experts were surprised by the extent of the computer and information technology revolution because they had not recognized that two hundred years of industrialization had fundamentally weakened the dialectically enfolded character of human life and society to bring it within reach of computers.[66] Their sales far outstripped predictions. What began as the re-engineering of the corporation has now spread to all organizations and institutions, even to the fabric of individual and collective human life, to the detriment of freedom and human dignity.[67] Feminism ought to ask itself whether women will change technique or whether technique will change women the way it did men. Environmentalists should ask whether environmentalism can change technique or whether technique has already all but subverted it. Pacifists need to question whether people can affect the techniques of making war or whether these techniques have already transformed human life and society. The same question should be asked by each and every political tradition, and certainly by every religious tradition.[68] As we are busy struggling against each other on all these fronts, mathematics-based technique is changing humanity far more than we have been able to change it. Even the most rapid and casual historical examination of any of the above traditions for the last century will tell the tale.

In sum, "technique changing people and society" has made our social world increasingly amenable to mathematics as a result of alienation and reification. In turn, this alienation and reification have greatly facilitated the further re-engineering of human life in a way that makes it more amenable to being treated as a game of some kind. A starting point and an end point can be defined as well as a way of dividing, separating, and measuring aspects of the activities between them in terms of ultimate criteria such as money (capital), efficiency, and power. All this is symbolically grounded in two successive generations of secular myths. Again, we need to ask what we as a so-called secular civilization are sacrificing to our gods in the name of our secular "fertility cults" of money, efficiency, and power.

No metaphysics of any kind is required for such an undertaking. Ever since technique began to graft itself onto culture and take over from us the organizing and adapting of our ways of life, we should have exercised our responsibilities and intervened accordingly. Doing so involves both a general and a specific understanding of how technique influences human life, society, and the biosphere, and using this understanding to adjust our participation in it so as to achieve a human purpose and at the same time prevent or greatly minimize harmful effects. We must consider how the domains of the disciplines we apply to a particular area technically mediate our relationships with it. We will quickly discover that this mediation is far from neutral; we must thus regard it as a kind of filter that allows certain aspects of the situation to pass through while blocking other aspects. For those aspects that pass through the filter, we need to examine how they are affected by it. As noted, what is entirely blocked by all disciplines is the enfolded and dialectically enfolded character of everything living. In addition, disciplines will also filter out anything that is not rational with respect to the goal we are imposing on an area of application.

Mathematics constitutes one of the primary components of this filter. To gain a comprehensive insight into what we are doing, we need to symbolize the situation in terms of the meaning and value of these activities for human life, society, and the biosphere by relying on the secondary frame of reference. On an entirely different level, we need to examine, analyse, and optimize the details of such a design exemplar by means of the analytical exemplars provided by the disciplines being used. Consequently, a responsible intervention by means of discipline-based approaches necessarily must occur on two levels: one associated with the secondary frame of reference and the other associated with the tertiary frame of reference. We can use our cultural bifocal lens to benefit from cultural approaches within their limits, and to transcend these limits by using discipline-based approaches within their own limits revealed by the cultural approaches. (Of course, such an undertaking would challenge our entire educational system, whose highest levels have been entirely surrendered to discipline-based approaches.)[69] We are then able to evaluate the use of game theory or the construction of a statistical "average person" for a particular category of significance to a large organization or institution. We will return to all this in a later chapter.

Mathematics and Education

When the university for which I worked during my entire career was compelled by our provincial government to absorb an independent institution created by an earlier government to advance public education, our president told a number of us that the only solution to its problems was an evacuation of the building in order to blow it up. When I told this story to a retired mathematics professor, he added that this would never have solved the problem because the occupants of the building would have congregated somewhere else to continue their ruining of mathematics education in the province. There is little doubt in the minds of many of us that it is not only our mathematics education that has been derailed in large measure by this institution and many others similar to it.

When I served two terms as president of an international academic association that focused on the interactions between science, technology, and society, I was always disturbed after my representing the organization at academic sessions organized by my colleagues in education. There was always a kind of zeal for the techniques they had developed to "doctor" the cultural souls of children and young people. Furthermore, during my thirteen years as editor-in-chief of an international journal devoted to science-technology-society interactions, I cannot recall a single article submitted by education professors that could be recommended for publication by its referees. I had some idea from a social and historical perspective as to what was going on in education, but to encounter it "in the flesh" was deeply troubling because the lives of so many children and young people were being played with in a kind of educational game.

As noted, the separation of knowing and doing from experience and culture and their reorganization by means of disciplines required a mass education system that could not be run effectively and efficiently without techniques of all kinds. The warnings of Jacques Ellul almost sixty years ago as to the consequences of this new education had, like many of his other predictions, come all too true.[70] Pedagogical techniques of all kinds do what all techniques seek to accomplish, namely, to make human activity more effective and efficient on its own terms, unfolded from everything else both in the lives of children and in society.[71] In the case of mathematics, this removes it from the kind of context described above and makes it next to impossible to ask the elementary questions. How can mathematics be taught in elementary schools without undermining the development of the secondary frame of reference essential for children's acquisition of language and culture? How can high school mathematics be taught to contribute to the development of a tertiary frame of reference required for the learning of any discipline without undermining the ongoing development of the secondary frame of reference? How can both these frames of reference flourish by teaching children and young people the limits of one relative to those of the other, and vice versa, in a way that is integral to the learning of mathematics? How can mathematics be taught in universities, where much of it occurs as an integral part of learning a wide variety of disciplines? No metaphysics, end-of-pipe ethics, or the latest educational fad is going to make any difference capable of addressing these kinds of questions. The evidence of this is all around us. It is the responsible doing of mathematics that education should be concerned about.

The obstacles faced by our schools and teachers as they deal with children and young people growing up in anti-societies participating in the building of a global technical order are almost insurmountable. The weakening of all social relationships and families as well as people's fascination with screen-based devices are helping to create two streams of experiences that significantly undermine one another in the development of children and young people. Learning disabilities and mental health issues are extensive; and schools cannot possibly compensate for what many children no longer receive from their families. It is thus not difficult to understand why, for decades, we have latched on to the hope that technology can save education and our schools. Have we forgotten the exaggerated expectations of promoting educational television, linking classrooms through information highways, putting a computer on the desk of every student, providing classroom access to the Internet, using educational software, and now the "Google classrooms" of this world? Nevertheless, these technologies have not delivered and will not deliver what they promise – for the kinds of reasons that were recognized earlier.[72] For example, before the emergence of the Internet, there was a leading professor of computer science who spent a great deal of time warning school boards to get computers away from children until their late teenage years. When I chaired an information technology ethics committee at my university in the mid-1980s, the chairs representing the key departments all stated that they would not have computers in their homes until their children were well into their teens. This kind of advice was essentially overtaken by the Internet. Why can we not recognize that when young children are encouraged to present their school work by means of PowerPoint, we are undermining the development of their secondary frame of reference by demanding that they unfold, into separate and distinct bullets, something that is dialectically enfolded? When all these kinds of developments are compared to how we prepare teachers, after their first degree, by a one- or two-year bachelor of education program, it is a miracle that the situation is not far worse.

In the case of mathematics education, what are we to do? Are we to trust the intuitions of mathematics professors as a guide for training mathematics teachers? We have noted that such intuitions are supported by a metaconscious knowledge of mathematics separated from experience and culture and acquired through a tertiary frame of reference. Are we to trust the intuitions of the children learning mathematics, given what we know about how their development is undermined by a premature stream of experiences related to screen-based devices? Should we explore ways that appear intuitive and sensible to them? Should we rely on the intuitions of their teachers as they, in their daily struggles, develop an extensive metaconscious knowledge of what appears to work for some children and not for others? Teachers used to be able to rely on a tradition, but technique has destroyed it along with all other traditions. We have replaced the role of tradition with one educational fad after another in a kind of Brownian movement that has gone nowhere.

In facing the challenge of teaching mathematics to children and teenagers, the point of departure surely ought to be that these unique activities must remain integral to the general relativity of the lives of teachers and students, as well as the relativity of their world. Any escape from this general relativity will involve absolutizing something by means of secular religious attitudes. In our contemporary anti-societies, two approaches to the teaching of mathematics are possible. Some parts of mathematics can be interpreted relative to human life and the world by means of the secondary frame of reference. Examples of this approach are furnished by all traditional cultures who proceeded as if they were building on the symbolization of numbers and space, without recognizing that this usually involved the imposition of a diametrically opposite architecture. In Western civilization, a well-known example is that of Euclidean geometry, believed to be the only possible geometry until non-Euclidean spaces were developed in mathematics. In the same vein, simple arithmetic was assumed to be intuitively self-evident. This approach, based on the secondary frame of reference, can still be used for the teaching of arithmetic and geometry to children in our societies. Doing so would have the considerable advantage of strengthening their secondary frame of reference without undermining the potential development of the tertiary frame of reference. This approach adopts the vantage point of the lives of children as they become persons of a time, place, and highly desymbolized culture within a technical order. It involves the use of culture-based reasoning.

The second approach teaches mathematics as a discipline, which involves doing so relative to each domain and its many subdomains. It intellectually brushes aside "intuitive mathematics" embedded in experience and culture in a manner analogous to the way the discipline of physics is taught in high school. It requires the adoption of the vantage point of a highly disembodied person whose participation and commitment are limited to a mathematical imagination, and advancing in this approach by means of logical reasoning separated from experience and culture in order to deal with a domain that exists nowhere in our world. All foundations of mathematics have proved to be illusory, however, with the result that there are no interfaces between the domain of mathematics and the general relativity of human life and the world. The only way in which this is possible is for this domain to be extended to infinity through one or more subdomains, in order to make it impossible that it could intellectually interfere with the domain of any other discipline, or with the general relativity of human life and the world.

Consequently, this second approach to the teaching of mathematics necessitates the prior establishment of an embryonic tertiary frame of reference derived from the mediation of experiences dealing with reality by screen-based devices. When this approach is adopted to teach simple arithmetic to young children in a manner founded on set theory, for example, we risk undermining the development of their secondary frame of reference, because it is not yet able to contribute to the development of their tertiary frame of reference. The following demonstrations can illustrate the potential problems and confusions.

Imagine a teacher who has laid out three parallel lines made up of tokens successively marked with the numbers from one to twenty, on a table at the front of the classroom. Assuming their appropriate ages, young children will have little difficulty recognizing that each of the three series has the same number of tokens and thus is equally long. Next, the teacher proceeds by removing all tokens marked with even numbers from the first series, all tokens with odd numbers from the second one, but leaving the third series unaltered. Again, the children will have little difficulty recognizing that now two of the series are shorter, which they can confirm by counting the tokens.

Next, the teacher can ask them to imagine repeating the same exercise by adding another twenty tokens marked with the numbers twenty-one to forty to each of the three series, and then removing the even numbered tokens from the first, the odd numbered tokens from the second, and again leaving the third unaltered. When the children conclude that once again two of the series are shorter than the third, they can be asked: If the whole procedure is repeated once again, and again, and as often as they like, will the results be any different? Most children will be able to recognize that the length of the first two series will always be shorter than the third no matter how often the procedure is repeated. This conclusion is reached by the use of the secondary frame of reference.

Next, imagine the teacher attempting to use the approach of set theory. The introduction of infinite series leads to something that is contradictory; and if it is at all intuited by the children, it will lead to no end of confusion because they have as yet no tertiary frame of reference to fall back on. No matter how long we make the three series, two of them will always be shorter than the third. However, the moment we attempt to imagine these series going on to infinity, the lengths of two of the series relative to that of the third, and the length of the third relative to those of the other two, have become equal. In other words, endlessly lengthening the three series in the world of the children will never approach a situation such as the three equally long infinite series in the domain of mathematics. This contradiction is implicit in Gödel's theorems.

Next, imagine another exercise designed to teach children that Euclidean geometry has nothing in common with the space in which they are embodied as part of the general relativity of their lives and the world. If the teacher takes a classroom globe of the earth and places a small figure on it at a location where it is vertical relative to the classroom space, the children can be asked to imagine what happens when this figure goes on a very long walking tour. The tour can then be simulated by moving the figure along the globe, causing it to tilt to a horizontal position relative to the classroom, and still later, to an upside-down position. The children can then be asked if this is what happened to them when they went on a long car ride or a plane trip. They will recognize that when they got out of the car or the plane, they were as upright as when they started the journey. It may be more difficult to help children recognize that what they observed with the figure on the globe is not a good model for what happens in daily life. The model may be further illustrated by means of a curved line drawn on the blackboard with the outline of the figure drawn on it at regular intervals, but perpendicular to the line. This geometrical model does not conform to their daily life experience either. When they become a little older, they may be ready to understand that the space in which they are embodied, and which is part of the general relativity of human life and the world, is completely different from Euclidean geometry.

With a great deal of hindsight, it is possible to reinterpret the development of mathematics in Western civilization as the gradual addition of the infinite to complement the finite. The limitations of the secondary frame of reference embedded in experience and culture were gradually overcome by the reorganization of all human knowing and doing by means of disciplines. As a result, "intuitive mathematics" gradually gave way to "school mathematics." Comparisons of the two have revealed all manner of contradictions between them as well as contradictions internal to the domain of mathematics.

For a long time, mathematicians and philosophers believed that geometry was the most reliable form of human knowledge. When this turned out to be untrue, attempts were made to found geometry on arithmetic, and arithmetic on set theory. This necessitated the introduction of infinite series and resulted in many contradictions. The descriptions of daily-life activities by means of the secondary frame of reference involved intuitions sustained by the metaconscious knowledge derived from differentiating and integrating the corresponding experiences, but these could not be a foundation for anything. It turned out that foundationalism in philosophy could not do so either. By means of the tertiary frame of reference, daily-life activities could also be described relative to the domain of mathematics, which imposed a diametrically opposite architecture on them. For example, the trajectory traced by someone falling from a tower can be described in terms of the remaining distance to the ground being constantly cut in half – with the result that the person never touches the ground! This contradiction with daily life can be avoided, of course, by clarifying whether the description is relative to the general relativity of human life and the world, or to the domain of mathematics. Doing one or the other places the activity in different contexts with diametrically opposite architectures. One involves the secondary frame of reference and culture-based reasoning, and the other, the tertiary frame of reference and logical mathematical reasoning. One involves an embodied vantage point and the other an imagined one. During their discussions of these kinds of demonstrations, it is essential for teachers to avoid the creation of a "hidden curriculum" that devalues the experiences of the students to the benefit of "objective" mathematics and science.

The argument that mathematics is the science of the infinite corresponds to its domain being infinite, in the sense that it cannot overlap with any other discipline or with the general relativity of human life and the world. Mathematics is supposed to be founded on the undoubtable principle of non-contradiction as the key to its architecture. However, from the perspective of the general relativity of human life and the world, everything casts doubt on this self-evidence. The architectures of enfolding and dialectical enfolding encountered in all life are based on characteristics that are the diametrical opposite of that of non-contradiction. I am not in the least suggesting that we attempt to explain this to children or even to teenagers, but we should stop teaching mathematics as if everyone is expected to become a pure mathematician.

It may well be more fruitful to adopt an orientation akin to applied mathematics. A subdomain may then be regarded as a finite model delimited by the boundary conditions imposed by the general relativity of the world. Such an approach can be fruitfully applied to all constructed artifacts, including classical and information machines and their interconnections. However, any application of such a delimited subdomain to the general relativity of human life and the world is scientifically not valid, and thus can intellectually and practically destroy what is modelled by dealing with it as if it were already dead. Such applications have caused the scientific and technical externalities of a theoretical and practical kind that have accumulated into the crises that plague all life and the planet. We must not deceive ourselves about the way we have transformed all dimensions of cultural mediation into constructs of the global technical order, because "underneath" them lie human activities and lives stretched on the rack of the principle of non-contradiction. We have hidden all these problems by means of our secular religious attitudes, which have allowed all discipline-based approaches (including mathematics) to escape the general relativity of human life and the world. Our dependence on infinite series in set theory is but one symptom of this escape.

In a world where everything is related to and evolves in relation to everything else, the principle of non-contradiction is radically impossible. There can be no repetition in this world, but only an evolution in time. Nothing can grow out of control to become infinite because it is held back by its interconnectedness to everything else. Infinity represents a kind of secular modern magic without which pure mathematics would be unthinkable. What are the implications of the use of mathematics, which depends extensively on infinity, to a world in which everything is finite? Is this the secular way of reintroducing something god-like into our world in order to gain power over it?

We ought to give serious consideration to what may be learned from a genuine history of mathematics. Does it imply that geometry is its most intuitive form, and thus potentially most accessible to children with a partly developed secondary frame of reference and only an embryonic tertiary frame of reference? Does the development of Euclidean geometries, and the attempt to "found" geometry on arithmetic, and arithmetic on "intuitive numbers," suggest that it may be useful to take seriously humanity's evolving mathematical intuitions as a possible inspiration for teaching mathematics to children? Does this also imply that teenagers can most easily move from geometry to algebra via an analytical approach using Cartesian coordinates? Can this geometry also become the stepping stone into calculus and other branches of mathematics?

We need to remember that the closest thing to the principle of non-contradiction in the general relativity of human life and the world is associated with those experiences that place what is visually perceived in the foreground, and everything else in the background. Mathematical approaches may thus be interpreted as extensions of these visual approaches under the control of the secondary frame of reference, and thus not in opposition to symbolic approaches. If we add to this interpretation the fact that the philosophy of mathematics has never been able to uncover any foundations, as well as the collapse of foundationalism in philosophy, I believe we have a strong argument that our current approaches to the teaching of mathematics to children and teenagers are about as pedagogically unsound as they can possibly be. We ought to explore geometry as the exemplar for arithmetic, algebra, calculus, and other branches of mathematics, as a gradual extension of the visual approach to the world under the control of the symbolic one through the secondary frame of reference.

The development of mathematics within cultures shows that it is entirely possible to teach the beginnings of mathematics to children and young teenagers as extensions of their visual approach to the general relativity of their lives and the world, much like the way in which the members of traditional societies approached mathematics. For example, the experiences of children learning intuitive arithmetics would focus on counting the items they see, with everything else they experience at that moment forming the background. Similarly, the learning of intuitive geometry by children and young teenagers would be through experiences resulting from visually focusing their attention on diagrams and associated theorems, with everything else remaining in the background. The structure of these experiences would be no different from other daily-life experiences with a visual focus of attention, with the result that they would remain under the control of the secondary frame of reference to the greatest extent possible. Proceeding in this way would have the considerable benefit that at some point (likely at the beginning of high school) it becomes possible to show them that when mathematics is applied to anything in the general relativity of human life and the world through engineering, architecture, medicine, and other practices, a small portion of this general relativity is represented as a distinct and separate domain. In this domain they will be playing a game of mathematics, which is extremely useful and powerful as an extension of what we can intellectually grasp through symbolization, language, and culture. If my engineering students had such a preparation, they would have a much better intellectual grip on how the boundary conditions to any application of mathematics ultimately derive from the general relativity of human life and the world. They would realize that they must be on the alert for the possibility that the application of mathematics to this relativity would distort its architecture, producing non-trivial consequences. At present, even my best students require an entire course before they can clearly conceptualize the intellectual structure of the preventive approaches that have become necessary to transcend the limitations of conventional ones. We will return to this subject in a later chapter. For now, the teaching of mathematics still proceeds as if most students will become mathematicians, as opposed to teaching it as a powerful modelling game that has a vast unforeseen potential of intensifying all our current crises.

It should be noted that "natural numbers" embedded in experience and culture and thus accessible by the secondary frame of reference had to be recreated by means of a "linear continuum" of the real number system, but a way of doing so without contradictions has never been found. Possibly the most pedagogically sound manner of teaching children and teenagers mathematics is to follow exactly the same pathway that turned Euclidean geometry, as the exemplar of the most certain and reliable human knowledge, into something that eventually had to be separated from experience and culture, and thus from the general relativity of human life and the world, to take on the diametrically opposite architecture of the domain of mathematics.

Consider the parallel development of teenagers learning the discipline of physics in high school. These teenagers now possess a highly developed secondary frame of reference in tension with an embryonic form of the tertiary frame of reference, the latter derived from the differentiation and integration of those experiences mostly mediated by screen-based devices as their access windows on reality. Consequently, at this age it has become quite feasible to disregard everything these teenagers have learned about "daily-life physics" and focus exclusively on "domain-physics." It is now possible to further develop their tertiary frame of reference without undermining their secondary frame of reference, provided that we do not confuse the differences by means of secular religious attitudes; that is, by making scientific activities into unique activities entirely open to sociological and historical analyses. In the same way, teenagers will now be ready to shift to learning "domain-mathematics" and to sever their connections to what they have learned about "symbolic mathematics" embedded in experience and culture.

The closest but less responsible alternative to teaching of mathematics is to teach children to play it as a game, much like the other games they learn to play. They know very well that a game has to be played by its own rules, and nothing of their daily lives must interfere with them. The only interference of their daily lives permitted in a game is the decision to start or stop playing the game. Consequently, the game can never become a micro-world entirely separated from experience and culture. In the same vein, we can bring children to the point where they can exit the game of mathematics and, as teenagers enter into the domain of mathematics, teach them how to sever all ties between the two, much like the way school physics became separated from intuitive physics. However, this separation relies far too heavily on secular religious approaches; and we will therefore examine this problem more closely in the next chapter.

We also gradually need to tie the teaching of certain subdomains of mathematics to the use our society makes of them, in order to uncover the collateral damage that results when we apply mathematics to anything other than the world of constructed artifacts. Doing so will go well beyond the destruction caused by our weapons of math destruction. It is a structural issue that is all-pervasive and exceedingly damaging because it squeezes the life out of everything as a result of a will to efficiency and power (and thus a will to death) by imposing an architecture suitable to these destructive endeavours. The application of mathematics to anything living is integral to technique and the building of the technical order, with the exception of anything that has an architecture in the image of constructed artifacts (such as classical and information machines).

It is essential to recognize that there are three parallel schools of thoughts in today's philosophy of mathematics: the Platonists, and the constructivist and formalist schools. There is no agreement between them and thus no foundation to be found anywhere. There is no common sense from which you can depart in the domain of mathematics. In order that this domain can be separated from everything living, it must go to infinity. This necessity produces all manner of contradictions that cannot be eliminated, according to Gödel's theorem. Nevertheless, in building the technological order, we have entrusted everything to mathematics as the non-language of all disciplines. Hence, we have embarked on a fantastic gamble with our lives and our planet, one that wagers that the collateral damage imposed on them will always be outpaced by new constructions advancing the technical order. It is the bargain we have made with technique against culture, and against our being a symbolic species.

What we have brought into focus is what I do not hesitate to call the "spiritual" significance of pure mathematics, provided that we understand this as being related to the forces ruling over human life, to the point of having the power to enslave it. The necessary introduction of infinity has progressively recreated mathematics into the means for essentially making a new beginning for humanity. When mathematics began to embed itself into an infinite domain, it simultaneously began to exclude the general relativity of human life and the world from this domain. It no longer had any place for the living world in a universe constructed by means of the principle of non-contradiction. When mathematics became the non-language of all disciplines, this exclusion was transferred into all its applications. Thus began the process of remaking humanity in the image of the principle of non-contradiction, which is the principle dominating the architecture of everything non-living. Until now, the history of humanity had grown out of a first genesis in which language and culture had led to its enslavement (alienation). The separation of knowing and doing from experience and culture, and their reorganization by means of disciplines, marked a kind of second genesis, in which technique began to displace culture to the greatest extent possible without ever being able to do so entirely.

The implications are far-reaching: an enslaved humanity can be liberated, but a humanity reconstituted as non-living can only be resurrected. For this latter situation, all our political, ideological, humanitarian, and religious endeavours are entirely irrelevant. They become the great illusions that prevent us from understanding what we are doing to ourselves and our planet, as we seek to submerge ourselves into a non-living reality to the greatest extent possible. Our secular political religion of democracy thus possibly becomes the greatest illusion of them all. It is from this ultimate perspective that we can begin to understand the necessity of introducing secular religious attitudes toward mathematics and all the disciplines it helps to sustain. These attitudes are clearly manifested in the so-called hidden curricula. With this interpretation it may perhaps be easier to identify them in order to minimize our dependence on them when we transmit mathematics and all other disciplines from generation to generation.

It has been argued that what is irrational about rationality is the rationalization of human life by technique. The unlimited rationality represented by technique must not be contributed to by our teaching of mathematics. Somehow we need to teach the beauty, power, and usefulness of mathematics within its limits, and simultaneously show its potential for destruction beyond these limits. If we proceed on the basis of this premise by destroying all secular religious attitudes toward anything that appears to have no limits, we may be able to introduce mathematics into the lives of children in such a way that it does not interfere with the further development of their secondary frame of reference, and also prepares them for the tertiary frame of reference required for the learning of all disciplines. Hence, it is a respecting of the limits of the discipline of mathematics relative to those of cultural approaches, and vice versa. I do not believe that this is an impossible task.

Children know very well that playing games is different from all their other activities. We may be able to reinforce this by showing that in the game of mathematics, four minus one equals three; but that in daily life, for example, subtracting one member from a family is to end that portion of the lives of the three survivors that was completely intertwined with the person who is no longer with them. In other words, the three surviving members are completely altered by this subtraction, with the result that representing a death in a family by means of arithmetic is so absurd that every child can get the point. We can then move forward to where arithmetic does apply to daily life: when, for example, children keep track of their pocket money and other financial gifts as well as their spending of the money. Surely these kinds of examples can be elaborated to ensure that mathematics education genuinely contributes to the development of children and teenagers in a manner that equips them for a more responsible use of mathematics than what is currently the case in our civilization.

When teaching mathematics to children and teenagers, here the teacher will have to be extremely sensitive to the extending of their visual apprehension of the world at the expense of their access based on language and culture. As noted, it is not difficult to show that the space in which children live does not correspond to two-dimensional and three-dimensional geometry. If we are going to be teaching children a body of knowledge that we know is separated from experience and culture because it cannot be founded on them, this ought to become clear through our teaching. The child will then be not confused between life and this new game of mathematics. It is especially important not to confuse a child's vocabulary by mistaking what is exact in the domain of mathematics with what is true in life. At some point, it can be shown that the spatial aspects of a house can be described by means of a three-dimensional geometry that is exact for all its dimensions and spatial relationships. However, this three-dimensional space must not be extrapolated to apply to Australia, because it would no longer be exact: people would walk upside-down and rain would rise. In sum, it may well turn out that an important part of the preparation of a mathematics teacher is to get a grip on the limits of mathematics, which would imply a corresponding destruction of all secular religious attitudes towards it. I recognize that this will contribute to undermining our present symbolic grounding in secular myths; but this will simultaneously contribute to weakening our alienation and reification by them to open up human history to other possibilities.

From this perspective, a genuine history of mathematics that would encompass the human, cultural, and the chaotic aspects of its development could furnish good examples to make the above points. There are good models for this in the sciences, especially the previously noted works of Fleck and Kuhn.[73] They restore scientific activities to their status as human activities, be it of a very unique kind, but nevertheless open to analysis by the social sciences and humanities. Since the majority of people using mathematics in their professional lives do so in the context of a scientific or technical discipline, we ought to stop teaching mathematics as if it is a preparation for becoming a pure mathematician. The lack of a foundation for mathematics ought not to prevent us from teaching what is intuitively plausible to children as it was supposed to be intuitively obvious to mathematicians until it was shown that it had no basis in human experience. We can thus begin with the teaching of arithmetic in relation to objects only, and its inapplicability to living entities. We can then move into geometry as an extension of the intuitions children have derived from visually apprehending the world. From this geometry we can move into calculus and algebra. It is not until children have learned a great deal of mathematics that we can officially sever it from all their early intuitions based on the secondary frame of reference, which by now will be relatively distinct from the tertiary frame of reference, especially if the limits of mathematics have always been kept in the background.

Such a strategy would correspond to an acknowledgment of our being a symbolic species who, after growing up as members of a culture, are able to learn disciplines such as mathematics. Pure mathematicians act as if we can learn mathematics by skipping over all of this, as if we do not need to acquire a secondary frame of reference before we can acquire the tertiary frame of reference necessary for the learning of mathematics. It seems to me that this is the strongest argument for insisting that we build on the intuitions of children when we teach them arithmetic and geometry. We can then respect them as human subjects by avoiding any metaphysics about objective or detached mathematicians that denies this subjectivity.

It is time that we really take seriously what is happening in the lives of these children from a psychological, social, cultural, and historical perspective, as best we can understand it at this point in time. As teachers, we should be able to back what we do with our lives so that our words are true and avoid any metaphysics and ideology to the greatest extent possible. It may be a cliché, but if we can be true to ourselves we will also be true to others. Playing the objective mathematician is a silly denial of our humanity.

It is preferable to explain all this in terms of a highly unique participation limited to our mathematical imagination and thus with minimal embodiment, participation, and commitment – an antidote to any illusions about objectivity. A symbolically and culturally effective pedagogical strategy will enfold something of the general relativity of human life and the world. Even in the teaching of mathematics, this is absolutely essential because of the distortions we have introduced by mathematizing the world through our discipline-based approaches and the apparently unlimited scope of mathematics.

Is Mathematics the Secular Religion of Technique?

We must appreciate the difficulties many of our fellow human beings face as they perform their technical roles in the organizations and institutions of our contemporary anti-societies participating in the building of a global technical order. These go well beyond what we commonly associate with the lack of stable social relationships and groups in mass societies.[74] It is simply impossible today to meet enough people in a socially viable manner that would make us capable of replacing what used to be provided by traditions and cultures of all kinds. As noted, this has led to the rise of public opinion replacing private opinions, and a statistical morality replacing a traditional morality. It is also impossible for large organizations and institutions to make the necessary decisions based on the alternatives that have replaced the culture-based resources offered by a traditional society. For example: How is an economist supposed to understand the normal economic behaviour of a particular group of people, or how can a judge assess what constitutes a reasonable person and what such a person would have done under particular circumstances? Because of the limitations of our experience and our highly desymbolized cultures, they, as well as all of us, will have to transcend these limits; and mathematics is frequently an integral part of doing so.

The social sciences play a necessary role in providing us with profiles of an average person, one that is constructed from the statistics of real persons belonging to a particular category. It is one of the primary reasons why many of the social sciences have transformed themselves into disciplines relying extensively on mathematical and statistical procedures. Those who continue to insist on more qualitative alternatives based on an intellectual tradition established by founders, for example, are regarded as being essentially unscientific. The result is that the social sciences have uncritically enslaved themselves to technique by refusing to develop a symbolic and qualitative understanding of what is happening to human life and society to act as the intellectual guide for quantitative research.

We have now entered a stage where dealing with others as "average" persons in some category is becoming a near substitute for the way societies used to rely on experience and culture. With the growing use of computers and making our cars, homes, and offices "intelligent," almost anything we do can be tracked. The data that is collected can then be used to categorize us, systems can then be designed to put this to use, and we will be on the receiving end of weapons of math destruction, with considerable collateral damage. This partial technical substitute for culture creates an all-pervasive omnipresent power of persuasion and manipulation. The celebration of the possibilities of big data is simply another symptom of our enslavement to the second generation of our secular myths. It is important to recognize that without mathematics none of this would be possible. We are so riveted to our screen-based devices that we simply do not appear to appreciate what we are doing to ourselves and the future of our children. To claim that pure mathematics is exempt from this development cannot stand up to any careful historical scrutiny. What constitutes the pure mathematics of today will very likely be the applied mathematics of tomorrow. It is important, therefore, that we teach mathematics accordingly.

Our failure to adequately distinguish between our relating to each other and the world by means of the secondary frame of reference, and our relating to the domains of disciplines (including mathematics) by means of the tertiary frame of reference, has far-reaching implications. In the next two chapters, we will examine these implications for science and technique, but for now, we will focus on artificial intelligence and the deep learning of a subdomain sustained by mathematics. It is producing one of the greatest technological bluffs of our age with regard to what this application will accomplish. For example, what has actually been achieved when a computer is able to defeat a world champion in chess? It certainly is not the case that the computer now plays better chess in the way a human being does. What has been accomplished is the unfolding of the game of chess from the general relativity of individual and collective human life by translating it, by means of the non-language of mathematics, into a diametrically opposite architecture with the elements of a subdomain of artificial intelligence and deep learning in the image of reality. Doing so marks an impressive technical accomplishment, but it has nothing whatsoever in common with human beings playing chess by means of their secondary frame of reference. If such human beings imagined themselves as operating within this subdomain by means of their tertiary frame of reference derived from learning mathematics, they would achieve very little because the strengths and weaknesses of symbolic approaches are the diametrical opposite of those of discipline-based approaches. If, instead, these human beings continued to play chess the usual way against the computer, they would have little choice but to perfect their game in the image of the machine operating in the subdomain of the translated chess game. With a great deal of practice, they would eventually play chess more and more like a machine, and thus lose what it is to play chess as a human activity.

When artificial intelligence and deep learning are applied to human activities other than games, the difficulties of translation increase because such activities are more deeply dialectically enfolded into the relativity of individual and collective human life. A game is as close as we can come in human life to a micro-world. Consequently, translating other activities is mathematically more challenging and represents an even more impressive technical accomplishment. Nevertheless, what is truly being accomplished is the translation of the general relativity of human life and the world into the life-milieu and system of technique, one human activity at a time. Such an imposition of a technical order on human life and the world results in the imposition of non-sense on a life of highly desymbolized sense. Systematically proceeding with this re-engineering of human life and the planet will produce unfathomable and devastating consequences.

All this has been legitimated by a second generation of secular myths that one day will vanish like a mist under the sun. This is exactly what happened to all the myths of traditional civilizations. This founding of human life will continue to turn each human activity into a vanity in the image of these myths as the vanity of vanities. We are now re-engineering these activities into constructs of the technical order grounded in our present second generation of secular myths. As a result, the alienation of our highly desymbolized humanity is being turned into its partial reification. All this amounts to the antithesis of an enslaved life and a highly desymbolized symbolic species. With the hindsight of some two hundred years, we have been transforming enslaved lives into non-lives, beginning with the introduction of the technical division of labour and rapidly intensified by subsequent mechanization, automation, and computerization.

Consequently, the theoretical and practical externalities that flow from the use of discipline-based approaches must no longer be regarded as accidental collateral damage. Instead, these approaches must be regarded as the very core of the new megaproject that humanity has progressively imposed on itself ever since the beginning of industrialization, urbanization, and secularization. This megaproject is in part sustained by a new ideology: a technological bluff that a new humanity is about to be born that will live eternally on the net. Some imagine this as humanity playing a new game with itself and with the universe, a game that it will be able to abandon if things do not work out. However, such an abandonment, often identified with an escape from this planet, implies that when we stop playing a game we return to our lives. We forget that what would be left of the general relativity of human life and the world will have been so reified that such a return may no longer be possible.

For some two hundred years, we have been transforming enslaved lives into non-lives by disassembling them into their constituent activities and reorganizing them in the image of the machine one activity at a time, with the result that these "pieces" can never be reassembled into a dialectically enfolded life, any more than we can assemble an ecosystem from its constituent elements. Attempting to transform life on the basis of what is supposed to be the self-evident and undoubtable principle of non-contradiction is a venture that is unthinkable without the non-language and discipline of mathematics.

It would be somewhat naïve to assume that an effective change in mathematics education could make a significant dent in technique. The latter is simply too efficient, powerful, and profitable for the large institutions to pass up. Nevertheless, a complete overhaul of elementary and high school education, especially in the teaching of mathematics and science, could potentially create a growing appreciation of the limits of mathematics and thus weaken our secular religious attitudes toward it. In turn, such a development may increase public resistance to the growing and unconstrained use of ever more powerful and invasive "weapons of math destruction."

To dismiss these arguments as "mathematics bashing" amounts to claiming that we should get rid of all our thermostats because they are continually "bashing" our furnaces that keep us warm in winter, and our air conditioning systems that keep us cool in the summer. Without these thermostats constantly adjusting these devices, their effective way of controlling them by means of negative feedback would become impossible. An accusation of "mathematics bashing" makes no sense and is the equivalent of violating a taboo in a traditional society; and in our situation it implies that mathematics has been turned into an infinite good, that is, incapable of doing any harm. Consequently, if humanity becomes highly reified, the root cause of our problems could not be found in this infinite good, with the result that instead of controlling it, other "end-of-pipe" approaches would have to be invented. For example, in the case of our furnaces, thermal motors would have to be installed on all our windows in order to admit cold air in the winter whenever a room becomes too hot! In the same vein, given our present secular religious attitudes, whatever has become a secular sacred or a myth must not be touched – a taboo analogous to the religious taboos of the past. The danger of our current secular religious attitudes stems from making it next to impossible to attribute all our human, social, economic, political, environmental, and religious crises to what we have (collectively and metaconsciously) declared to be the absolute greatest good that can possibly exist in the universe for all times and places. Our greatest good is limitless and incapable of doing us the slightest bit of harm. If we genuinely desire to live secular lives by refusing anything sacred and by destroying secular religious attitudes toward it, a great deal of harm could be prevented or at least somewhat reduced.

In Western civilization this very possibility was supposed to have been the business of the Jewish and Christian communities, or at least of the faithful remnants outside of their establishments. We are referring to those people who refuse to bow their knees to any cultural gods or their secular equivalents. In a world where everything is related to and evolves in relation to everything else in a general relativity (understood as a creation), every element embedded in this relativity is delimited by it and thus cannot indefinitely grow, reach infinity, or become god-like. It is somewhat analogous to a healthy ecosystem whose populations of constituent species are related in such a way as to balance each other's populations so that none of them can grow out of control and destroy the ecosystem. In the same way, the infinite character of any discipline, as confirmed by its autonomy from all the others, is a manifestation of its no longer being delimited by its dependence on the others, and thus it will tend to become infinite. It will then require the non-language of mathematics to describe its domain. The entities and relationships of such a domain are then out of control, and their mathematical representations require that the infinite be added to what was once finite. Such a mathematical mapping is bound to be nonsensical relative to what it represents; and this would be impossible without a dependence on secular religious attitudes on our part. We thus unleash secular religious powers against which we must struggle. Mathematics itself is not the cause, provided that it is used within its own limits. Doing so requires the refusal of any secular religious attitudes. The same holds for all the other disciplines. In other words, we must struggle against any secular religious attitudes, either toward technique or toward culture. This is possible only by keeping each one within its limits to the best of our ability by being as honest as possible about what technique or culture can and cannot do for us. However, because of the threat of cultural chaos, we must engage in this struggle, even though we will not succeed. Nevertheless, the struggle is well worth the effort because it will always reduce the harm we do to ourselves and our planet.

From this perspective, minimizing our secular religious attitudes when teaching mathematics to children and teenagers represents the reinforcement of a kind of strategic beachhead from which it is possible to further weaken these attitudes in favour of our being a symbolic species and our having a more liveable and viable future on this planet. With the expansion of this beachhead, further transforming developments will come within reach in science and technique.

From a historical and social perspective, we may conclude that, following the great cultural divide, mathematics increasingly became not only the non-language of all the discipline-based approaches that comprise science and technique but also the primary non-culture of our symbolic species under the growing pressure of desymbolization. It has conferred an entirely new architecture on all knowing and doing as approaches for making sense of and living in a world that is being reorganized by technique. Ultimately, it separates this world into what is for technique and what is against it, in accordance with the secular religious attitudes generally adopted by all who build and live within its technical order. In terms of the history of our symbolic species, mathematics may thus be regarded as the non-language and non-culture of technique as well as the basis for the secular religious attitudes of those who serve technique. All this will also vanish, as did the traditional and secular myths of all earlier societies. By denying mathematics, science, and technique their unlimited status, we can help open up a more liveable and sustainable future.

In preparation for the chapters that follow, it is important to note that when we compare the architecture of the domain of mathematics with that of the general relativity of human life and the world, one of the most fundamental differences is that, in the latter, there can be no independent variables. This observation would be trivial were it not for our discipline-based science and technique mediating all relations through domains built up with mathematics. Consequently, this mediation imposes independent variables on the general relativity of human life and the world. Doing so in any application is the secular equivalent of traditional cultures projecting religious attitudes on certain entities, thereby transforming them into gods. The imposition of independent variables implies removing what they designate from it in order to put them above and out of reach of it. In terms of the general relativity of human life and the world, what these variables designate becomes autonomous and omnipotent. The resulting mini-gods take the form of facts that are established by scientific and technical disciplines as being absolute in character, that is, objective in themselves and on their own terms, no longer relative to a frame of reference.

It is also important to note that the specialists involved in the establishment of such facts do so by means of their tertiary frames of reference, and thus with their lives working minimally in the background. Through this involvement, they contribute to and confirm the secular myths that ground and orient our contemporary societies – as we will discover in this work. In the same vein, our entire scientific and technical intellectual division of labour hinges on our making a single category of phenomena into the equivalent of an independent variable relative to the general relativity of the world of which we are a part. Doing so imposes autonomous disciplinary domains that also contribute to and confirm these same myths. It is only by discovering and respecting the limits of discipline-based approaches that we are able to diminish our possession by these secular myths.

5 Human Knowing and Discipline-Based Science

Is Our Science Unlike All Others?

As members of our current global civilization, it is difficult for us to understand that for the greater part of human history, what came closest to our science took the form of a dimension of cultural mediation. It relied mostly on metaconscious knowledge embedded in experience and culture, which was dialectically enfolded into all other metaconscious knowledge by means of the organizations of people's brain-minds in a manner that characterized humanity as a symbolic species. For example: the totemic system for organizing clans into early societies was inseparable from a social division of labour that facilitated the discovery of the meaning and significance for human life and society of everything in what we refer to as nature. The members of any clan thus became the specialists who knew the most about how the plants and animals of their totem were enfolded into everything else natural. It produced a knowledge of them that was so comprehensive as to be able to sustain all the early civilizations that succeeded these totemic societies.

This totemic form of science had to be dialectically enfolded into all the other dimensions of cultural mediation, which makes it very difficult, if not impossible, for us to think of it as being scientific. The totemic conception of a nature populated and ruled over by spirits made it impossible for people to take for granted the regularity of any natural cycles or processes. Any regularity of nature could only be guaranteed by an agreement with these spirits that took a magic-religious-juridical form. The metaconscious knowledge derived from the experiences lived according to a totemic way of life sustained all human practices, including food gathering and hunting, early agricultural practices, the use of natural materials, food preparation and preservation, and the medicinal uses of plants. Many other activities were affected, as well as all their institutions. For example, before an animal could be hunted, a moral-religious-legal agreement had to be made with one or more spirits that ruled over it. The same was true for the domestication of any plants or animals. Any notion of private natural property thus had to be defended against the ruling spirits. All dimensions of cultural mediation were jointly grounded in a secular sacred and myths. These manifested themselves in natural moralities and religions that kept cultural chaos at bay.

The emergence of the early civilizations helped to diffuse tradition-based societies that interposed themselves between the group and the natural life-milieu, forming the new life-milieu of society. Their dimensions of cultural mediation were jointly grounded in a sacred and its sustaining myths, which blended the social with the natural. Occasionally, the scientific dimension of the cultural mediation of these traditional societies went further by attempting to ground scientific knowledge in additional ways, as in a divine order of the stars or a sacred calendar. In Western civilization, which gave birth to our current global one, the Greeks made attempts at scientifically grounding some or all cultural knowing and doing, by means of universal knowledge grounded in the principle of non-contradiction. Much later, Western civilization invented modern science, although its foundations became an endless source of debate. In the late nineteenth century, the limitations of culture-based approaches to knowing and doing became evident in the industries and economies of the industrializing societies, but these limitations were overcome by reorganizing this knowing and doing by means of autonomous disciplines. However, all knowing and doing thus separated from experience and culture cannot be *lived* directly, and can at best be used as principles *by* which people can choose to attempt to live.

Whatever form of science we may seek to understand, our approach would be incomplete without attempting to account for how the corresponding activities are carried out by members of a symbolic species who cannot but mediate all relationships primarily by means of symbolization, language, and culture, and only secondarily by means of constructs taken to be autonomous from this mediation. Nevertheless, such an autonomous mediation must ultimately account for the general relativity of human life and the world by means of secular religious attitudes. Once again, focusing on Western civilization because it gave birth to our present one, the possibility of an autonomous science became thinkable following the emergence of contradictions implicit in the ways of life of the culture of the societies in their attempt to create a Christian civilization referred to as Christendom. The medieval way of life brought into focus the irreconcilable contradictions between the three so-called perfections on which it was founded: Greek philosophy, Roman law, and the Christian revelation. The latter had been almost entirely transformed into something else by attempts to clarify it by means of Greek philosophy and by its accommodation to the moral and religious necessities of these medieval societies. One of the most decisive contradictions gave rise to attempts to discover the interactions between the material and the spiritual and the influence each one had on the other. It led to a variety of dead ends, since it was not possible to answer such questions as how many angels could dance on the head of a pin, where the philosopher's stone was to be found, and where the soul was located within the body.

Gradually, the conclusion imposed itself that the universe was entirely material, except for the human mind. It became self-evident that the organization of matter could best be imagined in terms of the most perfect organization of this matter known to Western civilization, namely, that of machines in general and the clock in particular. The Christian creation was thus turned into a gigantic clockwork, which imposed the image of God as a clockmaker who first created all the parts and then assembled them into an organization that could be described by means of laws. From the perspective of a symbolic species, such a mechanistic worldview implied that a dialectically enfolded individual and collective human life in an enfolded living world was essentially eliminated in favour of all manner of new relationships. These relationships were seen as having a diametrically opposite architecture expressible in terms of the principles of non-contradiction, separability, closed definitions, openness to measurement and quantification, and a simple complexity. Such an architecture not only implied that everything could potentially be described in terms of mathematics, but that such a description was more accurate and reliable than what could be understood by means of symbolization, experience, and culture. Such a mathematical revolution preceded a scientific revolution. In turn, the scientific revolution gradually paved the way for reorganizing scientific knowing by means of autonomous disciplines that could examine the clockwork universe one category of phenomena at a time. The door was also opened to a future scientism: a belief in an objective and autonomous science that could be constructed by means of mathematics and whose relativity had to be grounded in entirely new secular myths generated by highly desymbolized cultures. Without these developments, this objectivity and autonomy within a general relativity of human life and the world would have remained unthinkable, and in any case, unliveable.

Our current secular religious attitudes have created the illusion that we are no longer suspended in languages and cultures, as if a scientific dimension of cultural mediation is no longer required because this objective and autonomous science has an unmediated access to human life and the universe. This supposed unmediated access to the unknown required that the relativity of the corresponding human activities be made unknowable and unliveable by these secular myths.

Facts could thus be directly obtained from the unknown, and because of the supposed lack of any mediation influencing this procedure, they would be valid forever. Such facts could then be interpreted and given their theoretical significance by integrating them into theoretical constructs, best described by means of mathematics. The scientific observer of these facts regarding the universe-machine was no longer dialectically enfolded into these scientific activities, thus making the concept of a detached and objective observer thinkable and believable. Of course, all this was centuries in the making. The most decisive beginning was made by Isaac Newton, who replaced Aristotelean physics (based on the systematization of the human experience of the physical behaviour of objects) with a new physics that systematized observations by means of mathematics.

In sum, our discipline-based science owes as much to the cultural developments that gave rise to a mechanistic worldview as technique does to the creation of the technological and economic orders. Both have the same architecture within the cultural orders of the industrializing societies. Before highlighting a few important milestones in these developments, it is important to emphasize that what all these developments have in common is the underlying translation of one or more aspects of a dialectically enfolded cultural order into the architecture of a reality expressible in terms of the previously noted five principles. An adequate history of the emergence of our discipline-based science can as yet not be found in our libraries because its dependence on highly desymbolized cultures has been almost entirely overlooked, thus obscuring the necessary secular religious attitudes grounded in our secular myths. The same is true for an adequate history of our discipline-based technique, which includes, but is far from limited to, what is usually understood by technology.

The relationships of both discipline-based science and technique depend on their separation from experience and culture, which was made possible by the kinds of cultural developments briefly and inadequately noted above. Scientific and technical activities create a kaleidoscope of domains within which we cannot live and by means of which we are unable to conduct our lives. We are able to relate to these domains only by means of a tertiary frame of reference that must be grafted into a secondary frame of reference, in turn grafted into a primary frame of reference. The primary frame of reference is acquired when the organization of the brain develops into that of the brain-mind as a result of a person's first experiences of beginning to live a life relative to the world. The secondary frame of reference can be acquired only by members of a symbolic species learning a language and a culture. A culturally mediated relationship symbolizes everything in relation to everything else in the life of a person, as a unique individual embodiment of a community and culture. It thus makes the greatest possible use of context, delimited only by its being grounded in myths that exclude anything that is radically other than what is known and lived. In contrast, discipline-based approaches limit the context to the domain of the discipline that mediates a relationship and thus restricts the context to a single category of phenomena represented (in mathematical form to the greatest extent possible) in this domain. Consequently, this discipline-based mediation not only restricts the context taken into account but also translates it into the architecture of what we have referred to as reality.

Because the tertiary frame of reference is necessarily grafted into the secondary frame of reference, the relationship between the two is parasitic in character. Discipline-based approaches translate everything into reality to the greatest extent possible, which makes them desymbolizing relative to their cultural alternatives. These approaches cannot be acquired until a secondary frame of reference has formed, but their formation has a desymbolizing influence on its further development. It is not impossible that these desymbolizing pressures, resulting from giving preference to discipline-based approaches over culture-based approaches in contemporary ways of life, may reach a point at which a resymbolization of individual and collective human life becomes impossible and the collapse of a highly desymbolized culture becomes inevitable.

In the meantime, we must learn to live with our scientific and technical creations by avoiding the worst collateral damage they can exert on our lives. Doing so, however, has resulted in our ongoing remaking of ourselves in the image of science and technique. This re-engineering of our being as a symbolic species into an embodiment of our discipline-based approaches constitutes the very core of the most significant developments in the evolution of human life and history in the late twentieth and twenty-first centuries. Our embodiment of discipline-based science and technique has created a very different relationship between humanity on the one hand and its works of science and technique on the other. Human beings are now "internally" as well as "externally" connected to the transmission, development, and application of discipline-based knowing and doing within their institutional frameworks in their external world. Our embodiment in all this constitutes people as members of anti-societies, so named because their characteristics are the diametrical opposite of those of experience-based traditional societies.

It may be concluded, therefore, that discipline-based scientific activities have no common denominator with their culture-based precursors. They are no longer a dimension of cultural mediation because they have become autonomous relative to culture. Discipline-based science is incapable of extending our knowing based on symbolization and culture. Its unprecedented discipline-based intellectual division of labour imposes the rational at the expense of the symbolic character of scientific knowing, with each discipline in turn imposing its goal of extending its knowledge of a single category of phenomena. Its autonomy derives from either excluding all other categories of phenomena or isolating them in the background and barring them from participation. The general relativity of human life and the world is thus not merely distorted but disaggregated into the isolated domains of the constituent disciplines.

Disciplines and Daily-Life Knowing

We have noted that a discipline-based approach is scientific in character only if the situation at hand is dominated by the influence of the category of phenomena examined by the discipline, to the point that the influences of all other categories of phenomena may be neglected. Such situations appear to be almost absent in living entities, but are dominant in classical and information machines, as well as everything built up with them because of their simple complexity. Moreover, the absence of a "science of the sciences" capable of unifying the findings of all disciplines into a comprehensive overall understanding of human life and the world implies that science can never replace our culture-based knowledge of human life and the world.

From these rudimentary observations, it follows that our current scientisms essentially overlook the inability of science to improve our making sense of and living in the world by means of a culture. Discipline-based science can do no such thing. We begin to experience this during our first class in high school physics, where all our intuitive knowledge of physical phenomena, which is dialectically enfolded in our understanding of human life and the world, is kept at bay as we start afresh with the building of "school physics" in a domain that we are learning to imagine. All the physics classes in the world will never be able to contribute to the former kind of knowledge. People who have earned doctorates in physics know this very well, and they would not dream of turning to the equations of physics to improve their playing squash or hockey, for example. The reasons are obvious: the physical aspects are dialectically enfolded in their playing these games, and thus determined by the strategies being pursued via the boundary conditions for these equations. As a result, there is a parallel between "intuitive squash" and the school physics aspects of this game; and this is true for every other game as well. We learn to play games through a secondary frame of reference, while all our scientific knowing of these games is acquired through a tertiary frame of reference. Each involves a unique formation of metaconscious knowledge that is discontinuous relative to the other, since one is embedded in experience and culture and the other is separated from them. Any potential complementarity between our culture-based knowing and our scientific discipline-based knowing remains undeveloped because of our scientisms, ultimately anchored in our second generation of secular myths. These myths imply that discipline-based approaches are objective and reliable relative to their culture-based alternatives.

For example, Newton's great contribution was to go beyond interpolating and extrapolating astronomical observations in order to determine the orbits of the planets by proposing laws and principles that were (at least in part) implied in the empirical observations even when their physical basis was not understood, as in the case of gravitational attraction. In other words, he substantially improved the way astronomy worked mathematically without fully understanding the physical phenomena it described. It ushered in a more complex relationship between theory and empirical observations. This new approach was entirely dependent on the assumption that what was being scientifically investigated could have but one possible architecture, namely, the one that can be described in terms of mathematics. Such an implication caused Newtonian physics to break with Aristotelean physics, which sought to systematize the daily-life experiences of the behaviour of physical objects in the world.

We need to back away from the kinds of scientisms with which we have all been raised. It is too simplistic to believe that through experiments we discover the "facts" that either confirm scientific theories or compel us to modify and possibly replace them as we add more and more of the unknown to what we know. All human knowledge is relative and thus must be intellectually bound to its social and historical context.

As noted, industrialization and everything that accompanied it introduced an entirely new architecture into human life and the world. It first took the form of an economic order, which subsequently mutated into a technical order when knowing and doing became organized by means of disciplines. This reorganization of human knowing and doing began whenever their culture-based alternatives reached their limits in every branch of industry, first in relation to its technology and soon in relation to its large organizations. For example: Karl Marx placed science in the superstructure of a society and thus outside of the forces and relations of production that were supposed to make up the economic basis of a society.[1] During his lifetime, this essentially characterized the role of science in industrializing England. Later, Max Weber began to show that technological and economic rationality had begun to transcend some of the limits of culture encountered in those areas of human life involving a great deal of technology and in large organizations.[2] Still later, this rationality in its strongest form opened the doors to the discipline-based approaches pioneered in the natural sciences. In a little more than half a century, discipline-based approaches to human knowing and doing replaced their cultural counterparts in almost every area of human life and society.

Rationality and discipline-based approaches were to reality what symbolization, experience, and culture were to the relativity of human life and the world. Each of these two approaches is relative to a particular context with a different architecture. The assumption that they would be relevant beyond these contexts might have been questioned by the appearance of major difficulties (psychological, social, economic, political, legal, moral, and religious) whenever they were operated beyond their limits.[3] To this day, these kinds of difficulties are not associated with the transcendence of these limits as a result of two generations of secular myths. Humanity had always dealt with the general relativity of human life and the world by means of symbolization, language, and culture, which were capable of respecting this general relativity up to a point, but then having to symbolically ground all understanding and living with it in the form of myths. In roughly one century, humanity was behaving as if this general relativity of human life and the world had disappeared to make room for a simple complexity. This complexity could be dealt with by extending the approach to human life and the world by means of seeing rather than by listening to and acquiring a language and a culture.

The Known and the Unknown

We have noted that there is no clear boundary between the known and the unknown in human life and the biosphere as a consequence of their dialectically enfolded and enfolded architectures respectively. Everything we know and live enfolds something of the unknown and is either directly symbolically grounded in our secular myths via the secondary frame of reference, or indirectly grounded through the tertiary frame of reference that can only be grafted onto it.

From this perspective, it is more responsible to imagine the unknown relative to our human finitude as an inexhaustible complexity that almost certainly does not have a simple architecture, and thus does not resemble anything technological such as information machines and their information processing. When we undertake a new scientific experiment, we are creating a new way of interacting with this complexity that produces situations that did not exist before. How many of these situations remain unaffected by these interactions and how many are radically changed by them is unknowable. In other words, the results of such interactions are situated somewhere in between our gaining some insights into this inexhaustible complexity without having much of an effect on it and our modifying it into something else that remains also unknowable. It depends on the interpretation of the purpose of the experiment and its findings. Relative to our discipline-based approaches, possibly the only conclusion is the recognition of the vast ability of experiments to manipulate, modify, and dominate nature. We can say little, if anything, as to whether we know this nature any better than those who went before us, as Kuhn has clearly demonstrated for the discipline of physics.[4] Knowing more does not translate into knowing better other than though scientism and secular religious beliefs of certain kinds. All this would be little more than interesting speculation were it not for our planetary crisis.

How then are we to proceed? It may be helpful to examine what is happening in discipline-based science by critically applying both discipline-based approaches within their limits and culture-based approaches within theirs. We may then discover that, generally speaking, each one has diametrically opposite strengths and weaknesses, thus unveiling a potential limited complementarity that remains hidden by our secular religious attitudes, in turn sustained by our myths.[5] Since all human knowing can only be relative, we ought to spell out this relative character to the best of our ability. In the case of human knowing, we should therefore examine culture relative to science, and science relative to culture. At the same time, we need to consider culture relative to technique, and technique relative to culture, since our contemporary highly desymbolized cultures are unthinkable without technique. Technique is unthinkable without discipline-based science, and science in turn is unthinkable without the discipline of mathematics. Our analysis will continue my previous work on the relationship between technique and culture in order to better understand the dependence of the analyses of Thomas Kuhn[6] and Ludwik Fleck[7] on culture, which was begun in the previous chapter.

Culture and Discipline-Based Science

In the previous chapter, we suggested that culture is to a society what a paradigm is to a group of interacting practitioners of a scientific specialty. This analogy needs to be developed further in light of the recognition that a separate culture cannot exist, any more than a society can exist "out there." In the case of a society, we have noted that others help to constitute the society we live in, just as we help to constitute theirs. We are all both individual and society. A society can therefore not be reduced to a collection of its constituent members, even though the desymbolization of cultures has pushed us far in that direction.[8] Similarly, there is no distinct culture "out there." Both a society and its culture are inseparable from its members, because something of that society and culture is dialectically enfolded into all members in a way that is individually unique and yet culturally typical at the same time.

What the members of a society have in common and what makes them differ constitute the required dialectical tension for viable social relationships and groups.[9] In the same vein, it is the cultural differences between the members of a society and what they culturally share that form the dialectical tension that allows people to participate in the historical journey of their society.[10] Such a history is that of growth, decline, stagnation, possible regeneration, further decline, and so on until an eventual collapse occurs when the cultural unity becomes overwhelmed by individual diversity. At this point, a culture is no longer able to give a common meaning, purpose, and direction to a community.

The same dialectical tension that operates between the members of a society and between the ways they share their culture also functions in a group of experts who collaborate in the advancement of a subdomain of their discipline, be it in a highly desymbolized form. Their shared paradigm does not exist "out there." Each expert is an individually unique embodiment of the paradigm in his or her thought style, and this thought style is constrained by the unity of the paradigm. Since the participation of the experts in the group occurs on the basis of the their extensively developed tertiary frame of reference relative to the subdomain of their shared discipline, their embodiment, participation, and commitment are exceedingly constrained by it. Nevertheless, a dialectical tension between the members of the group can sustain an individual intellectual diversity within the unity of the paradigm, and can impose this unity on this diversity at the same time. From a sociological perspective, such a group constitutes a social entity operating by means of a tertiary frame of reference to make sense of a subdomain of their discipline. There is an almost complete lack of face-to-face communication, as technical mediation has now almost entirely displaced cultural mediation. The situation is somewhat comparable to an online "group" or "community." This does not preclude the importance of face-to-face communications at conferences, which may appreciably deepen patterns of collaboration when these become embedded in a more personal cultural mediation: meeting at receptions, going out for a drink, or sharing a lunch or dinner. Such contexts can then be strengthened through more personal telephone or video conversations.

The work of Ludwik Fleck contributes substantially to our understanding of what happens during a period in the history of a discipline that Kuhn has referred to as "normal science."[11] Normal science is essentially cumulative in character as the reigning paradigm is refined in detail and extended in scope. Fleck illustrates this with his hands-on experience of the process of how, over a period of decades, the disease of syphilis became increasingly differentiated from a number of others that had overlapping symptoms and other characteristics. He showed this to be a very "messy" process that was complicated by equally messy experimentation involving researchers in various laboratories in different institutions. Assumptions that for a time were influential turned out to be questionable later on. This did not prevent them from playing an essential role in this messy process. In the same vein, it involved experiments that later on turned out to be irreproducible, and yet they also played an essential role for a time. As a result, some valid discoveries occurred on the way, despite some details found to be erroneous relative to a later understanding. What made the whole process even more "messy" was the inability of the principal researchers and participants to provide an accurate account of their thoughts and actions, because (with hindsight) they could not avoid rationalizing and idealizing what had happened.

Such difficulties were unavoidable because the researchers' intervening experiences continued to modify their paradigm and thus their individual thought styles. These intervening experiences also affected their recollections of earlier events. If they attempted to repeat the same experiment ten years later, they could hardly avoid doing it differently; and they would certainly interpret each and every detail and finding in a somewhat different manner. In some cases, it made earlier experiments impossible to replicate.

All this confirms that a paradigm is an individually unique development within the brain-minds of the researchers, as shown by the way they worked in evolving the subdomain of a discipline. As a consequence of brain plasticity, they were unable to recreate anything in exactly the same manner at a later date because intervening experiences had modified the way their experiences were differentiated and integrated, thereby changing their scientific significance relative to one another. Furthermore, this was backed by metaconscious knowledge on many levels, some of which corresponded to the explicit entries in Kuhn's concept of a disciplinary matrix. Consequently, what was deemed plausible and implausible, self-evident or impossible, or worthy of follow-up or merely extraneous noise was constantly evolving along with the individual researcher's brain-mind. Apart from greater constraints and a development relative to a domain rather than to life in the world, a tertiary frame of reference depends on the same brain functions as does the secondary frame of reference. Because of the significantly constrained embodiment, participation, and commitment required by a tertiary frame of reference, it is not possible for a researcher's entire life to work in the background, but only the highly constrained portion of it associated with being a researcher. Both daily-life activities and research activities involve a person's semantic memory. In the case of the former, these are differentiated from each other and (jointly) from the research activities, and vice versa. This interdependence between the two kinds of activities can be understood relative to growing up in a culture and then, through a long formal education, grafting a tertiary frame of reference into a secondary frame of reference.

The above perspective is, of course, entirely out of step with most of contemporary psychology. Many of its experiments are asocial and ahistorical to the degree that their design, execution, and interpretation appear to require no reference to what is so unique to our contemporary societies, not the least of which is the coexistence of secondary and tertiary frames of reference. Since our lives work in the background of anything we undertake, the experimental design needs to carefully take into account which developments in people's brain-minds will be dominant, which will be peripheral, and which will be marginal, while at the same time recognizing their inevitable leading into one another. Without this social and historical context, the researchers will work the usual "garbage in, garbage out" procedures. Moreover, many experiments essentially eliminate what has made us into a symbolic species capable of respecting the general relativity of human life and the world, by stripping off anything that is enfolded or dialectically enfolded in order to reduce everything to information processing based on a simple architecture in the image of our information machines. This is accomplished by assigning tasks to subjects in which they have not previously acquired a high skill level based on metaconscious knowledge either embedded in or separated from experience and culture. The subjects are then operating on the lowest levels of skill acquisition, characterized by analysis and problem solving associated with different areas of the brain showing as being active on MRIs. It amounts to an implicit denial of our human history, but this is hardly surprising given the autonomy of all disciplines, including that of psychology. Our suspension in secular myths makes it possible to simply ignore alternative explanations, such as interpretations of what little we know of the functioning of our brain-minds.[12]

If we are at all interested in a unified science capable of developing an overall understanding of human life and the world in a non-piecemeal manner, we ought to at least attempt to establish a measure of compatibility between the findings of our many disciplines, allowing for the presence of the many ambiguities and contradictions associated with enfolding as well as dialectical enfolding. It would amount to an attempt at resymbolizing all these findings. The rather "messy" character of scientific research, even during a period of "normal" science, cannot be accounted for in the image of our information machines and processes, nor can anything else regarding human life be represented in this way. We need to recall why decades of knowledge engineering and early attempts to build a general artificial intelligence were complete and total failures, except where a set of human activities could be isolated from the rest of a person's life to the point of having something in common with the micro-worlds of games. Even in these cases, the separation from human life is incomplete because a life remains indivisible until mental health issues begin to interfere. We also need to be mindful of the re-engineering of our symbolic species by having discipline-based approaches all but displace their culture-based alternatives. It makes *homo informaticus* plausible and thinkable, and many disciplines uncritically take its inevitability for granted.

The perspective we are developing is also completely out of step with the efforts of epistemology that attempt to strip out any symbolic and cultural elements from scientific activities. Fleck was highly critical of such efforts, which he regarded as doomed to failure already during his time.[13] They represent an attempt at a complete reconstruction of what is happening in these activities, giving rise to the kinds of stereotypes and asocial and ahistorical images of science that feed our scientisms. It methodologically excludes any possible contributions from the psychology of invention, the sociology of research groups, the cultural anthropology of their societal embeddedness, their historical changes over time, the economics and politics of research funding, the technological state of the art of scientific apparatus, and much more. Any logical reconstruction of scientific activities squeezes out the messy, erroneous, accidental, ideological, and cultural dimensions. It is true that the tertiary frame of reference greatly constrains all this, but scientific research never comes anywhere close to their epistemological reconstructions.

Fleck also explained that a thought style is capable of taking on a mood.[14] I will slightly reinterpret this concept relative to different phases of normal science. Following a scientific revolution during which a paradigm change has taken place, the professional mood of researchers may exhibit an almost supreme confidence in the ability of the new paradigm to explain a great deal more than did its predecessor. This confidence translates into higher than usual levels of excitement, a greater tolerance toward ambiguities and the expectation that they will be short-lived since everything will soon fall into place, an openness to trying out hunches that are intellectually or experimentally "on the wild side," a greater than usual tolerance toward differences of opinion in the expectation that these also will be short-lived, and so on. However, as the paradigm continues to be developed in both detail and scope, the mood of the researchers is likely to become more conservative and cautious. A lot of work is now at stake; and anything that challenges the researchers' extensive accomplishments is likely to be met with a great deal of resistance, if not outright dismissal and even hostility. Surely all this work cannot just be wrong.

This mood may change once again when it becomes evident that, on the perimeter of the paradigm, there is an obvious accumulation of questioning of some details, experimental procedures, and interpretations of the theories. Sooner or later, this accumulation will weaken people's confidence in the paradigm. It may turn into a self-reinforcing cycle if some researchers begin to recognize that, hidden in all these developments, there are things that may soon turn out to be anomalies. Other researchers may become very defensive of the paradigm, to the point of a growing intolerance toward all this theoretical and experimental turbulence. Eventually, battle lines may even be drawn and thought styles polarized in one way or another. Nearby parallel interpretations of the paradigm may emerge.

Fleck likened this situation to one in which a number of artists paint the same subject from adjacent vantage points but with different styles.[15] It becomes impossible to take different parts from different paintings to create a composite picture of the subject that would sum up all their efforts. Even though the subject of the composite painting would be the same, the different styles would portray variations in the impression of the subject on every individual painter as well as differences of interpretation, as shown by the way all this is worked out in the details and highlighting, and so on. In the same vein, the developments relative to the findings of the researchers could become different enough to manifest a somewhat incoherent paradigm. An overhaul of the paradigm may be attempted; and if this does not prove satisfactory to most researchers, the stage may be set for a paradigm shift in the course of a scientific revolution.

All these moods will be affected by the extent to which the domain and subdomain of a discipline have been mathematically represented. If the level of mathematization is high, there will be much less tolerance for ambiguity. Moreover, anomalies will be easily recognizable and much more difficult to dismiss. In contrast, when the level of mathematization is low, there will be much more room for ambiguity and disagreement over what constitutes an anomaly. This situation used to characterize many of the social sciences before the quantitative approach based on survey methodology became dominant.

Fleck also noted the important relationship that exists between a paradigm, the group of scientific specialists that advance it, and the society(ies) in which these specialists participate.[16] For example, a group of researchers may become a scientific elite if the paradigm is receiving a great deal of attention from the traditional or new media, to the point of generating a public opinion about it. Such an elite can ignore public opinion at its peril because its status is deeply affected by it. It will be compelled to respond by giving interviews, lobbying politicians, and giving public lectures. Fleck noted that there was a time when syphilis was widely regarded as a much more pressing disease than tuberculosis; consequently, it was almost impossible to get research funding for the latter even though it was a very common problem.[17]

As in any human group, some researchers may be rugged individualists who are quite prepared to follow their intuitions even when this means being temporarily or permanently marginalized from the other researchers in the specialty. The motivation of such a researcher may be driven by the expectation that eventually recognition may come as being ahead of his or her time, the hope of achieving certain practical results related to personal convictions, or a determination to pursue a course even if it means being written off by other researchers as not "fitting in." As noted for the discipline of mathematics, when specialists in a field write research proposals, accept media interviews, teach courses to non-specialists, or undertake other such activities, their paradigm will necessarily "bleed" into the reality lived by the members of a society and portrayed by the media, and in turn this reality will bleed into their specialty. When specialists desire to communicate a message, they need to pay careful attention to public opinion and popular views of science in general, and their specialty in particular. Hence, there is a constant interpenetration as a result of these kinds of efforts, which will cause the developments in the brain-minds of the specialists associated with the secondary frame of reference, those related to the tertiary frame of reference (related to reality), and those associated with the tertiary frame of reference (related to the domain of a discipline) to bleed into one another. Much of this will be in conformity to and sanctioned by the second generation of secular myths, which symbolically ground the relative status of these three developments in the lives of scientific specialists.

In many respects, the perspective Fleck has on science converges with this analysis in that both refuse any absolutization of science. For Fleck, science comprises a diversity of human activities that can be interpreted in terms of thought styles and what he referred to as "thought collectives" (the research group as a social entity), ultimately embedded in culture.[18] He thus opposed the perspective and approaches of the Vienna Circle and its many iterations that sought to liberate science from human and cultural influences in order to establish it on an objective and absolute basis. Unfortunately, attempts at relativizing science have often gone too far in an age of ideology, as if it were possible to develop a science in the service of national socialism or communism, for example. Fleck noted that in his day, science was generally regarded as an inappropriate subject for sociology. He also noted the dependence of science on culture, but he did not elaborate the implications of this dependence.

We can go beyond the works of Fleck and Kuhn by exploring the relationships between our contemporary discipline-based science and our highly desymbolized cultures. Scientists, like everyone else, grow up in a society and become members of a symbolic species through the acquisition of a language and a culture, desymbolized as these may be. Consequently, their learning of science is necessarily grafted into and thus relative to their culture. We have shown that by the time they become working scientists, their semantic memories contain three "universes" of differentiated and integrated constellations of symbolized experiences. These universes are each uniquely differentiated from the other two, and vice versa. The first resulted from the acquisition of language through the development of the secondary frame of reference. It permitted those experiences with a language foreground to be directly differentiated from each other, and jointly differentiated from all others whose foregrounds constituted the focus of attention and were embedded into the backgrounds that together interpreted the situation at hand. This latter "universe" of symbolized experiences is associated with a person's episodic rather than semantic memory. The two memories are distinct since, in rare cases, one can be lost without the other being much affected.

Into the first development within the semantic memory is grafted a second universe of symbolized experiences. It becomes progressively differentiated from the first because the foregrounds and primary backgrounds of its symbolized experiences are constituted of image-stories sustained by image-words, as opposed to language foregrounds. The architecture of the former is that of a simple complexity while that of the latter is a dialectically enfolded one. In the one, the undivided background and in the other, the secondary background of the symbolized experiences have the architecture of a dialectically enfolded complexity. This second universe within the semantic memory results from a stream of experiences frequently associated with screen-based devices, which undermines the usual stream of experiences that contribute to people's becoming members of a symbolic species through the acquisition of a language and a culture. This second universe of highly desymbolized experiences portrays the "world" in which the members of an anti-society live. It is the reality that is grafted onto what little had developed of what, in a traditional society, would have become a symbolic universe. It is the reality engineered by the original and new media and by the Googles, Facebooks, Apples, and Amazons of this world. Living on our screens enables powerful interests to re-engineer us as a symbolic species on historically unprecedented levels, without any significant oversight.

Into this second "universe" of highly desymbolized experiences is grafted a third. It derives from the learning of discipline-based approaches in high school, which may or may not be continued by higher education. The foregrounds of these highly desymbolized experiences do not deal with the reality in which people live but with the domains of disciplines that no one can inhabit. As a result, it is not until the acquisition of discipline-based approaches that the tertiary frame of reference becomes fully developed. Such a further development is essential because the domains are separated from experience and culture and must thus be interpreted relative to themselves. This development is externalized in human behaviour, implying the coexistence of parallel forms of knowing and doing (one embedded in and the other separated from experience and culture). Every discipline-based approach focused on a human activity will go beyond the existing "intuitive" form to begin a "school" form of it. The former is learned by doing, while the latter is learned abstractly in a classroom and from textbooks.

Within each of these three developments, metaconscious knowledge plays a different but fundamental role. The metaconscious knowledges associated with the differentiation and integration of the experiences of the first two "universes" will directly "bleed" into one another. However, the metaconscious knowledge associated with the full development of the tertiary frame of reference is discontinuous with the other two forms of metaconscious knowledge, because it is separated from experience and culture and associated with a domain of a discipline in which we cannot dwell. Despite these differences, the essential role of metaconscious knowledge has been noted by Michael Polanyi as personal knowledge and tacit knowledge,[19] by Ludwik Fleck as a thought-style,[20] by Thomas S. Kuhn in his original concept of a paradigm,[21] and by Hubert and Stuart Dreyfus as the intuitive knowledge required for reaching the two highest levels of human skill referred to as proficiency and expertise,[22] as well as the metaconscious knowledge essential for the functioning of any culture.[23] The scope of this latter metaconscious knowledge encompasses those of the other kinds, which are particular expressions of it.

The interpretation of the second universe of desymbolized experiences within people's semantic memories is ambivalent because of its dependence on how a person grows up and on the kind of society in which this happens. This universe as a whole can be more strongly differentiated from the first than from the third, or vice versa. The research of Sherry Turkle provided a fascinating window into this development at a very unique time, because the process of desymbolization had not yet advanced to anything resembling today's levels.[24] When children began to encounter computers in their lives, they were struck by how similar they were to these machines in some ways and how different they were in other ways. It presented them with very different ways of living their lives, with far-reaching consequences that they were incapable of understanding. Nevertheless, they were obliged to make choices. They could spend more or less time playing with other children through face-to-face contexts, to experience the rich complexity and ambiguities of the social world. They also had to decide how much time to devote to their machines, which offered solitude in a much simpler and more controllable world.[25] A preference for the social world would more strongly develop their secondary frame of reference, while a preference for the computer world would more strongly prepare for the development of their tertiary frame of reference. These choices would also profoundly influence the building of metaconscious knowledge that would be working in the background of their future behaviour. These choices would also significantly shape their metaconscious images of their social selves, of the social selves of others, and of being alive in either the social or the computer world. All such developments would have been significantly affected by peer group pressure, parental guidance, and their growing up in mass societies with highly desymbolized cultures. When, a decade or so later, the levels of desymbolization began to increase enormously as a result of a growing dependence on smartphones and the Internet, children had no choice but to more strongly develop their tertiary frame of reference. The choices children make and the influences they need to deal with build up a kind of "cultural inertia" that will affect their lives for years to come. Their metaconscious knowledge helps to integrate their universes of experiences; and this in turn is externalized in their behaviour as personality traits and lifestyle patterns that eventually become symbolically grounded in even deeper metaconscious knowledge.

Because so much of our culture is metaconscious and thus partly or wholly hidden from critical scrutiny, millions of people, who on the surface appear to disagree about almost everything, can nevertheless participate in the history of their society. If all this metaconscious knowledge could readily be made explicit, everything about the role of culture would be stripped down to its rational and technical elements. This would make it impossible for a culture to function because we would all end up disagreeing about a lot more than we do already. No group or society would survive very long under such conditions. Individual differences would grow with few constraints, since all metaconscious knowledge, including its deepest forms, would have disappeared. Whether this knowledge had been built up mainly through the secondary or tertiary frame of reference does not alter this very much, but the levels of desymbolization may weaken the scope and depth of the new metaconscious knowledge that must be built up by the organizations of our brain-minds. When this happens the integrality and integrity of our lives is greatly weakened.

The disappearance of all metaconscious knowledge would all but eliminate the dialectical tensions that are essential for our communicating through a shared language, in order to sustain relationships, to participate in viable groups and communities, and to make an essential contribution to our society's ability to adapt and evolve.[26] We have suggested that when we talk to one another, there needs to be a dialectical tension between our similarities and differences for genuine communication to take place.[27] The same is true for sustaining intimate relationships. Groups and societies would no longer be able to draw on the dialectical tensions between individual differences and what these people culturally share. The result would be the emergence of a hell on earth, in the sense of making the sharing of any lives impossible.

The same kinds of observations can be made regarding the functioning of a paradigm relative to the group of researchers or practitioners associated with its development and application. Consequently, any attempt to intellectually "squeeze" the cultural elements out of science is misguided and possibly nihilistic. It is another reason why, when Kuhn abandoned his concept of a paradigm for that of a disciplinary matrix, it constituted an intellectual wrong turn – although he was under enormous pressure to do so.[28] His original concept of a paradigm included an intellectual life related to a scientific specialty working in the background through its values, beliefs, best practices, model solutions, and experimental techniques. This included whatever bled into them as a consequence of the tertiary frame of reference having been grafted onto the secondary frame of reference in a particular social and historical context.

The relationship between science and culture can be further illuminated by examining the shared relative character of the human activities associated with them. No human activities of any kind have an unmediated access to the unknown; thus they cannot deal with the threats of cultural chaos other than through myths. Unacceptable as this must appear to many readers, the behaviour of scientists in a laboratory makes this clear. The difficulties they face are related to their having no fixed reference point in relation to which they can access the unknown, whether or not they are strengthening their intellectual grip on a situation. What they face is not unlike the difficulties dealt with by toddlers and children in their attempts to get an intellectual grip on the world, as discussed in a previous chapter.

It is as if scientists were attempting to put together an intellectual jigsaw puzzle of which the shapes of the pieces are constantly changing, because the experiences of their laboratory work are endlessly modified by their ongoing differentiation and integration in their brain-minds. They are constantly faced with details that, experimentally or theoretically, do not quite fit. Getting a handle on this requires a constant going back and forth between what the situation appears to be from many different perspectives: the paradigm, the experimental design, the limits of the instrumentation, difficulties with the experimental techniques, what appears to be implied in the experimental data, the hunches derived from an evolving metaconscious knowledge, their interpretation and possible follow-up by trying something else, the opinions of fellow researchers, time constraints, and so on. Much of this greatly depends on their intuitions of how well all these "puzzle pieces" appear to fit together: whether or not there is much significance in the "noise" of a lack of fit; the likelihood of success in adjusting certain details to discover whether the fit will hold or possibly improve; or, if there is something not quite right about any of it. All the scientisms assuming that our experimental designs confirm or challenge our theoretical frameworks and that we are the detached and objective observers of all of this are just plain foolish. No one who has ever worked in a laboratory can escape the "messiness" that derives from the lack of a point of reference relative to which we can clearly mark the adequacy of our intellectual grip on a situation. There is no getting away from a constant going back and forth that will not stop until we become convinced that nothing more significant is likely to be gained. It corresponds to a kind of intuitive sense of having reached a dynamic equilibrium between the likely shapes of the individual puzzle pieces and the overall fit we can achieve with them. Minor or more significant anomalies may occur on the way, and we must decide either to face or ignore them.

In addition to this going back and forth on the level of individual researchers, a similar movement is usually present in the research group to which they belong. There is an important collective dimension of obtaining an intellectual grip on things in the absence of any clear reference point. The same kind of going back and forth may spill over into the body of researchers associated with the advancement and application of a paradigm. At the same time, there is a trade-off between this going back and forth and the attempts on the part of individual researchers to be the first to submit the results for publication, in the hope of an article and possibly getting one's name on a discovery. No one has the luxury to investigate all possible hunches, doubts, and theoretical or experimental possibilities. This highlights the crucial but unacknowledged role of metaconscious knowledge in expert researchers who follow hunches as to what strategically may have the greatest likelihood of delivering results in a reasonable time, in the knowledge that only with hindsight will it be possible to theoretically and experimentally justify it. There is thus an ever-present pervading messiness even in the best research, even during a period of normal science. As is the case for a culture, there are remnants of a past and the embryonic beginnings of future alternatives, which coexist with the present paradigm. All this adds to the importance of the intuitions of expert researchers, which may translate into their willingness to take their hunches seriously as being indicative of evolving metaconscious knowledge that will require more explicit follow-up. There are always many questions: Is the researcher dealing with limitations of the instrumentation or the experimental design? Is there a weakness in the intellectual framework associated with the paradigm itself, or even if all the details are deemed to be plausible, are they not necessarily equally promising? Of course, there are many constraints: the individual's thought style; the "mood" of the researcher and colleagues; the "mood" of the researchers associated with the paradigm as expressed at conferences, through publications, peer reviews, and so on; a willingness of the researcher and colleagues to go against the flow; and so on. All of this and more cannot be avoided because of a lack of a reference point.

The paradigm internalized by the members of a research community associated with a particular scientific specialty directly constrains the thought styles of the members as a consequence of its being embodied in the organizations of their brain-minds. It also externally constrains their behaviour because any applications for research funding, submissions of abstracts for conference presentations, submissions of articles or books for publication, considerations for tenure and promotion, and more, all involve a peer review process by other researchers, who have internalized the same paradigm. Consequently, the individual diversity rendered possible by the tertiary frame of reference related to the paradigm will be constrained by what is thinkable or unthinkable, plausible or implausible, potentially fruitful or lacking promise. The internal constraints will complement the external constraints as individual researchers externalize their understanding of the paradigm in a variety of ways. As is the case in any culture or society, individual diversity will thus be constrained by the unity of the paradigm.

The possibility of gaining a deeper understanding of our scientific activities is also barred by our present intellectual division of labour, centred on autonomous disciplines and a university organization built around it. For example, the sociology, history, and philosophy of science are stunted in their development because their practitioners cannot understand what they are examining; they do not have the necessary development of the tertiary frame of reference that corresponds to the domain of the discipline that they are studying. Without the education required to develop such a tertiary frame of reference, they will be in the same position as knowledge engineers were when they attempted to have experts explain the rules that (supposedly) underlie their work.

With rare exceptions, most research scientists have little or no understanding of how the tertiary frame of reference they have acquired through an extensive education and practice works in the background of all their activities. They are unable to appreciate how any explanations they provide of their work are intellectual reconstructions of what is really happening, and how this involves a great deal of metaconscious knowledge to which they have no access. We suppress this obvious impasse by all kinds of scientisms at an enormous cost to humanity and this planet. If we all possessed a better intellectual grip on our discipline-based scientific activities, we would have a much better knowledge of the limits of such activities and know when we could count on them and when we should reach for alternatives.

For example, after obtaining a doctorate in engineering, I continued my studies of technology on the post-doctoral level via the social sciences and humanities. Analyses by three different authors had shown me how much I could learn about engineering. First, there was the work of Thomas Kuhn, which I found immensely helpful when I faced a crisis in the doctoral research I was undertaking in fluid mechanics.[29] Second, there was John Kenneth Galbraith, who helped me make sense of ever-greater scientific and technical specialization and how this changed the workings of technology and the economy.[30] All this came to a head when I read Jacques Ellul's study of technique, which implied the best description of how my engineering mindset worked.[31] The practical implications of this have kept me busy for my entire professional life. I cannot understand how we can spend fortunes on scientific and technical research without undertaking the above kind of research to gain a critical insight into how this discipline-based knowledge is transmitted, applied, and developed within our institutions and how this is surely related to our successes and failures. From this point on, it does not require rocket science to begin to redesign this knowledge system to ensure a more socially viable and environmentally sustainable future.

On the way, I made some mistakes. For example, I wrote the first draft of my first book as a NATO postdoctoral fellow, but decided not to present it as a second doctoral thesis. It was a terrible mistake. For forty years I have been teaching engineering, sociology, and environmental studies. My engineering colleagues never shied away from reminding me that, even though I had a doctorate in engineering, I was not really teaching or doing research in "real" engineering. In sociology and environmental studies, my colleagues recognized that I had made significant and award-winning work; but since I did not have an official doctorate in these disciplines, I was only able to hold cross-appointments, which did not permit me to supervise graduate students, thereby cutting me off from research assistants. Since my doctoral students had to register in engineering, the kinds of research projects they were able to undertake did not suit their interests or mine, and I graduated only two doctoral students in my entire career.

During those forty years, I discovered that home departments generally regard what people do through cross-appointments as a kind of extra-curricular activity that is not taken very seriously, or is regarded as time away from what really counts. As president of an international society that sought to sustain interdisciplinary studies in society, technology, science, and the biosphere as well as the practical application of this understanding, I sat through endless discussions in which colleagues explained how their departments only fully counted work that would eventually end up in publications in the journals related to its disciplines. As the competition for tenure, promotion, and research grants sharply increased, many said they could no longer afford to spend time and travel money on these activities. I was unable to save the society as I served two terms on this sinking ship. Similarly, as the editor-in-chief of an international journal publishing this kind of work, my editorial board and I resigned after thirteen years because we could not sustain it without seriously compromising our standards. Nevertheless, my research has pointed to ways in which we could restructure technical education and practice, thereby improving the economic use we would make of it – to the point that my provincial government became interested in having me prepare a detailed study. Some years later, the federal government used these initiatives to attempt to reorganize research in Canadian universities. Unfortunately, both endeavours were halted by conservative governments who believed that the market could do all this much better. Nevertheless, the Canada Foundation for Innovation recognized my development of preventive approaches as one of twenty-five leading Canadian innovations.[32]

The following analogy may illustrate my point. As academics, we all face situations where poor students come to complain about low grades they have received because some of their classmates' papers they had read received much higher grades, and yet they could not find any significant differences between these papers and their own. Of course, it is always possible that an error has been made; but in many cases, the problem arises from their inability to adequately differentiate between the two papers – which they are unable to do because their background in the subject is weak. It involves the kind of metaconscious knowledge that they have not developed.

An understanding of the relative character of our scientific activities will necessarily lead to a better understanding of these activities. They are human activities that, within their limits, exhibit substantial advantages over their symbolic and cultural alternatives. However, beyond their limits, they are turning out to be much more destructive. They cannot provide us with the kind of knowledge we require to deal with our various crises. For example: How can reasonable people aware of their finitude believe that, one day, the branches of physics will be unified in a set of equations that will explain everything about everything in our universe? Even if we had such equations on a piece of paper in our pockets as it were, what difference would it truly make? All it would tell us is that we have finally succeeded in squeezing everything we know about this universe into an architecture corresponding to the simple complexity of our mathematics. It would also mean that anything beyond its physical mass-energy aspects, and thus anything beyond the discipline of physics, would not be capable of challenging these equations and the simple complexity they represent. In other words, the question whether or not this universe is able to sustain life as a consequence of this "materiality" cannot possibly matter. The existence or non-existence of humanity changes nothing in such equations. I hope the reader can forgive me if I find this an absurd form of nihilism. If we refuse the disastrous body-mind dualism, as many have done, then it would appear that there is something about matter and energy that is capable of sustaining an enfolded biological life and a dialectically enfolded cultural life. This potentiality can never be injected into the kind of architecture that our current mathematics is able to represent. If the branches of physics could ever be mathematically unified, we would only have a scientism backed by secular religious attitudes that proclaim that life within this universe is of no consequence whatsoever. In the meantime, this scientism provides a kind of hope to some people – that the discipline of physics has no real limits; and that, as the leading discipline, this means that the potential of our discipline-based science as a whole is unlimited. Let us all bend our knees to this new omnipotent science!

Scientism does not stop there. We now encounter people who believe that this universe is a kind of mathematical video game. After all, if it can be represented in mathematical terms, why not? Consequently, if we destroy this planet, it is our responsibility to move to another one.

If we run out of planets, we will soon be capable of stepping out of this video game, starting another one with a different architecture, and so on. To maintain some kind of sanity in response to this folly, all we need to accept is the general relativity of human life and the world. The autonomous discipline of physics is our creation with its own unique limitations, which will become plain as day if we question our secular religious attitudes toward it. Doing so is as simple as putting our relative knowledge in its context.

These kinds of beliefs belong to Western civilization, which spent much of its so-called Middle Ages trying to find where the soul, spirit, and mind were located, in the creation accounts of a God as revealed in the Jewish and Christian Bibles. When these could not be found, the "stuff" of the universe became differentiated into matter and mind. What was understood as matter was relative to what was understood by mind, and vice versa. The structure of this matter was then compared to what, at that time, was regarded as humanity's most perfect organization of this matter – the clock – as if no architecture other than that of our own works was thinkable. It eventually became the cornerstone for the technological and economic orders with their simple complexity built within a dialectically enfolded cultural order of each industrializing society. By this time, it had become entirely self-evident that the "super clockwork" universe could be tinkered with just like any other machine, and was soon explained in terms of natural laws best described by mathematics. The relativity of all this kind of human knowing thus became obscured by a scientism that endowed us with the power to know the ultimate architecture of our universe, and soon everything else within it. The unknown no longer had any ultimate mystery in terms of its architecture. The gods had come to earth, and the god of human knowledge no longer had any possible limits. All that remained for it to do was to fill in the details. The Vienna School was possibly the summit of this divination of science. However, it was permanently called into question by the works of Fleck and Kuhn, which dared to suggest that science was a unique kind of activity open to psychological, social, and historical investigation. If anyone thinks I am exaggerating, I remember very well that when Kuhn came to the university where I worked, the controversy over his work had become such an intellectual tempest that we had to meet behind closed doors in order to have a fruitful and reasonable discussion. As noted, I had found his book immensely helpful when confronted with a minor crisis in my doctoral research in fluid mechanics.

One day, three doctoral students, who were all doing large wind tunnel experiments, happened to exchange the instruments they were using, each of which was an accepted version of the standard reference instruments for calibrating many fluid mechanics measurements. It turned into a disastrous twist in their work, because it became obvious that these instruments were not nearly as accurate as they were supposed to be. They had limitations that had become embarrassingly obvious. Our doctoral supervisors had no answers. In the library, all we could find was a laboratory manual from a good research university that gave us some hints but no answers. With our work being in the final stages and thus with everything at stake, we decided that it would take the equivalent of several doctoral research projects to sort this out; and that we had better get on with our degrees. Kuhn's work helped me to sort out this problem, and to conclude that reference probes were not absolute but relative. They had limits that needed to be determined.

Returning to physics: When Albert Einstein developed his paradigm, commonly referred to as the general theory of relativity, he may well have introduced a fundamental contradiction into physics. The meaning of time was now relative to the meaning of space, and vice versa. As work proceeded in filling in this paradigm, it may be argued that the meaning of matter underwent a fundamental transformation. The mathematical significance of mass needed to be understood relative to the mathematical significance of time and space, and vice versa. Did this mean that the assumption regarding the architecture of the universe being a simple complexity also needed to be questioned – along the lines of David Bohm,[33] for example? Does general relativity imply the impossibility of independent variables? If there are no independent variables, can mathematics be used to describe the architecture of the universe? Are the difficulties related to the mathematical and intellectual reintegration of the discipline of physics due to having assumed the wrong underlying architecture? I understand very well that such questions are not seriously being entertained. What if we discovered that a truly general theory of relativity requires an interpretation of matter relative to life, and vice versa? In such a case, our mathematics cannot apply. Everything in physics would have to be rethought.

The present work has struggled with similar issues. Language and culture may be regarded as implicit *lived theories* of the general relativity of human life and the world, which end up being completely distorted as a consequence of being symbolically grounded in myths. As far as I am able to understand, it is as close to as we have ever been able to come to living with the general relativity of our lives and the world. The Greek invention of universal knowledge based on the principle of non-contradiction did not in the least overcome the limitations of language and culture, but set out to replace them with rules, algorithms, mathematical theories, and much else that we cannot *live*. If, instead, we accept the finitude and relativity of human activities, the brain-mind can no longer be the eternal "society" of universal information processors that much of psychology and other disciplines believe it to be. Of course, I am not about to "stuff" something non-material back into matter. What I am arguing is that if we are serious about a truly general theory of relativity, the relativity of matter cannot stop with time, energy, and space. I am simply continuing the perspective that underlies all my work, namely, that of the general relativity of human life and the world.

In contrast, our global civilization has gone to the other extreme of simply declaring the unlimited abilities of discipline-based approaches, which implies that the unknown cannot possibly have any other architecture than the simple complexity we can represent with our mathematics. This architecture has to describe everything, including what has grown out of its introduction into our societies, beginning with an economic order that paved the way for the present technical order of this civilization. After our Taylorizing of our hands and our brains, we have somehow convinced ourselves that this is how human life really *is*. When I began my postdoctoral research in the social sciences, I was astounded that this simple architecture was taken for granted as underlying even everything living. My first book shows some of the astonishment that arose in my mind from the incredible differences between living and non-living entities, of which engineering had taught me a great deal.[34]

The above must not be interpreted as another attempt at suggesting that the so-called Big Bang began a process of unfolding the potential of this universe, including that of all life on our planet. I will not join the perfectly silly debate about the misreading of the opening chapters of the book of Genesis in the Jewish and Christian Bibles. Their meaning cannot be forced relative to the meaning of the work of Charles Darwin, or vice versa. Throughout Western civilization, Jews and Christians have always fashioned images of their God that satisfied their religious needs. These arose from an unwillingness to accept our finitude and relativity. All this can be well understood and explained in terms of the history of religion, the sociology of religion, cultural anthropology, depth psychology, and much else, and thus has no need for a revelation from a transcendent God. Presumably, this revelation is about something else; and for believing Jews and Christians, the whole creation-evolution debate should have been a non-starter from the very beginning. All this is amply confirmed by the fact that the opening chapters of the book of Genesis were always read very differently for as long as they have been in existence.

Science, Reality, and Our Life-Milieu

Niels Bohr believed that we are suspended in language to a point of being unable to know what is ultimately up or down.[35] In other words, he believed that all our relationships were mediated by language (and thus, via it, through culture), and that this leaves us without a fixed reference point. I have gone further by showing that in order to make this liveable, humanity has always interposed a symbolic universe (in the case of a traditional society) or a reality (in the case of our contemporary societies) between itself and the unknown. By means of myths, the symbolic universe and the reality were always turned into embryonic representations of the unknown in the image of the known that societies were busy filling in, through their discoveries and the development of these discoveries. The symbolic universes took the forms of a natural life-milieu in the case of food-gathering and hunting groups, and a social life-milieu with a secondary natural life-milieu for the early societies.

Our global civilization has altered the suspension of our symbolic species in an ultimately unknowable universe in one significant way. It has mostly replaced a culture-based mediation with a discipline-based technical mediation, one with a secondary reliance on culture-based mediation for bringing up each new generation as members of a symbolic species and for sustaining personal relationships. As a result, what we commonly refer to as reality may be interpreted as the life-milieu of technique, with a secondary life-milieu (of what little remains) of society, and a tertiary life-milieu of a (highly modified) biosphere.[36] The reality in relation to which we live and evolve our lives is no longer the result of symbolizing our lives and the world, and thus of respecting their relativity (at least up to a point). The reason for this change is that our reality is portrayed by the old and new media and through the findings of all discipline-based approaches. These jointly constitute the life-milieu of technique that mediates (and thus filters) almost all our experiences.[37] In the past, people participated in the evolution and adaptation of their symbolic universe by means of their secondary frame of reference, which permitted the symbolization of everything relative to everything else as constrained only by the culture's myths. In the present global civilization, however, our participation in evolving and adapting what we refer to as reality is much more constrained by the tertiary frame of reference, which substantially limits our embodiment, participation, and commitment to it to a degree that we can no longer symbolize it as before.[38] We have transformed our being persons of a time, place, and culture.

Our reality is now separated from experience and culture in a way that is unprecedented in human history. It represents a kind of disenchantment of the world, first described by Max Weber.[39] There is no longer a *lived* time, space, and a social, but a distant and cold reality that we no longer symbolize in terms of the meaning and value it has for individual and collective human life.[40]

This reality derived from technically mediating almost all our relationships has all but eliminated our awareness of the dialectically enfolded character of our talking to one another and our forming relationships, groups, communities, and societies in their contemporary highly desymbolized forms. The building of this independent and apparently autonomous reality begins with the second stream of highly desymbolized experiences that help to build a tertiary frame of reference in children – one that cannot fully develop until they begin to learn discipline-based approaches. These approaches involve image-words and image-language because our access to the domains of disciplines depends on the extension of our visual apprehension of human life and the world. We recognize this when we say that a picture is worth a thousand words; and in this immediate context, that is true. However, in a broader context, we forget that we continue to be members of a symbolic species, who are first and foremost dependent on mediating all our relationships by means of a language and a culture before we can develop a tertiary frame of reference. Nevertheless, we have essentially surrendered our *lived* time, space, and the social for a reality that is separated from our experience and culture. Its simple complexity excludes the very possibility of enfolded and dialectically enfolded relationships, and thus excludes life as we know it. Our reality becomes endowed with its distant and remote character, making humanity into a kind of "accidental happening" within it. We are no longer "existentially rooted" in this reality in the way earlier societies were rooted in their symbolic universes. It is as if we had existentially removed ourselves from it by means of our second generation of secular myths.

From a psychological, social, and cultural anthropological perspective, these developments have completely transformed the socialization of the members of each new generation, although we continue to formally educate them as if it were "business as usual." The concepts of a symbol and a language have been appropriated by discipline-based approaches, even though they cannot have any meaning in a domain separated from experience and culture other than a purely technical one. Nevertheless, much of the literature on mathematics, the natural sciences, and artificial intelligence implies that these disciplines are capable of adding symbols to our experience, and that these symbols can even constitute a language. There are thus "natural" symbols and languages as well as "non-natural" symbols and languages. Ever since the invention of Boolean algebra, particularly during the early decades of seeking to establish a general artificial intelligence, mathematics became regarded as a language capable of expressing anything that could be symbolized through the language and culture of a society.

Such a perspective overlooks two fundamental issues. First, there are no natural symbols or languages, since all cultures are artificial.[41] Second, culture-based approaches to knowing and doing use symbols and a language whose meanings are dialectically enfolded into each other. Consequently, they cannot be divided, separated, defined in closed terms, measured, or mathematically represented, and contribute to a non-simple complexity. The universality of cultural approaches all throughout human history until very recently may be interpreted as attempts to deal with a general relativity that cannot be captured by a simple complexity based on the principle of non-contradiction. Consequently, mathematics, information theory, and computer language have a fundamentally different architecture that is incommensurate with the architecture of natural and cultural life.

The implications are far-reaching. We cannot behave as if teenagers are simply acquiring a few more symbols and a new language in high school in the way they might learn a foreign language. Nevertheless, this is essentially what we are doing. I recognize that our current secular myths make almost any alternative unthinkable or impossible to take seriously, for the simple reason that doing so would require a complete overhaul of primary, secondary, and higher education. It would make the limits of discipline-based approaches thinkable; and this in turn would make room for alternative approaches that could transcend these limits. If such alternative approaches began to be used to complement discipline-based approaches, thus holding each within its limits, all our institutions would gradually be transformed, beginning with the university.[42]

Since all human knowing and doing are relative, these kinds of arguments are inevitably circular. Hence, it is essential that we obtain the best possible intellectual grip on what are fundamentally different kinds of symbols and languages. We will attempt this, beginning with what happens in our learning of high school physics.

Physics as a Mathematical Game?

In their physics classes, students enter the domain of the discipline of physics, which has nothing in common with their experiences of the world acquired through their secondary frame of reference. It does share some features of what they have come to know as reality through the second stream of highly desymbolized experiences, which they learned to make sense of by developing the beginning of a tertiary frame of reference. Initially, all this is rather confusing because they are not explicitly told that they must use their imaginations to enter a domain governed by Newton's laws of motion and the principle of gravitational attraction. All this is introduced in a mathematical form that limits all interactions to purely external causal relationships between distinct and separate entities that can be defined in closed terms. It involves image-words and image-language that are somewhat similar to the image-words and image-stories of the reality they have encountered through their screen-based devices. The similarity is due to the priority given to free-body diagrams over language, while the differences derived from these free-body diagrams are relative to the domain of physics and not to the world in which we live. It may be argued that this is temporary; but the domain of physics will never resemble the lived-in world because it is always limited to physical phenomena, to the exclusion of all others.

The relativity of all physical entities represented in this domain manifests itself by the assumption that everything in this domain is entirely disembodied. For example, forces, velocities, and acceleration may be represented by arrows. Masses could be represented as points, but it is more practical to represent them as small rectangles in order to make them more visible. The planes with which some of the masses are in contact are non-material, given that they are all the same, besides lacking in friction and having an infinite extension. The surrounding air is disembodied in the same way. It offers no resistance and does not exhibit any of the phenomena that occur when gases pass over solid objects, such as the formation of boundary layers and vortices. Pulleys are mere geometry, without any inertia. Strings resemble mathematical lines, since they are weightless and incapable of being affected by gravity or subject to inertia. We could continue the examples, but the picture is sufficiently clear. The students are entering the world of a kind of game in which they are going to "play" with entities that exist nowhere in the world of their daily lives. In any case, all this has nothing in common with a detached and objective observer examining physical phenomena in the world. It reminds us of a kind of secular magic: entities and their domains appear out of nowhere and then are supposed to be dealt with as if they were "real." Moreover, there can never be any life in this domain, since all enfolded and dialectically enfolded relationships have been ruled out by a simple architecture beyond which mathematics cannot go.

Without some kind of scientific ideology, most students would probably be completely lost. Many of us were probably rescued by our growing fascination by, and immersion in, solving the puzzles that our teachers referred to as physics problems. Their structure is identical to that of any game. The domain is a micro-world in which certain entities are permitted to interact exclusively on the basis of certain rules that are referred to as laws. Students must then discover what happens in a situation that is partly specified by a problem statement. Initially, success may be more or less difficult to achieve depending on students' relying more strongly on their secondary frame of reference than on their embryonic tertiary frame of reference, or vice versa. This dependence will have been greatly influenced by the way they were brought up and the choices they have made in their lives.[43]

If teachers had been able to remove all scientisms and secular religious attitudes by restoring the relative status of learning physics, they would surely have followed a very different approach. They might have explained to their students that the world in which they live is so complex that, from their daily-life experiences, they can learn only so much about it. In order to go beyond these limits, they will learn physics by means of an alternative approach – one that temporarily forgets about the world in order to teach them a number of scientific "games" that will eventually permit them to obtain another kind of intellectual grip on the behaviour of physical objects in the world, in a manner that is very different from how they have gone about this in daily life. With some simple examples, teachers could explain to their students why their daily-life experiences of the behaviour of physical objects could lead them to the wrong conclusions, as did Aristotelean physics. Teachers could also explain that, even though the Newtonian physics they are learning has been replaced by the general theory of relativity and that of quantum mechanics, it is nevertheless very useful because it mathematically "works" for many applications. For example, whether or not Cartesian geometry describes the space in which we live, it is what all engineers use when they design buildings, airplanes, machines, or anything else of which they need to work out all the spatial relationships and dimensions. They would never be able to do this if they placed what they are designing in the space we have come to know through experience, or the one described in the general theory of relativity.

In the same vein, gravitational attraction "works" mathematically when engineers are determining the kinds of footings necessary to support a building or the structural steel frame that must carry all the loads imposed on it. If they attempted to do these calculations in terms of the concept of inertia as described in the general theory of relativity, they would make life extremely difficult for themselves.

Regarding the concept of time relative to the domain of physics, teachers might explain to their students that, in response to the need to synchronize ever more human activities in our globally integrated civilization, we have invented a standard time that is the same within each time zone but different from the time in all other time zones into which the planet has been divided. This abstract time has no correspondence to the *lived* time of their lives because it has been separated from experience and culture.

Teachers may also find it necessary to explain that concepts such as mass and energy have different physical interpretations depending on the domain of physics relative to which they operate. Nevertheless, they will be able to solve their physics problems by accepting the Newtonian definitions and the laws that govern their behaviour. Doing so will allow them to solve many physics problems, and this in turn will have many practical applications. Students can be told that Newton used the principle of gravitational attraction even though he did not understand its physical basis. Nevertheless, he was able to describe the motions of the planets better than anyone else had been able to do.

Arguments against this pedagogical strategy will have more to do with our scientisms than with pedagogy. Do we really wish to confuse these young teenagers by suggesting that the architecture of our universe can perhaps not be expressed in terms of our mathematics? Do we wish to unintentionally diminish the prestige of physics as the leading discipline and, along with it, possibly undermine the confidence of our students in this and other disciplines, and thus in science itself? These are questions that must be seriously considered; but from the perspective of this work, our current way of going about the teaching of physics amounts to trading short-term gains for long-term pain. It makes the entry into discipline-based science more difficult for teenagers than it needs to be, and possibly unattractive to far too many of them. It may well be better to "come clean" from the beginning: to acknowledge that the relativity of human life and the world implies that living entities have diametrically opposite architectures from those of non-living entities. For example, in a living world, the entities can be internally as well as externally related to each other; therefore no simple causality can occur as a result. It will then be possible to explain that mathematics, like all other human inventions, has its own unique strengths and weaknesses – with far-reaching implications for human life and our planet. It would allow us to help students understand the limits of what they are learning, right from the beginning.

Is this any different from teaching teenagers to use power tools in a workshop? Surely, to use these tools safely they need to learn to use them within their limits, and when to put one tool down to exchange it for another. Why should intellectual tools be any different, and why should the learning of intellectual skills not be accompanied by learning their limits as well? We would then begin to undermine our current scientisms, which are barring us from gaining the kind of knowledge we need to create a more socially viable and environmentally sustainable future. It would lay the foundations for creating the kind of knowledge infrastructure we need for such a future – one that will require a complete restructuring of primary, secondary, and higher education.[44]

Our nanotechnology, biotechnology, and information technology have the same structure and organization as all other disciplines. How is it possible, then, that university students are told that these disciplines may hold the key to solving many of our structural crises? As disciplines, these endeavours will continue to produce the same kinds of dazzling successes and equally spectacular failures that plague all other disciplines when they are used beyond their limits, where they have no scientific or technical validity. In the same vein: in engineering it appears impossible to recognize that discipline-based approaches are a necessary but entirely insufficient means for the design of anything.

What we require are design exemplars that operate on the level of experience and culture, and analytical exemplars that operate relative to the domains of disciplines. When they are used together in a way that each approach is used within its limits, their complementarity can achieve things that could radically change our future.[45] Moreover, such a restructuring of engineering may also help to address its inability to represent half of humanity adequately among its practitioners, for reasons I have outlined elsewhere.[46] Imposing quotas for the admission of women students (as my faculty has now done) amounts to a refusal to recognize that no other profession continues to have these problems, and thus demonstrates a failure to address the deeper underlying structural issues in the curriculum and the profession.[47] We cannot avoid the difficulties posed by the fact that all human knowing and doing is relative, while the associated practices lack a reference point. It will undoubtedly lead to the destruction of many of our scientisms, and may go a long way toward a scientific and technical professionalism capable of acting in the public interest by its actions rather than by ideology. Our present situation, where most scientists and engineers have no idea what discipline-based knowing and doing will never be able to accomplish for humanity, is a sign of a very dangerous secular religious blindness.

Our Metaphors for Space, Time, Matter, and Numbers

The discipline of physics has always had an enormous impact on how the members of our contemporary anti-societies live within the world of their daily lives. We will illustrate this with the metaphors for space, time, matter, and numbers. We will attempt to show that when physics became a discipline, its knowledge became separated from experience and culture, thereby excluding human beings from space, time, matter, and numbers. This was necessary and unavoidable because no human being can dwell within the domain of physics. Its concepts of space, time, matter, and numbers require the development of a tertiary frame of reference, while their daily-life counterparts rely on the secondary frame of reference. Consequently, when these concepts become metaphors for making sense of our lives in the world and are thus confused with their *lived* equivalents arrived at through experience and secondary frame of reference, significant existential issues occur. Such issues would have been impossible if mathematics were just another language.

For the following discussion dealing with the concepts of space, time, matter, and numbers, I will reinterpret the findings of Roger S. Jones,[48] which are indebted to the work of Owen Barfield.[49] Like most works on this topic, it does not consider the possibility of there being a fundamental difference between the "symbols" and "language" of physics and mathematics and their culture-based alternatives. I will therefore reach very different conclusions. However, the work stands out for its intellectual honesty and clarity. In the following discussions, I will limit myself to Western civilization, since it gave birth to our present global civilization, which would be unthinkable without it.

Beginning with the concept of space, the Greeks developed geometry largely in relation to astronomy. Euclid attempted to extend what his culture knew about space by turning it into a universal knowledge built up with the principle of non-contradiction. Doing so required an explanation of how the space of the stars and planets of the universe reached down to the space in which the Greeks lived on this planet. The answers came from Greek astrology and a form of popular stoicism that took this one step further. The former explained how the relationships between the planets were related to and affected those between people. The latter taught people how to live within the order of the stars. The unity between astronomical space and the world was thus affirmed. Euclidean geometry remained unchallenged until after the Middle Ages. It did not challenge the concepts of space in the cultures of medieval societies, because these were radically different. Simply put: their concepts of space resembled what we mean when we speak of our mental and emotional space, that is, as the felt connections between everything in the lives of the members of the societies. It was as if this space were directly differentiated from what we would refer to as psychological space, and vice versa, within a dialectically enfolded life-space. It was thus the diametrical opposite of our "empty" space, which we fill with everything around us. Medieval space was alive with meaning, knowledge, and wisdom and thus endowed with the ability to nurture human life, which was dialectically enfolded into it. A separate and distinct "inner" space and an "outer" space were entirely unthinkable.

Cartesian space emerged along with analytical geometry as a combination of algebra and geometry. It permitted any geometrical figure to be described by means of an algebraic equation and thus by a logical mathematical statement. The reverse became equally possible. Hence, algebra and geometry became interpreted as transformations of one another, to constitute a new logical metaphor of space. "Space" became separated from experience and culture, to become a lifeless reality apart from us. It was conceptually turned into a vast void at an extremely low temperature – a kind of empty locus where matter functions and from which life and consciousness emerged. It is a metaphor of space that can be expressed in terms of non-spatial mathematical statements, as a mathematical abstraction.

Initially, the reintroduction of the concept of infinity into mathematics by means of the calculus developed by Newton and Leibniz did not pose any serious difficulty because of the operational definitions of taking a limit to zero or infinity. However, in the case of analytical geometry, the three axes extend to infinity and, along with them, many figures (such as straight lines or parabolas). This space eliminates any possible locus for a *lived* space; and with its disappearance, we disappeared as observers of that space and as human beings. Hence, we must intellectually annihilate ourselves as persons engaged in doing geometry. Such a result is clearly impossible with any lived space interpreted through the secondary frame of reference.

To conceptualize this mathematical abstraction of space requires a tertiary frame of reference. It is important to recognize that the separation of analytical geometry from experience and culture cannot be absolute, because this tertiary frame of reference is grafted into the secondary frame of reference in the course of learning this geometry. The separation appears absolute to us because of the second generation of secular myths.

The above contradiction is rooted in the development of Boolean algebra. Such an algebra cannot be turned into a way of representing any culture's symbols or language. It cannot be interpreted as the development of another language that is independent from all other cultures. Hence, we do not have a choice of speaking about space in one language or another: as a logical statement, a geometrical statement, or a statement of a symbolic language embedded in a culture, desymbolized or not. As previously noted, what mathematicians and scientists refer to as symbols are always relative to the domain of a discipline, while the symbols of a human language are relative to our living in the world. Living in the world involves interposing a symbolic universe or a reality between ourselves and the unknown. Doing so is very different from interposing one or more domains of disciplines because this requires a fully developed tertiary frame of reference grafted into a secondary one. As we will continue to show, this technical mediation is fundamentally different from any culture-based mediation – with far-reaching consequences, including the spectacular successes and equally disastrous failures of our global civilization.

These differences can in turn be related to those between a secondary and a tertiary frame of reference. As noted, all languages associated with the culture of food-gathering and hunting groups, as well as the societies that succeeded them, appeared to be able to deal with the non-simple architectures of human life and the world. In these architectures, everything is differentiated from everything else as a negative expression of what something is *not* relative to everything else. Integration is the positive expression of what something *is* relative to everything else within this interrelatedness. This (negative and positive) expression of what something is within an interrelatedness is what we externalize in our behaviour as the symbolization of its meaning and value relative to everything else in individual and collective human life. This symbolization in its turn is symbolically grounded in traditional or secular myths.

In the case of the tertiary frame of reference, this differentiation and integration, and consequently the symbolization of anything, are extraordinarily constrained by an embodiment limited to our imagination of the domain of mathematics or a scientific discipline, our participation in a purely abstract intellectual "game," and a commitment that would exclude our very life were it not for our secular myths. It is the metaconscious knowledge of these constraints that has made the concept of an objective detached observer somewhat intuitively plausible. Hence, our participation in the domain of mathematics and those of the disciplines of science and technique is not merely separated from experience and culture. It would also be a denial of the very possibility of human life itself were it not for our secular myths. These require that all domains are organized in terms of a simple complexity that excludes all life.

It is becoming apparent that symbolization through differentiation and integration within a secondary frame of reference seeks to respect and maintain the general relativity of human life and the world. In contrast, the tertiary frame of reference is a denial of that general relativity, which is thus excluded from all domains. Ironically, in the leading scientific discipline of physics, a different kind of relativity has become accepted regarding space-time-matter, to which more hyphenated concepts may have to be added in the future. Even if this occurs, it will model nothing more than the greatest possible relativity within the architecture of the simple complexity of mathematics.

We will briefly make a similar argument for the metaphor of time as implied in physics. From the beginning, Western civilization was torn between two irreconcilable concepts of lived time. One was derived from the influence of Greek culture and may be thought of as time conceptualized relative to the cosmos. The other was derived from the influence of Judaism – a culture that was probably the first to conceptualize time relative to human history. Doing so continues to resemble our contemporary way of organizing human life relative to time. For example: From very early on, Jewish commentaries have interpreted the creation of light in the opening chapter of the Jewish Bible as being related to the creation of time. Moreover, the earth was seen as a series of events in time.[50] We all continue to acquire a metaconscious knowledge of how even our highly desymbolized cultures still arrange human life relative to time. This knowledge is so fundamental and vast in scope that it does not fully develop until we are well into our teenage years.

Once again, our metaconscious knowledge of lived time developed through the secondary frame of reference is very different from the one used in the domains of mathematics and physics developed through the tertiary frame of reference. The latter appears to more closely resemble the concept of time implicit in the behaviour of people who have symbolized periods of time as solid time-objects in accordance with the spatial orientation of the tertiary frame of reference in particular and that of our cultures in general. We seem to have accepted a four-dimensional space-time that we are able to imagine mathematically or physically, but that certainly cannot exist in any symbolic universe or in the reality of any contemporary desymbolized culture. What can happen in this space-time is limited by the speed of light. Two events within this space cannot be connected in any causal manner if the time interval between them did not permit light to travel from one to the other. In any case, this space-time of physics is a space emptied out of all human life and thus exists outside of us. It is the space of the domain of physics; and this is how it will remain regardless of how this domain evolves mathematically. Such developments can have substantial effects on this time and on this space, as will become evident when we examine the concept of matter.

As noted, the development of physics in Western civilization began with Aristotelean physics, which sought to systematize our daily-life experiences of the physical behaviour of objects. Also, when we grow up, we have already developed a great deal of "intuitive physics." This physics is dialectically enfolded into the experiences of our daily-life activities and in the metaconscious knowledge that was built up with them. Hence, attempting to systematize this knowledge initially appeared to be the best way of developing a "school physics."

In our daily-life world, every object soon comes to rest unless something moves it along. This was entirely satisfactory until Galileo postulated the opposite: all bodies have an inertia that keeps them moving in a straight line at a uniform speed unless they strike another object or are subjected to friction, air resistance, gravitation, and so on. This synthesis of the behaviour of objects was counterintuitive precisely because there is no object in our world that does not encounter resistance from one source or another. The concepts of matter and motion were thus separated from experience and culture. They were placed in the kind of space encountered in what was beginning to emerge as a domain of a discipline, a space in which matter occupied different positions that could be marked by time intervals.

Newton was able to significantly advance the understanding of the orbits of the planets by means of his three laws of motion and the principle of gravitational attraction, for which he had no physical basis other than that these planets appeared to "mathematically" work this way. It marked another decisive step toward physics taking on the form of a discipline separated from experience and culture. Its powers of prediction were thus improved, but the physics itself did not derive from them. Nor was any connection made between inertia and gravity.

Einstein wondered why both the inertia and the gravitational attraction of a body were characterized by its mass. What was it about inertia and gravity that made this possible? He refused to believe that this was a mere coincidence. He decided to give it a physical interpretation, which completely changed what had been understood by space, time, and matter. It lead to the reorganization of the domain of physics by means of the general theory of relativity, which held that the mass that had been associated with inertia and the mass that had been associated with gravity were identical. The space-time continuum became curved, thus eliminating Euclidean geometry from this domain. Similarly, gravity was displaced by inertia, marking the end of Newtonian physics. It is now the mass-energy present in space that can mathematically determine its geometry and its inertial characteristics. Mathematically speaking, the greater the concentration, the greater will be the curvature of the associated space.

The theory of general relativity provided a description of the universe that worked better than the Newtonian description, but it goes so far beyond our experience of the world that we are unable to make sense of it. The equations now predict the readings we can make of this universe even more exactly, but we have no idea of the physical basis of the equations. It is as if the domain of physics no longer has anything physical within it of a kind that we can directly or indirectly experience and interpret. Of course, all this has developed a great deal further since the days of Einstein, but it does not in the least alter the fundamental point: physics improves how well it works mathematically; but very little, if any, of this is based on improvements in our physical understanding of the universe.

This brings us to numbers, quantification, and measurement. Jones has a wonderful detailed analysis of what all this involves.[51] It should be noted that until recently, the significance of numbers in all cultures was first and foremost symbolic rather than quantitative. Again, any logical extension of these symbolic meanings by having some of them function as postulates led to insurmountable contradictions. Gödel showed that arithmetic and geometry were either inconsistent or incomplete.[52] Hence, if we attempt to derive arithmetic or geometry from these postulates, we will encounter contradictions or find that some results known to be exact cannot be derived from these postulates. In other words, when we transfer numbers from human life and the world into a rational domain separated from experience and culture, this domain will not find any common-sense basis. It is likely indicative of the architecture of one being incommensurate with that of the other, and vice versa.

All this was further confirmed by Georg Cantor, who proved a variety of theorems regarding infinite sets that are completely counterintuitive and non-sensical.[53] In other words, what may be exact in the domain of mathematics has no foundation in common sense. As noted, any search for the foundations of mathematics has long been abandoned.

Let us return to the question as to whether or not contemporary science provides us with metaphors that can improve the intellectual grip we have on our lives and the world. What we have attempted to show is that we have transformed symbolized lived time into reality-time, symbolized lived space into reality-space, symbolized matter into reality-matter, and so on. These kinds of transformations are not limited to the above examples. Jointly they constitute a transformation of our symbolized surroundings, composed of entities endowed with meanings and values for human life, into a detached, separate technical life-milieu that exists "out there" because we no longer symbolize it. This transformation has been accomplished by the piecemeal playing of our discipline-based mathematical games, which continue to transform human life and the world one category of phenomena at a time. The resulting concepts are no longer designed to improve the intellectual grip we have on our lives in the world, but to make everything work mathematically in order to extend our power. Everything in our lives and the world is dealt with as if all its aspects are reducible to mathematical equations in ways that exclude all life and responsibility, other than trying to repair the "collateral damage" in an after-the-fact and end-of-pipe fashion.

The findings of our discipline-based science bleed into the constructed reality of our daily lives mostly through the original and the new media. Nevertheless, the scientific metaphors (such as time, space, matter, and numbers) have to be made liveable by means of scientisms and secular religious attitudes. The greater the extent to which the domain of a discipline is expressible in mathematical terms, the more this discipline follows the path of mathematization, thus turning everything into a game – with one important difference. In these mathematical games, we move from statement to statement according to the principle of consistency; but this now occurs within an overall constraint imposed by the results derived from experiments. Nevertheless, playing this game is to act as if the category of phenomena examined by the discipline is fully comprehensible in terms of the architecture of mathematics. Whatever cannot be captured by this architecture (and thus mathematically represented) is literally eliminated from consideration, excluded from our way of life and thus from our history.

We are placing ourselves into a self-reinforcing cycle in which our history is increasingly made by playing these kinds of games (especially those of technique, which we will deal with in the next chapter). It has thrust us into a dark tunnel of our own making.

Can we conclude that discipline-based science as a whole has become a metaphor for anything being reducible to a gigantic mathematical game? Can this assumption be responsible for so many people giving credibility to those who in one way or another reduce human life to the playing of a kind of video game? This game can be changed for another simply by changing its underlying architecture. According to some people, turning individual and collective human life over to the cognitivist version of these games will soon permit us to live eternally on the Internet. These developments are indicative of the kinds of metaconscious knowledge that we derive from living in our contemporary anti-societies. It makes reified life "normal." Nevertheless, what appears plausible and, to some observers, self-evident is that these are manifestations of our being possessed by our secular myths, as opposed to our having a grip on what is happening to human life and the planet. We help sustain these developments through the scientisms dispensed by our education systems. All this seems to be leading to a secular religious global frenzy that is oblivious to the way we are reifying all life. Will this frenzy intensify the service of our secular gods that we have made for ourselves in the knowledge that they will eventually destroy us, but that they will also provide us with the means to save ourselves?

The historical tunnel into which we have thrust ourselves is constituted by our heedlessly playing the games of mathematics, discipline-based science, and discipline-based technique. We utterly trust these games because they "move us forward." We cannot conceive of anything different or radically other. Have we in this way already declared ourselves as dead, as if all life has already been potentially eliminated through the unlimited use of mathematics? Will we be left with a gigantic system of information flows that we serve with all our hearts, minds, and strength?

As noted, we are obscuring these developments by using the same words for living and non-living entities. These include cognition, information, knowledge, expertise, intelligence, memory, and a great deal more. It is further evidence that scientific and technical concepts, whose significance is entirely relative to the domain of a discipline, have bled into the daily-life reality, providing us with a range of metaphors by which we are attempting to obtain an intellectual grip on our lives and the world.

We must be as clear as we can be about playing the above games. There is nothing the matter with playing a game. We know very well that when we stop playing it, we return to something entirely different, namely, our lives. However, it appears that our civilization is systematically exchanging the playing of the games of mathematics, discipline-based science, and discipline-based technique for the living of our lives. Worse, our lives are deemed to be incapable of contributing anything essential to these games. Again, we appear to be oblivious to the limits of what science will never be able to know, what technique will never be able to do, and what the state-nation will never be able to politically deliver. There is a general lack of awareness that all these games cannot help us with getting on with living meaningful and viable lives that can be sustained by our planet. Discovering exactly where these limits lie could help to deliver us from our scientisms and secular religious attitudes. It could also help our young people not to be drawn into these dangerous addictions. We could explain to them that if they would like to study mathematics, for example, they should know that it is a fascinating game, it works, it is very powerful, but do not let it take over your life. It will never be anything more than a game played within a domain. Hence, we need to teach our young people when to play it, and when to get on with their lives. We will then be able to help them to resymbolize their deeper metaconscious knowledge and thus begin to weaken our common secular religious attitudes and myths. Collectively, we need to learn to have mathematics serve us and to unlearn being its slaves. The same is true for discipline-based science and technique.

Science, Religion, and Christianity

We have already ruled out the classical interpretation of the relationship between science and religion. Science cannot be understood as pushing back the boundary between the known and the unknown, to the detriment of religion. Such a relationship has become untenable ever since the discovery of paradigm shifts or scientific revolutions in the course of the historical development of a discipline.[54] These events could not have occurred in the way we currently understand them without the unknown having been enfolded into the known. There is no evidence whatsoever that the spectacular growth of discipline-based science has in any way diminished the need of our anti-societies to ground their ways of life in secular myths to avoid cultural chaos. Those forms of atheism that celebrate the triumph of science over religion have failed to recognize the clear and obvious limitations of discipline-based science, thereby creating a scientism that sustains secular religious attitudes based on a supposed omnipotence of science as the ultimate authority over all human knowing. Any association of science with truth has long been abandoned with the collapse of philosophical foundationalism. Like mathematics, science has no absolute foundations.

Both science and religion are internally divided. Science presents humanity with a kaleidoscope of domains produced by disciplines that are all autonomous from one another. Consequently, a synthesis capable of producing a scientific equivalent of a cultural order or a reality is impossible. Interdisciplinarity of any kind as a path toward a unified science has escaped us.

The religious phenomenon is also internally divided. Prior to the great cultural divide, the nations that constituted humanity were divided by the unique and total commitments they needed to make to their myths to hold cultural chaos at bay. These commitments manifested themselves in the religions and moralities through which they served these myths. Some time after the great cultural divide, however, the anti-societies of our present global civilization became united in their ultimate commitments to the same second generation of secular myths. Prior to the great cultural divide, what distinguished the member societies of any civilization from the member societies of all other civilizations was their ultimate commitment to traditional myths. Today, anti-societies are distinguished from one another only by their highly desymbolized cultures for the bringing up of each new generation and the intimate predominantly face-to-face relationships among their members. They are also divided according to their unequal success in serving the second generation of secular myths, and thus the power they have attained in the global technical-economic race, which helps to create the relationships of power and exploitation among these state-nations. As we will examine shortly, this new situation is affecting the forms of religions everywhere.

We first need to examine the secular religious role played by discipline-based science in our global civilization. Our scientific disciplines furnish us with the facts by which we seek to understand human life and the world. We have transformed these facts into absolute entities by denying their status of being entirely relative to the domain of their corresponding disciplines. We have thus provided ourselves with absolute information. However, from the perspective of the general relativity of human life and the world, our scientific facts have become the secular mini-equivalent of the traditional gods of the past. We have left ourselves no room for the interpretation of these facts in accordance with their meaning and value for human life and the world. These are the facts of non-sense; and their secular religious objectivity cannot be challenged. The result is that we live as if science had commanded us to obey its findings without question. Of course, science can do no such thing. We have thus adopted a secular religious attitude toward science that provides it with the authority we slavishly obey.

We will show in the next chapter that we act on our economic data, for example, as if their status as facts cannot be questioned. However, there is nothing more questionable than this economic information, given that we have not had an adequate economic framework since the discipline of economics replaced political economy. As noted, if we adopted a different intellectual framework, the "fact" that economies produce wealth would almost certainly turn into the opposite conclusion, that they actually produce debt. Consequently, our economic behaviour is akin to serving our economic mini-gods – with the same disastrous consequences that have always been associated with any culture serving its gods in order to fend off cultural chaos.

Without our secular religious attitudes toward science, facts would have a status relative to the intellectual framework of the discipline through which they were obtained, and that of this intellectual framework would be relative to the facts gathered by it. This circularity of scientific authority would be broken when an anomaly is discovered. At this point, the intellectual framework would have to be completely reorganized to make the anomaly into one of its first "normal" facts, after which some earlier facts could be reinstated as such, others rejected, but most retranslated into the new intellectual framework. Although this is historically and socially recognized by a number of scholars, we live as if facts are objective and thus must be obeyed.

Our attitudes toward scientific facts thus introduce an entirely new kind of authority into our lives and the world. We rely on our scientific disciplines to collect all the facts we need in order to act. If we discover that some facts are lacking, research funding is usually quickly directed toward producing them. Because these are facts of non-sense that lie outside of the sphere of sense, however, they must be introduced into the reality of our daily lives by the original and new media – which turn them into image-stories that in turn are transformed into "myth-information" by their participation in the mythology that organizes almost everything in our lives.

This mythology assures us that despite the many terrible things we see and hear on the media, we can be confident that we are moving forward in the direction of history. Democracy will triumph over tyranny. Our politicians will prevail on our behalf by imposing our values and political wish lists on the technical order and the state-nation. Free trade and global markets will produce the economic growth on which we all depend for our livelihood and for our governments' ability to provide us with the services we expect. Consumption is our ritual participation in the greatest good and the deepest meaning of life we know, namely, that the technical order, through its constant technical advances, will materially produce a socially better world with a happier humanity. Medical science will save us from disease. Education will keep us in the global race. Democracy will ensure that our taxes are well spent to create the kind of state-nation to which we aspire and for which we will lay down our lives if necessary, as we commemorate others who have done so in the past. We celebrate sports as the ultimate spectacle of all-out performance, efficiency, and power. This is the life that we either love with our hearts, minds, and strength – or out of anxiety and depression because there appears to be no meaningful alternative, even though it seems clear that something else is happening, other than what the media portray. It is the credo we practise and live by through our national symbols and the bath of images into which the media immerses us.

It is important to emphasize that this daily-life reality has a simple complexity. Consequently, events on our newscasts can come out of nowhere and disappear into nowhere without affecting anything else, or in any way undermining the mythology by which we live. It is this mythology that gives this simple complexity its only coherence, being capable of turning anything that happens into image-stories and myth-information. We cooperate in all this by reconstructing our lives on the social media, turning fragmented images into our own image-stories and myth-information. It is this reality that mediates our relationships with our discipline-based approaches in science, technique, and politics.

As noted, humanity has lived in three primary life-milieus by means of unique myths served by equally unique religions and moralities. Our technical life-milieu stands out, having been constructed by our global civilization as a separate environment "out there." We do not symbolize it because we do not relate to it by means of our secondary frame of reference. Its role is essentially restricted to the bringing up of each new generation and to our intimate, mostly face-to-face personal lives. We relate to our technical life-milieu by means of our tertiary frame of reference, which scans it for details considered to be essential for the situation at hand. Since it is grafted into the secondary frame of reference, it is essentially operated by a highly desymbolized sociocultural self related to our metaconscious knowledge embedded in experience and (our also highly desymbolized) culture. To make this possible, the use of the tertiary frame of reference is limited to relationships involving minimal embodiment, participation, and commitment from our highly desymbolized sociocultural self. The reality into which we are immersed by the media is separate and distinct from our lives as well. As a result, our global civilization is constructing a technical order "out there" in which we can only minimally participate, thereby shutting out much of our lives – which we must seek to live somewhere else. It raises the question as to what the purpose of this order is if it essentially shuts us out, making us stand at a distance from it.

Following the Second World War, social scientists intuited our changing relationships with our societies and our primary life-milieu by means of a series of "intellectual fashions" that included functionalism, structuralism, and systems theory. These were eventually displaced by intuitions derived from metaconscious knowledge about the implications of discipline-based approaches for human knowing. It helped to turn social sciences into disciplines using a variety of techniques such as survey methodology and statistical procedures.

Our new life-milieu appeared with the second generation of secular myths. It has put all traditional religions, inherited from a time when societies constituted the primary life-milieus of humanity, into a terrible predicament. These religions had to affiliate themselves with what little was left in people's lives of their highly desymbolized culture and secondary frame of reference, which thus became the "personal business" related to their life events: birth, growing up, entering a partnership, having children, living through difficulties, and dealing with disease and death. On the one hand, people had to protect themselves from cultural chaos by rooting their participation in an anti-society with its secular myths; and on the other hand, they had to continue to root their religious and spiritual lives in a traditional religion through a highly desymbolized culture. Doing so required a major reconstruction of all traditional religions.

For believing Jews and Christians (as the major traditions of the civilization that gave birth to our present one), living in an anti-society also presented them with a terrible dilemma. They could not avoid participating during the week in a world that they believed to be lost, and celebrate their religious life through personal devotion and participation in a synagogue or church organization. After the Second World War, this resulted in Protestant, Catholic, and Jewish forms of the American way of life.[55] From a sociological perspective, this was the tension lived by the mainstreams of these three religious traditions, which were losing their way in the face of everything that had destroyed all traditions. Many young people left because they found it impossible to live this tension. Those who stayed learned to use their tertiary frame of reference to scan in those details from their Bibles, communities, and anti-societies that would produce a satisfactory religious construct. It led to a religious and spiritual divide in which, underneath a surface veneer, is hidden a terribly divided existence. It resulted in the kinds of vanities and contradictions so well described by Qohelet (Ecclesiastes) in the Jewish and Christian Bibles.[56] These difficulties are glossed over by the conservative and liberal branches of these communities, all accusing one another of betrayal and having lost the way.

6 Human Doing, Technique, and the Living of Our Lives

Naming What We Have Lost

Because of our human finitude and our being a symbolic species, we appear to be able to come to terms with the general relativity of our lives and our world by proceeding as if what we know and live constitutes the one and only vantage point for making sense of and living in the world, and the one direction for its historical journey in time. We thus essentially ground ourselves in an ultimately unknowable universe by symbolizing this general relativity as more of what is known and lived, thereby ruling out any other possible vantage points and historical orientations and keeping the threat of cultural chaos at bay – at least, most of the time. A community essentially names itself by everything it has achieved, metaconsciously summed up in what cultural anthropology regards as its myths. However, because all neighbouring communities are obliged to make the same ultimate commitments to the way they have symbolized the general relativity of their lives in the world in different ways, they pose a potential threat of opening the door to cultural chaos. Generally speaking, each and every community thus loses the ability to "playfully" live with the endless wonders of the inexhaustible general relativity of its life and the world, the way its children do – until they begin to acquire the myths that ground the community in this relativity. At that point, the general relativity of their lives in the world is progressively distorted: they learn to live with something that is integral to this general relativity as if it was absolute and thus godlike in traditional societies, and without limits in secular societies. It gradually plunges their lives into endless contradictions, distortions, and ruptures of the general relativity of their lives and the world, with the result that they become alienated from all other people, themselves, and their world. It is the price any community pays for symbolically appropriating the entire universe as more of what their own vantage point and historical orientation can make knowable and liveable. The past thus becomes a preparation for the present, and the present will mark the future, either by continuing a tradition or by accepting the inevitability of having to "move forward." This makes the historical journeys of all other communities ultimately unknowable and unliveable from the absolutized vantage point and historical orientation of a person's community.

It is from this general perspective on our history as a symbolic species depending on the biosphere of our planet to sustain us that we need to assess what has been gained and lost during one of the great divides of this history: the shift from a primary reliance on culture to make sense of and live in the world to a primary reliance on highly desymbolized discipline-based approaches to knowing and doing, without being able to entirely leave behind a secondary reliance on highly desymbolized cultures for our myths. This shift had four primary dimensions: industrialization, urbanization, rationalization, and secularization. In all this we need to understand the role of technique relative to culture, and the roles of our highly desymbolized cultures relative to technique.

As a consequence of our developing both a secondary and a tertiary frame of reference, living our lives in contemporary anti-societies involves us in establishing, maintaining, and evolving two kinds of relationships. One kind contributes to what little is left of the culture-based connectedness of human life and society, and the other to the technique-based connectedness of our global technical order. The culture-based connectedness can be understood as the fabric of relationships established and lived on the basis of their relative meaning and value to each other, and jointly to human lives and society as grounded in the second generation of secular myths. As a consequence of the influence of these myths, the culture-based connectedness is limited in its ability to respect the general relativity of human life and the world by means of the secondary frame of reference associated with symbolization, language, and culture. As noted, this development is severely stunted in children by the stream of experiences involving screen-based devices and discipline-based approaches.

The resulting stunted development of the secondary frame of reference and the corresponding culture-based connectedness of individual and collective human life is confirmed by a dramatic transformation in the way people use the organizations of their brain-minds to make sense of and to live in the world. The work of David Riesman may be reinterpreted as more directly explaining the breakdown of tradition-based ways of life, their transformation into the ways of life of the first generation of industrializing societies, and their subsequent transformation into those of the so-called mass consumer societies, as opposed to only explaining the effects of a growing demographic density.[1] Such a more direct explanation focuses on the joint effects of industrialization, urbanization, rationalization, and secularization on individual and collective human life (summed up by what we have referred to as "technology changing people"). The introduction of the technical division of labour involved the translation of the activity of human work that was dialectically enfolded within individual and collective human life into a construct in the image of the machine. This meant that it now had an architecture characterized by the principles of non-contradiction, separability, closed definitions, openness to measurement (and thus to quantification and mathematical representation), and a simple complexity. This translation opened the door to subsequent mechanization, industrialization, and, later, automation and computerization. Increasingly, the technology-based connectedness was no longer related to satisfying local needs and thus had to be separated from the culture-based connectedness of individual and collective human life. It was the introduction of a technological order initially embedded into an economic order within the cultural orders of these industrializing societies. These constructed orders had to be dealt with by rational, technological, and economic approaches that began to desymbolize the culture-based approaches that remained dominant in the remainder of these societies.

These transformations were so fundamental and so different from anything that had preceded them in human history that prior experience was of little value, and thus all tradition-based ways of life began to break down, including their constituent traditions that had guided particular areas of human life. Consequently, it became impossible for people living these upheavals to continue to use the organizations of their brain-minds as symbolic mental maps of tradition-based ways of life. For a few generations, people could cling to those elements of these symbolic mental maps that endured (namely, the deeper metaconscious knowledge embedded in them) as a compass or gyroscope to orient their lives. It was not long before so little of these maps remained that people had to evolve an entirely new approach to having their lives work in the background by means of the organizations of their brain-minds.

Their cultures had become so highly desymbolized that they could no longer provide people with the meaning, direction, and purpose for their lives. An external supplement was required, which Jacques Ellul referred to as integration propaganda of a social kind.[2] In order to make effective use of this external guidance to supplement a highly desymbolized culture, people found it necessary to rely on the organizations of their brain-minds as a kind of radar. By doing so, they acquired public opinions to complement the private opinions they could no longer form in a great many cases, a statistical morality to acquire moral guidance from scanning what everyone did as being normal and normative, and much other guidance obtained from the bath of images diffused by the old and new media – all of which collectively began to constitute the reality relative to which people conducted their lives. What tradition-based cultures had accomplished by means of customs and traditions, integration propaganda accomplished by means of images, image-words, and image-stories. In other words, what little remained of the highly desymbolized cultural "world" had to be complemented by a reality with an architecture that was the diametrical opposite of the architecture of the culture-based connectedness of any earlier symbolic universe.

The dialectically enfolded character of the organizations of people's brain-minds was largely banished from collective human life, only remaining as potentially useful in individual life where it was also under pressure from desymbolization. As a result, the weakened secondary frame of reference gradually learned to use the tertiary frame of reference in which people used the corresponding portion of the organizations of their brain-minds as a radar that scans reality for those aspects necessary for the provision of external guidance for the situation at hand. Doing so was possible because this reality had an architecture with a simple complexity that was thus unaffected by the image-stories added and subtracted from it each and every day (mostly through the media), and the personal use they made of it by having their lives work in the background (including their sociocultural selves) to guide those portions of the organizations of their brain-minds constructed by, and acting as, a tertiary frame of reference. On the collective level, this extension of their visual approaches provided the essential details for constructing the "building blocks" for the economic, social, legal, political, moral, religious, and aesthetic constructs that in traditional societies had been the dimensions of cultural mediation. The resulting reification of human life was mostly veiled as being of no consequence to our lives, since the second generation of secular myths implied that every element unfolded from a dialectically enfolded human life in the world constituted a "moving forward" in the "history" of a technical order constructed by means of discipline-based approaches and the political order of our state-nation.[3] What was thus hidden from our individual and collective awareness was that the dialectically enfolded character of individual and collective human life was being put on the rack of discipline-based approaches, to be slowly stretched and distorted into an entirely new architecture with a simple complexity.

These developments are negatively affecting the ability to live our lives, because these lives cannot fully and integrally work in the background of each moment as a consequence of the transformation of their architecture. Does this mean that we will have to live our lives with a greater measure of cultural chaos? Does this mean that we are losing our "common sense" as the shared body of meanings and values that bound the dialectically enfolded communities of the past? Can a bath of images, image-words, and image-stories adequately take over from what had been accomplished by means of symbolization, language, and culture? Are these the hallmarks of the transformation of the mass societies of the late twentieth century into the anti-societies of the early twenty-first century? Parts of the organizations of our brain-minds may well have become a very effective radar picking up from the horizon of our experience the details that matter for our lives, but what is being lost in terms of how all these details are integral to the lives of others and of a world in which we all live? Are we conducting a war on ourselves as a symbolic species, and does this involve a battle for our human spirit?[4]

As noted, our lives also contribute to the technique-based connectedness of our global technical order. This connectedness may be understood as the universal order of relationships constructed and lived because of their absolute efficiency. This *absolute* efficiency is the diametrical opposite of the relative efficiency of everything that characterized the culture-based connectedness of all traditional societies. The relative efficiency of any human activity, or of anything we invent, organize, and build, may be interpreted as the contribution it makes to the general relativity of human life and the world lived by a community and thus grounded in myths. This relative effectiveness of efficiency was symbolized as the meaning and value of something relative to everything else, and thus in relation to human life and society as established by symbolization, language, and culture. In contrast, the absolute efficiency of something is established relative to that something having been abstracted from the general relativity of human life and the world, and relative only to itself. The absolute efficiency thus refers to how something internally transforms whatever inputs it requires from the general relativity into the desired outputs to be returned to this general relativity, as measured by ratios of the latter to the former. It is absolute in a double sense if this internal transformation of required inputs into desired outputs is also as great as it can become, given the state of development of the discipline-based approaches, used for this purpose. All undesired outputs are externalized by these approaches as well as how any internal technical rearrangement to obtain the absolute efficiency will affect the relative efficiency. Consequently, the technical approaches for achieving absolute efficiency are the diametrical opposite of the culture-based approaches that would have ensured the best possible relative efficiency.

What we are showing is that a symbolic culture is a way in which individual and collective human life makes sense of and lives in the world, whereas a technique constructed from technique-based approaches is a system of methods rationally arrived at to produce the greatest possible absolute efficiency, given the current state of advancement in sustaining and adapting an entire way of life one constituent technique at a time. Consequently, our participation in the advancement and adaptation of technique requires a tertiary frame of reference that is grafted into a secondary frame of reference, and whose limits are veiled by the second generation of secular myths, to the detriment of any symbolic culture.

The architecture of the culture-based connectedness of human life and society can be expressed in terms of the principles of enfolding and dialectical enfolding. In contrast, the architecture of technique-based connectedness of the universal technical order can be expressed in terms of the principles of non-contradiction, separability, closed definitions, openness to measurement (and thus to quantification and mathematical representation), and a simple complexity as encountered in all constructed entities, of which classical and information machines are the exemplars.

Both discipline-based science and technique make sense of and manipulate the world one category of phenomena at a time, the results of which are mathematically mapped in the domain of a corresponding discipline. In our discipline-based science, the findings of all disciplines, mathematically mapped in their corresponding domains, cannot be assembled into a scientific representation of human life and the world because the autonomy of the disciplines makes a corresponding scientific approach for doing so impossible. Any concept of a *scientific* interdisciplinarity, multidisciplinarity, transdisciplinarity, or systems thinking amounts to wishful thinking. This is because no *scientific* equivalent of anything like the culture-based connectedness of contemporary life or of the cultural orders of traditional societies is possible. In contrast, a universal technical order can be constructed out of those relationships that have been reorganized by means of techniques in order to achieve an absolute efficiency, in accordance with the technical significance of the desired outputs for this technical order.

The loss of people's dialectically enfolded lives working in the background as symbolic maps of the experiences of their lives lived in accordance with the culture-based connectedness of their way of life and society was veiled as being entirely redundant by the first generation of secular myths. These myths created an alternative, and supposedly more desirable, way for people to live their lives, based on the unlimited advancement of the technology-based connectedness of human life and society. According to the myth of progress, doing so would inevitably bring about social and spiritual betterment sustained by a new and more successful culture-based connectedness. This technology-based connectedness was organized and advanced by means of an economic order that distributed goods and services according to the supply and demand in their own markets – a process that is entirely piecemeal as compared to distributing them in accordance with their meaning and value for human life and society as established by an economic dimension of cultural mediation, as in traditional societies. This Market order was only half the story: the prerequisite reification and commoditization of everything before it could be traded in markets produced all manner of market externalities that accumulated in people's lives and in society. They were turned into Market forces, of which the great "invisible hand" is the best known positive one, and the endless economic, social, and environmental crises are the most prominent negative ones. Cultural orders produced no such externalities other than through the way their traditional myths distorted the general relativity of human life and society, which made it next to impossible to regulate anything that a culture had declared to be very good or the absolute good by means of these myths.

In the same vein, the second generation of secular myths that grounded the transition of the first generation of industrial societies into the consumer mass societies of the second half of the twentieth century, and then into contemporary anti-societies, oriented human life in a direction that implied that everything could be gained by advancing the technique-based connectedness. The already deteriorating culture-based connectedness was devalued, since it supposedly could not interfere with the enormous benefit of moving forward in human history. Moreover, this even more highly desymbolized culture-based connectedness was regarded as being of no interest to the political orders established by the state-nations when everything became political. What little remained of the culture-based connectedness could at best somewhat mitigate the effects of desymbolization: the symbolic development of the members of each new generation before they acquired their tertiary frame of reference, the reification of people's personal relationships, and what remained of people's private traditional religious commitments shared in these relationships. All this at best constituted a kind of private sphere that was not to interfere with the public one.

The progressive deterioration of the culture-based connectedness of human life and society has been examined in detail in my previous works.[5] The first generation of secular myths legitimated the use of an economic rationality as the only responsible way for people to use the organizations of their brain-minds, as opposed to using culture-based approaches and their secondary frame of reference. The second generation of secular myths gave full rein to the use of the tertiary frame of reference by creating a kind of "bifocal mental lens" for dealing either with the public sphere or with the private one.

These developments are confirmed by the uneasy coexistence of two forms of metaconscious knowledge capable of sustaining creativity, imagination, intuition, and human skills in two separate and distinct ways. One form of metaconscious knowledge is built up by integrating the experiences of activities embedded in experience and culture, while the other kind is built up by integrating the experiences of activities separated from experience and culture. We have discussed the coexistence of intuitive and school-based forms of the knowledge of the grammar of our mother tongue, physics, and stress analysis; and these examples can be multiplied almost indefinitely. Both forms of metaconscious knowledge are jointly sustained by the deepest metaconscious knowledge corresponding to our present secular myths, with the result that the creativity, imagination, intuition, and skills sustained by the metaconscious knowledge separated from experience and culture are considered to be reliable and important, while those sustained by metaconscious knowledge embedded in experience and culture are devalued and fit only for our personal and private lives. This latter kind is built up through the primary and secondary frames of reference, while the other kind is built up through the tertiary frame of reference.

The consequences for human life and society are far-reaching. For example, in engineering, design has essentially become associated with discipline-based analysis and optimization.[6] Managing is closely tied to the application of techniques taught in the disciplines of our business schools.[7] Medicine has become less of an art and more of a discipline-based applied science.[8] In other words, consider what we have lost: the art of engineering design practised relative to the general relativity of human life and society; the art of managing as a comprehensive strategy that interrelates the significant issues faced by a business within the general relativity of society and the biosphere; and the medical arts as the practice of healthcare rather than disease care, within the general relativity of human life and society. Genuine healthcare would be interpreted as placing demands on our resources to maintain our physical, social, and mental wellbeing during all daily-life activities: eating and drinking, working, socially sustaining and being sustained by others, relaxing, and sleeping. Such medical arts would include the prevention of disease as well as end-of-pipe disease care, where prevention has either failed or turned out to be impossible. They would consider the production of disease by agribusiness, food production, workplaces, the urban habitat, screen-based devices, and much else that may affect our ability to fend off disease by means of our resources. The art of any human activity is lost when it cannot depend on metaconscious knowledge embedded in experience and culture, which is eliminated when the tertiary frame of reference is employed. In such cases all intuitions that may occur are sustained exclusively by metaconscious knowledge separated from experience and culture because the practitioner of any technique is suspended in a triple abstraction. Because of the second generation of secular myths, there is little or no concern for the loss of engineering design that once dealt with the relative efficiency of a technology and was thus concerned with the culture-based connectedness of human life and the world. Nor is there much concern about the loss of the art of management or medicine within these contexts, relative to which businesses exist and relative to which health must be understood. These are but three examples of professions that have almost entirely fallen into the grip of discipline-based approaches that are entirely piecemeal in their dealing with everything one category of phenomena at a time.

The consequences are well illustrated by the differences between the technologies of traditional societies and those of contemporary anti-societies. The constituent technologies of traditional societies were created, applied, and evolved by people who were sustained by their dialectically enfolded lives working in the background. Consequently, these technologies were mostly compatible with the general relativity embodied in their ways of life and cultures and with local ecosystems. They bore the "imprint" of a people of a time, place, and culture to such a degree that archaeology could do its work. It was not until knowing and doing separated themselves from experience and culture, especially following the Second World War, to be reorganized on the basis of disciplines, that this context-compatibility could no longer be taken for granted. It had to be named – which led to the concepts of an appropriate technology and sustainable development.

These concepts have far from exhausted what was lost when the organization, adaptation, and evolution of ways of life were handed over to discipline-based approaches. Contemporary technology as a branch of technique has become so inappropriate and unsustainable that Ralph Estes estimated that, in 1993, the social costs passed on to the United States by its corporations were roughly twice the size of the budget of the federal government and half the GDP, which was equivalent to five times their profits.[9] The most significant factor that made the American way of life increasingly unsustainable during the decades after the Second World War was not the result of the combined effects of a growing population resulting from the baby boom and a rising level of affluence as the war economy was turned toward consumer goods, but the technology factor – which may be expressed as the load imposed on local ecosystems and the biosphere per unit of throughput of matter and energy.[10] Herman Daly and John Cobb estimated that the costs incurred in the production of the GDP have resulted in net wealth not having increased at all for many decades, but having declined by nearly half for the lifestyles of the American middle class, depending on the assumptions used in the calculations.[11] These are but a few indicators of the tremendous costs resulting from the collisions between technique on the one hand and human lives, societies, and the biosphere on the other hand. We can confidently assume that our corporations and national economies have become wealth extractors that are making all us poorer, except for a tiny global minority that is becoming wealthy beyond what could be imagined in the past, thanks to technique.

The commoditization and reification of all life that began when the economic dimension of cultural mediation was turned into a constructed Market system of the first generation of industrializing societies has been greatly accelerated by technique and its desymbolization of all cultures. As human beings internalized this reification, it began to desymbolize the dialectically enfolded character of their lives and thus the ability of these lives to work in the background in the way they did prior to their desymbolization. As a result, all daily-life activities became less and less a particular expression of the general relativity of human lives lived as individually unique expressions of a society, as symbolized by a culture. This desymbolization opened the door first to rational approaches and then to technical ones based on disciplines. Since the application of all techniques is necessarily suspended in a triple abstraction from the general relativity of human life and the world as symbolized by a culture, their application is relative to the domain of a technique and no longer to the general relativity of human life in this universe, symbolized and thus restricted by a culture because of its having to ground itself in myths to fend off cultural chaos. The resulting commoditization and reification of all life produces a comprehensive pattern of collisions between technique-based connectedness on the one hand and what remains of the culture-based connectedness of human life and the biology-based connectedness of all life on our planet on the other. Our highly desymbolized lives working as a kind of radar in the background of all our activities have become a self-reinforcing and self-sustaining desymbolization of the human species.

The only way in which we can put an end to this self-reinforcing destruction of our being a symbolic species is to resymbolize technique in order to gain an intellectual and practical grip on what it can and cannot do for our lives, our societies, and our planet. As long as we live with technique as if it has no limits (and thus as the secular equivalent of the traditional gods), there is no way of dealing with the root of all the crises that our global civilization has unleashed on itself by its secular religious attitudes toward its most powerful creations. There is nothing new under the sun in this, other than the magnitude of the power and scope of the collisions we have unleashed between all life and technique. All this is increasingly slipping beyond our awareness because we fail to recognize the radically opposite architectures of all life, on the one hand, and the global technical order we are imposing on it, on the other hand. The more we think and live life in the image of non-life, the more we weaken our capacity to come to terms with technique. Naming what we have lost is our ability as a symbolic species to live our lives with these dialectically enfolded but alienated lives working in the background of all our daily-life activities. However, our ability to do so has been decisively weakened because our lives are now working in the background in a reified form.

Recognizing the Symptoms of What We Have Lost

From the perspective of our daily lives, what is happening to human life and the world is made sense of in entirely different ways, judging by public opinion, political beliefs, and the bath of images diffused by the traditional and the new media. Our contemporary anti-societies appear to be divided according to three essentially political interpretations. The first may be referred to as an economic worldview. It holds that our economies are at the centre of everything. They produce the jobs essential for our full participation in society and the wealth required by governments to pay for the services on which we all depend. Without a sound economy, many people would be left out of meaningful participation, and the entire social fabric would be deeply and negatively affected. Moreover, there would be no resources for the necessary environmental protection measures. Hence, according to this worldview, all our aspirations and hopes for the future must be constantly re-evaluated in terms of their being economically realistic.

A second interpretation puts a society at the centre, and may thus be thought of as a social worldview. Our wellbeing affects all our daily-life activities, and thus helps to advance and adapt a way of life that gives meaning, direction, and purpose to all of us. As we move through our daily routines, our energy levels are depleted and our physical, social, and mental resources are degraded. Hence, we all depend on healthy nutrition, meaningful work, a need to sustain and to be sustained by others, while filling gaps in our lives with recreational activities that meet unfulfilled aspirations, and ending the day with a good night's sleep that restores us. In sum, we all need a sustaining social and natural life-milieu. We must all be capable of participating in the raising of each new generation and preparing each new member for a meaningful and responsible involvement in our society. All this includes, but is far from limited to, ensuring a well-educated, productive, and creative workforce. As we go about our daily lives, it is essential to take into account the many constraints that are imposed on us, but we must not allow ourselves to serve these constraints and become the slaves of our society.

A third interpretation is centred on the biosphere, and is commonly referred to as the deep ecology worldview. It essentially regards the other two worldviews as "well and good," but overlooking the fact that, without a sustaining biosphere, there would be no life on this planet. Hence, the protection of this biosphere, which we are weakening in so many ways, ought to be of the utmost importance. We must ensure that all life can go on and that this planet has a future. In other words, this worldview almost amounts to a reversal of the usual relationships between humanity and the biosphere, in which humanity must accept our being a constituent species of the biosphere.

What is fundamental, essential, and nevertheless missing from the above three worldviews can perhaps be illustrated by means of the following analogy. Imagine three people sitting side by side composing a description of the landscape before them. The landscape has a number of interesting features, including a grand old farmhouse, a stand of trees in full fall colours, and an extensive wetland. Because of different interests, each person takes one of these features and makes it the focus of their description, with the other features complementing it as part of the background. If at a later time and in a different place, we read the three descriptions, we would be unable to tell what the original landscape was like, unless we had "lived" it as a visitor. There is no possibility of arriving at a master description by synthesizing the three descriptions in some way. What is missing from each description is the landscape itself.

In the same vein, the above three worldviews cannot be describing three different worlds, since their adherents live in one and the same world. It is this world that is missing from all of them. During elections, people vote for political parties that essentially behave as if one of these worldviews is the only true one. In this way, we are all caught up in a political illusion that fails to recognize that all three worldviews are a fundamental distortion of the world in which we live. We need to question the soundness of all political commitments, beginning with our own, and get on with it by participating in round tables to jointly discover how we all distort the world. We will then quickly discover that we no longer share a "common sense" as a body of meanings and values by which we make sense of and live in the world. We have lost a common background that, in traditional societies, would have corresponded to the culture-based connectedness of a shared tradition-based way of life. With it we could have adapted and evolved our daily-life activities because that life never quite repeats itself in the same way; and we could do so without losing the dialectically enfolded character of what we had in common. With the beginning of industrialization, urbanization, rationalization, and secularization, this culture-based connectedness was steadily and relentlessly weakened, with the result that it brought on an age of ideology to supplement the diminished capacity of a culture to give meaning, direction, and purpose to the lives of the members of industrializing societies.

The only way we can jointly discover the world in which we all live is to retrace our steps by drawing a kind of dynamic intellectual map of how the present relationships between our lives, the economy, society, local ecosystems, and the biosphere came into existence as a consequence of a long and complex transformation whose primary dimensions are those of industrialization, urbanization, rationalization, and secularization. Prior to this transformation, the members of traditional societies did share a culture, which permitted the organizations of their brain-minds to function as a kind of mental base map of a tradition that was elaborated in detail according to the way the individual participated in a shared way of life. We no longer share a common culture-based connectedness, with the result that, if we wish to obtain a common intellectual and practical grip on what is happening to human life, society, and the biosphere, we will need to develop such a shared intellectual base map of these transformations.

The construction of such a base map is what I have attempted to do in previous volumes dealing with the relationships between technique and culture as cited throughout this work. This construction involved a resymbolization of the findings of as many relevant disciplines as possible in an attempt to recover something of the greatly weakened dialectically enfolded character of individual and collective human life. It uncovers the illusion of contemporary politics and the danger of our secular political religion of democracy.

The complexity of this undertaking may be illustrated by another analogy. The previous one, that of the three people describing a landscape, cannot sufficiently explain how our anti-societies are contributing to the building of a global technical order supported by a national political order. In our situation, attempting to describe what is happening to human life and the world by means of discipline-based approaches results in something even more fundamental going missing. If the three people describing the landscape were limited to discipline-based approaches, they would have to depict the landscape one category of phenomena at a time. A number of these categories would have been greatly restricted by their scientific or technical specialties, assuming that they had acquired them. The following analogy may help us to appreciate the implications.

Consider a slightly elaborated account of an exploratory mission undertaken by a team of experts to investigate a hunger problem in a particular region, for the purpose of recommending what should be done about it.[12] Following their individual explorations of what was happening, let us imagine them gathering around a round table to put their findings together. The nutritionist on the team told the others that what was happening was all too clear. The people in the area did not have a nutritious diet, therefore steps had to be taken to supplement this diet to improve their health. The expert in public health strongly disagreed: even if they had such a nutritious diet, they would be unable to digest it because the contamination of their drinking water; and also, the lack of sewage disposal and treatment was causing widespread diarrhea. Hence, the first thing that had to happen was to treat their drinking water and improve sanitation. The economist began to laugh. The reason the people did not have adequate nutrition, clean drinking water, and proper sewage disposal was simple: they could not afford them. Hence, the solution resided in economic development by bringing in some industry, or by initially encouraging some people to go out of the area to work and send home what they could save. The agronomist looked perplexed. These people were farmers, hence it would make much more sense to teach them modern agricultural methods to produce enough food to feed their families and sell the surplus for a modest income. Impossible, according to the political scientist on the team. None of these proposals had any chance of working because the government, army, and police were controlled by a small group of large landowners, with the result that the people did not have a political voice. They would have to politically mobilize to participate in making the decisions that affected their lives in the area. According to the sociologist, all this was well and good, but all the proposals thus far failed to take into account that, according to the cultural customs of the people, the land held by a family was split when the oldest son married to make it possible for him to raise a family. This custom had resulted in the parcels of land having become far too small to be farmed with modern agricultural practices. To change this, groups of families would probably have to form cooperatives that would permit them to jointly farm much larger fields. An emphatic no came from the demographer. This solution would not work any better than all the others, for the simple reason that after the introduction of some vaccinations and some elementary healthcare thanks to earlier foreign aid, infant mortality had dropped significantly, and the population of the area had become too large to be sustained by the local ecosystem. Some people would simply have to move somewhere else. Finally, the ethicist on the team told its members that they faced the most fundamental problem imaginable. If they could not agree on a common diagnosis, they could not possibly recommend an intervention with the kind of authority, responsibility, and accountability to justify telling others how to live. It was as if they were speaking different languages and describing incommensurate situations. Perhaps they ought to ask the people how they saw their situation and how they, according to their way of life and culture, would address it if they had the necessary resources.

How are we to interpret this analogy of our contemporary situation? It is clear that each and every specialist mediated the situation of the people with the hunger problem by the constituent disciplines of their expertise. Consequently, they could do nothing else but put those aspects of the situation that corresponded to these disciplines in the foreground as well as in the background. They were all suspended in a triple abstraction, which made it impossible for them to take the next step: interpreting this scientifically and technically mediated diagnosis into a kind of foreground against a background of how they would have made sense of the situation by means of their primary and secondary frames of reference if these had not been so desymbolized and stunted in their development. If this desymbolization, especially of their secondary frame of reference, had not occurred, would the members of the team have shared a symbolic interpretation equivalent to the metaconscious knowledge of the culture-based connectedness of a way of life encountered in the members of a society exposed to little or no desymbolization? If such a hypothesis is at all plausible, might it then be feasible to develop a kind of knowledge on the conscious level equivalent to the metaconscious knowledge of the culture-based connectedness as an intellectual base map of the situation? The members of the team would then be able to elaborate this base map in terms of those aspects that corresponded to their expertise. Such an elaboration would essentially constitute the beginning of a resymbolization of their expertise and would give the team of experts a good chance of arriving at a common diagnosis, and possibly even a shared proposal for an intervention. The intellectual base map would also contribute to the resymbolization of the findings of all the disciplines used by the team relative to one another.

Proposing this kind of hypothesis is bound to meet with a great deal of scepticism. First, it goes against our current secular myths, which rule out the need for and the validity of subjecting the findings of discipline-based research to resymbolization and their integration into an intellectual base map – which we will never be able to establish by means of science because of its discipline-based organization. Second, if each of the numbers of the team represented a university, consulting firm, or government department of their country, it is difficult to avoid the suspicion that everyone is acting on the basis of personal and institutional self-interest. How convenient that the core of the hunger issue happens to coincide with their own area of expertise, and how beneficial this is for their careers and their personal and institutional self-interests.

The above problem is a perfect exemplar for what occurs whenever our civilization applies discipline-based approaches to a human situation that inevitably has at least traces of a dialectically enfolded general relativity. Is it even possible that a group of specialists could rise above this, and for the sake of the people facing their hunger problem, take an intellectual journey as far as they can in order to establish this common base map, to which they can then all make their unique contributions according to their discipline-based approaches? Will self-interest and institutional factors not make this possibility highly improbable and likely to be dismissed as too naïve and idealistic to merit a time commitment? Moreover, even if the members of the team could make this commitment, how far could they intellectually journey down the road toward a base map before becoming bogged down in irreconcilable differences because of their ideological, political, or religious positions?

In other words, does culture and everything built up with symbolization, experience, and language have any chance whatsoever against our discipline-based science and discipline-based technique, which our secular myths have declared to be without limits? Have we not already surrendered our universities to these discipline-based approaches, leaving no room whatsoever for anything else – other than polite discussions about the need for interdisciplinarity, sustainability, entrepreneurship, or whatever the current academic fashion happens to be? We appear to have surrendered our limitless politics to an economic worldview; and when the results become too extreme, we resort to letting in a bit of a social worldview. The graduates from our universities are now playing key roles in all our institutions – roles that are inaccessible to people who have not been trained in discipline-based approaches of one kind or another. Do all discipline-based approaches not externalize the dialectically enfolded character of individual and collective human life and the biological enfolded character of all life? Using much of the organizations of our brain-minds as a radar to complement the highly truncated portion built up by our primary and secondary frames of reference means obtaining much or all that we need from what we have referred to as reality, with its diametrically opposite architecture to that of everything arrived at through symbolization, experience, and culture. How long will it be before (in both our private and professional lives) the possibility of an architecture different from that presupposed by our other-directed personalities and our discipline-based approaches becomes completely unthinkable and certainly implausible? Is this not implied by the widespread acceptance of the above three worldviews and the reality of our daily lives constructed by the original and the new media? Is this not borne out by our growing use of a vocabulary that we apply equally to ourselves and our machines – implied by cognitive psychology, artificial intelligence, and deep learning?

Since we are suspended in our current secular myths, we have no access to the general relativity of human life and the world, with the result that we are so trapped in our reality and the domains of our disciplines that it is next to impossible to imagine that for most of human history, what we take to be self-evident would have been completely unthinkable, unknowable, and unliveable. We have re-engineered ourselves as a consequence of "technology changing people" followed by "technique changing people" from a symbolic species into, first, a *homo economicus* and then a *homo informaticus*. We have remade ourselves in the image of our most important and valuable works, and have thus renamed who we are.

All this leads us to even more troubling questions. We seem to have created a global civilization that suffers from the collective equivalent of short-term memory loss because we appear to have fragmented the living of our lives by scanning in separate and distinct moments by means of the radar of our other-directed personalities, thereby constructing a reified life superimposed on a dialectically enfolded life built up with a highly desymbolized primary and secondary frames of reference. This could explain our creating a new niche for humanity that provokes all manner of new forms of anxiety and depression, new kinds of mental illness as well as new physical diseases such as immunological disorders.

Have we created a civilization that tends toward dealing with time as a continuum of space by dividing it into time-objects that cannot evolve and adapt? Our civilization appears to treat language as images and thus has trouble symbolically situating life in time. This development would imply that for a non-trivial number of people, life is lived with an architecture associated with seeing, images, and discipline-based approaches rather than an architecture developed through symbolization, language, and culture. Is this the beginning of a development that will result in our becoming *homo informaticus*? In any case, we have lost a body of common sense or a culture that functions as a shared symbolic way of making sense of and living in the world. Consequently, we no longer live in societies but in their diametrical opposites – what we have referred to as anti-societies. Does this mean that we are heading for living as Robinson Crusoes in the "filter bubbles" that the new media help us to create and adapt for ourselves to match our other-directed personalities?[13] Is the struggle between technique and culture all but over in our daily lives? Will these developments plunge us into a kind of secular hell: the impossibility of viable and sustaining intimate relationships? Are freedom and love to be the collateral damage of technique?

Absolute and Relative Efficiency

The above questions are not at all designed to trap us into a paralysis of hopelessness. On the contrary, they should encourage us to get involved in the struggle between technique and culture that rages in all our lives. To illustrate what is involved, let us return to our previous observation: that the technological dimension of cultural mediation of traditional societies was generally appropriate to their cultures and also sustainable by their local ecosystems. Today, technology is constructed as a branch of technique. What this discipline-based technology can and cannot achieve for us can be induced from what future engineers learn about the way technology influences human life, society, and the biosphere, and how this understanding must be put to use in order to design and build technologies that achieve their goals, while at the same time preventing or greatly minimizing collisions with everything non-technological. As noted, the answer was established by an exhaustive investigation of undergraduate engineering education, which scored all its components by means of two research instruments exploring how well engineers learn to take responsibility for the consequences of their design and decision making.[14] The scores imply that learning engineering is like learning to drive a car with all its windows covered over with paper, to ensure that the driver's undivided attention is directed toward its performance as indicated by the gauges on the dashboard. The undergraduate curriculum all but prevents future engineers from learning to look out of the windows of their profession to observe what is moving down the road of human history along with technology. We have noted the costs we incur in the many collisions with our economies, societies, and the biosphere that technology produces.

The results of this comprehensive study of undergraduate engineering education were entirely predictable by the theory of technique as advanced by Jacques Ellul decades earlier.[15] The curriculum is almost entirely devoted to discipline-based approaches, both within the technical core and in the so-called complementary studies component that deals mostly with the social sciences (with an occasional humanities course). The consequences for the personal and professional lives of future engineers have been thoroughly examined by Benson Snyder.[16] These findings also confirm Ellul's theory of technique. The curriculum is dedicated to the pursuit of absolute efficiency, and almost entirely ignores a potentially complementary pursuit of relative efficiency. The above two questions posed of undergraduate engineering education may be reinterpreted as measuring the likely ability of future engineers to secure the best relative efficiency of everything they design, build, or manage. Relative efficiency evaluates the contribution made by a technology to human life, society, and the biosphere. An investigation of the best achievable relative efficiency involves the establishment of a design exemplar of a technology that deals with the way this technology fits into, interacts with, and contributes to its contexts.[17] It deals with its "external" excellence or lack thereof in the appropriate architecture – that of the general relativity of human life and the world. In contrast, the discipline-based approaches provide the analytical exemplars by which the details of the internal architecture of the technology can be analysed and optimized to produce the greatest absolute efficiency within the boundary conditions specified by the design exemplars. Consequently, the interdependence between this relative efficiency and the absolute efficiency is established by the design exemplar, which provides the boundary conditions for solving the equations involved in the analysis and optimization based on the analytical exemplars. This complementarity ought to have been clear and obvious, but it is made unthinkable, and in any case impractical, by our current secular myths. According to these myths, the discipline-based approaches for analysing and optimizing the technology have no limits; thus the technology is good in itself. Any collisions it creates must therefore be dealt with by end-of-pipe approaches that add mitigation devices and services to deal with the collisions after they have taken place. Of course, such end-of-pipe approaches will endlessly add additional costs to the technology, even with our minimal labour, social, health, and environmental standards. Consequently, competition is increasingly based on the option of moving to those of the parts of the world where these costs can be externalized to the greatest extent possible. In sum, our secular myths, like all the traditional myths before us, condemn us to a path of ultimate self-destruction.

Can we generalize this brief discussion of technology to apply to technique? It may be argued that the disciplines of technology are to technique what the discipline of technique is to discipline-based science: the model and gold standard.[18] As a result, undergraduate engineering education may well be an excellent window into what we may expect from all discipline-based professional education, and university education in general.[19] The pursuit of absolute efficiency will do nothing else but multiply the collisions between our global technical order and our national political orders on the one hand, and human life and the biosphere on the other hand. We may also confidently predict that as long as we bow to our current secular myths, even as the collisions grow in number, scope, and intensity, we will interpret this as evidence that we need to pursue absolute efficiency with even greater diligence. Any symbolic equivalent to negative feedback is excluded by the same secular myths: What they declare to be our ultimate good could not possibly bring about all these collisions; and we must therefore use this good more diligently to get ourselves out of our predicament.

We must never forget that myths, whether traditional or our current secular ones, are self-evident only during a historical epoch, after which they vanish like the mist under the sun, to the point that they can never be lived again. Consequently, in principle it is not very difficult to transform undergraduate engineering education and thus all other discipline-based education. I may speak from forty-one years of experience. As I approach the end of my involvement (well beyond my official retirement), it is difficult to know if anything will come of it. It is true that the development of preventive approaches to complement our discipline-based approaches has received a great deal of recognition, but this is possibly the easiest way in which the "system" is able to claim to be on top of things while at the same time changing nothing fundamental. However, tens of thousands of students have been presented with an alternative to the conventional approaches to engineering. Roughly a quarter of them referred to this experience, on course evaluations and other communications, as life-changing. Another quarter found the material very stimulating, but dozens of other courses going in the opposite direction soon washed away any enduring effects. For the remaining half, it was an interesting spectacle that was neither good nor bad but something you simply had to pass to graduate. In other words, the situation is far from hopeless as long as we recognize (as a former colleague and friend put it) that we can sow new seed but we cannot make it grow. We are engaged in a kind of spiritual battle because we are dealing with forces and powers capable of alienating and reifying our lives. We can respond to these forces the way all religions did and continue to do: sacralize them as the traditional gods or the limitless entities that rule our lives. Alternatively, we can accept that in human life and the world everything is related to and evolves in relation to everything else, with the result that nothing can be absolute or limitless.

If we take the latter position, the most effective stand we can take against the rule of our secular myths over our lives is to attempt a genuine realism – by symbolizing our discipline-based approaches in science and technique and our limitless politics of the state-nation, in order to discern what these can accomplish for us and what they will never be able to achieve. Doing so will weaken the hold our secular myths have on our lives, with the result that the fruits of our labour will be a measure of freedom relative to the forces and powers that rule our lives and our planet.[20]

If such efforts spread and become mainstream, we would usher in the beginning of an entirely new epoch in our history. It would result in the reorganization of our universities to make room for intellectual and practical approaches based on mapmaking with a preventive orientation, as opposed to the conventional ones based on optimization and absolute efficiency.[21] As the graduates of these universities diffuse such approaches into all our institutions, we would be initiating an entirely new kind of "people changing technique" and thus a new "technique changing people" in a self-reinforcing way characterized by a continuing and sharp reduction in all the collisions that provoke our many crises.[22] It is essential to point out that doing so will not lead to any utopia of the kind promised by Karl Marx, where all alienation would cease.[23] All I can be certain of is that our secular myths would be dethroned by others, which would assume the throne during a new epoch of our history. However, it is difficult to imagine a force capable of systematically reifying human life and society that would be more powerful than our discipline-based approaches.

It is essential, therefore, that we recover the world we have lost according to our three political visions of what is happening to us. It is urgent that we establish something equivalent to the culture-based connectedness, which could act as a shared intellectual base map to replace the one we lost by the introduction of the technological, economic, and (later) the technical order, as a consequence of their having the diametrically opposite architecture to that of all prior cultural orders. We need to engage whatever symbolic capacities we have left by beginning a major endeavour to resymbolize our lives and our world.

We may then be able to understand the many contradictions in our lives, including the one between our economic worldview and our lives, that we ignore at our peril. Imagine a typical suburban couple with two children who are deeply concerned about the possibility of one or both partners losing their jobs as a consequence of so-called free trade agreements. These amount to the "forced trade" of their jobs because their employers are unable to compete with others who can externalize their social costs. This applies to an even greater extent in low-wage countries with authoritarian governments that are willing to sacrifice their people in return for technical and economic powers. In order to lower the chance of being laid off, both partners are voluntarily working longer hours. Their fatigue levels increase to the point that they are unable to recover their resources to a sufficient degree.

One night, one of the partners is involved in a car accident. He needs a great deal of medical attention, and the car is a write-off and needs to be replaced. As a result of such incidents, the GDP increases. With this anxiety and tension, the two partners get more easily embroiled in arguments over money and how they will deal with one or both of them being laid off. The financial worries undermine their relationship to the point that they decide to go to counselling to save it, and once again, the GDP goes up.

Because of the tensions at home, one of the children begins to "act up," and it is not long before she "hangs out" with the wrong crowd and gets into trouble. She no longer listens to her parents, who suspect she is also taking drugs. Feeling completely impotent, they ask their daughter to accompany them to counselling, and again the GDP goes up. The son insists that he wants to drop out of school and live on his own. He also ignores his parents, who feel that they have no recourse other than to enroll him in a private boarding school, since he is still underage, and they have the legal power to do so. Once again, the GDP goes up. The partners themselves are deeply affected by the home situation. One of them needs anti-depressant drugs and the other decides to try sleeping pills to get a better night's rest – and again the GDP goes up.

Eventually, the situation becomes so desperate that the parents decide that a major intervention is required in order to save the relationship between them and between them and their children. They resign their jobs, purchase a small property just outside of town, and begin to grow organic vegetables for the local market. To their surprise, the kids respond really well: but now the GDP takes a big hit because they only need one car, require no counsellors and less medication, and grow much of their own food. The message is obvious, and I am certainly not the first to explain this.[24] Collisions of all kinds are a boost to the GDP but a power of destruction of our lives, our families, and our communities.

This raises the question of how we *live* these kinds of contradictions. In a pre-industrial traditional society, such contradictions might have given rise to rebellions or even revolutions. For the last hundred years, the rise of technique has led to very different kinds of phenomena. As noted, it put enormous desymbolizing pressures on all cultures, to the point that they needed to be complemented by what we have referred to as integration propaganda, diffused by the traditional media and later reinforced by the social media. These media constructed an almost complete substitute for the symbolized culture-based connectedness of a society, in the form of a reality that was scanned by the dominant other-directed personality to deal with the situations of daily life. Doing so required that a tertiary frame of reference be developed and that it would overshadow the roles of the primary and secondary frames of reference. Nevertheless, there remained a fundamental incompatibility between this reality and what little remained of the culture-based connectedness of our personal lives. The former is built up by means of all manner of techniques that are separated from experience and culture and practised by specialists of all kinds, who are suspended in a triple abstraction. In contrast, what little remains of the culture-based connectedness is embedded in experience and culture and thus has a diametrically opposite architecture leading to irreconcilable differences that were taken care of by the first and (later) the second generations of secular myths in favour of technique and everything that led up to it.

All this has been explained by the theory of technique, which includes the techniques of integration propaganda to complement any highly desymbolized culture.[25]

However, for most of the last hundred years, the contradictions between economic reality and the economic experiences of people's daily lives have been so great as to require additional political techniques. In western Europe and North America, except for a few decades following the Second World War, democracy has been constantly endangered by the political Right, which, in one way or another, gave itself the ultimate and exclusive legitimation based on these secular myths of these societies. Have we forgotten how, in Germany, the leading industrial power of that time, unemployment and inflation had created such a crisis that Hitler could take power, and that many captains of industry as well as politicians in a great many countries admired his "economic miracle" of putting Germany back to work and bringing inflation under control within two years? Have we forgotten the rise of fascism in other countries? Have we forgotten what happened after the failure of post-war Keynesian economic policies because of the proliferation of collisions as a result of technique? Although the economic anomie was certainly less extreme, the contradictions were strong enough that the proposals of some politicians became thinkable, believable, and politically decisive. First, there was the political economic fundamentalism that swept through Western civilization. It was pioneered by Margaret Thatcher and a whole string of faithful disciples that included a number of American presidents and a rather resilient Canadian prime minister. Have we forgotten the "axis of evil," Star Wars, and a host of wars fought for political reasons?

By the time this political economic fundamentalism had run its course, it had paved the way for something else that was even more dangerous. The "cultural soil" was prepared for a new populism that grounded itself in our secular myths by the extreme self-justification of a "might is right" approach to dealing with everything, and by dismissing all opposition as "false news" and the actions of the enemies of the nation. Complete political polarization results from these secular religious attitudes, which divide "believers" from "non-believers." This neo-fascism is a strange combination of a pursuit of absolute efficiency and power on the one hand, and a lip service to family values that are no longer *lived* by most people on the other. Democracy has become a secular political religion based on the secular myths of our society, with the result that no informed debate is necessary. Everything is true if you are on one side, and false if you are on the other side. There is no middle ground. The complexity of the kinds of issues we face today vanishes – if you are not for us you are against us. McCarthy did it all before.

We must remember that politicians must constantly legitimate themselves, and that the only ultimate legitimation can come from our present secular myths. They must intuitively bridge the gap between these myths and the life of the nation. When this becomes very difficult, extreme interpretations and solutions become almost irresistible, but it has been consistently the extreme Right, not the Left, that has threatened democracy from the "inside." There appear to be more and more politicians who cannot resist the temptation to take advantage of our current "cultural soil" in order to gain power. They are apparently unaware of the secular religious political forces with which they are playing and to which they will sooner or later fall victim, like all the other extremists before them.

The moment a movement denies the voice of everyone else in a nation by essentially turning them into enemies, everything is lost. There is an ultimate division and separation between parents, between parents and their children, between friends and neighbours. Dividing everyone from everyone else in this way unleashes a spirit of division, separation, and accusation that, in the Christian tradition, is referred to as demonic and Satanic. Many Christians have the strangest political bedfellows today.

Economics as Technique

How is it possible to resist this secular religious use of democracy and politics? If we refuse to fight fire with fire, we must question the legitimacy of what is happening as best we can through resymbolization. Our highly unique way of conducting our daily lives can in part be understood by the fact that every dimension of cultural mediation has been transformed into a construct organized and adapted on the basis of technique. The economy is but a microcosm of what technique is doing to individual and collective human life. Nevertheless, because of its importance, the economic construct is illustrative. Like all other discipline-based approaches, economic knowing and doing proceed as if the general relativity of human life and the world is of no consequence. It is this problem that makes economics structurally unscientific.

For economics to take on the form of an autonomous discipline, the category of economic phenomena it investigates must be autonomous relative to the categories of phenomena examined by all other disciplines. The absurdity of this possibility will be briefly demonstrated by reviewing its historical transformation with the four primary dimensions of industrialization, urbanization, rationalization, and secularization in terms of how the economic phenomena affected and were affected by almost everything else.

Since an economy can neither create nor destroy the matter and energy on which it depends, it participates within the biosphere as a subsystem with a radically different architecture. As this subsystem grew to an unprecedented level, it soon began to interfere with the services provided to all life by the biosphere, with the result that these could no longer be assumed to be infinitely available and had to be priced. This assumption led to a variety of market externalities that accumulated into Market forces that pushed the economy into a form that over-consumed the biosphere and under-"consumed" people, thus producing unemployment and underemployment. Entrepreneurs sought to reduce their production costs by primarily economizing labour, with much less attention given to matter and energy. The same Market force produced the environmental crisis. Since the discipline of economics cannot take these dependencies into account without surrendering its autonomy as a discipline, these developments were, and continue to be, ignored.[26]

The process of industrialization completely changed the dependence of economic phenomena on human knowing and doing. As every tradition for evolving a way of life began to break down, culture-based approaches dependent on the metaconscious knowledge accumulated through past experience were of little use in the face of the radically new conditions created by the comprehensive transformation of both technology and society through industrialization. There was thus a shift away from the dependence of an economy on culture-based approaches toward rational economic approaches, and eventually toward discipline-based approaches. Changes in this fundamental dependency went almost unnoticed, with a few important exceptions including Max Weber, John Kenneth Galbraith, and Radovan Richta, especially when we interpret them through the lens of the concept of technique.[27] The growing integration of more and more national economies into a single global technical order is unimaginable without their joint dependence on discipline-based approaches; but once again, this cannot be accounted for by an autonomous discipline of economics.[28] Nevertheless, discipline-based approaches to knowing and doing now constitute what is surely the most influential factor of production, and certainly the one that mostly governs how capital is allocated. For example: the corporation cannot apply this discipline-based knowing and doing to the analysis and optimization of the design exemplar of any new product without recovering its huge investment from a large volume of units, thus necessitating mass production and mass consumption. The simultaneous planning of the design of the product, the design and construction of a dedicated production facility, and a sales strategy that included extensive consumer testing created an even greater reliance on discipline-based approaches as well as the impossibility of relying on markets on both the "input" and the "output" sides of the corporation. The Market economy was thus transformed by a great deal of planning and a minimal reliance on markets. In order to stabilize these new economies, governments had to get involved. They did so by means of Keynesian economics that prevented market failures for decades, for the first time in the history of industrial civilization. However, Keynesian economics was unable to deal with the effects of technique on the economy. This reopened the door to neoclassical economics and monetarism, which could not deal with it either – as is evident from our present situation.

One particular way in which modern economies rely on discipline-based approaches has also been overlooked. By using very sophisticated computer algorithms, banks and other large investors have created an entirely new category of economic activities that "produce money from money" without the usual intervening economic activities. For years, observers such as Noam Chomsky, Herman Daly, and Bernard Lietaer issued warnings about the tremendous growth in the speculative bubble constituted by these new activities, which have vastly increased the money supply *ex nihilo*. These were estimated to account for some 97 per cent of all financial flows produced around the world each and every day – a fact that even the monetarists have entirely overlooked.[29] Wealth can now be produced from wealth, which is multiplying inequality beyond anything seen before.

Nor did the discipline of economics take note of the sharply rising social costs resulting from the use of technique that the new economies imposed on their societies. Keynesian economic policies failed because governments were no longer able to achieve budget surpluses during times of economic growth – money that was required to finance the expansions of the public sector to offset the contractions in the private sector during difficult economic times. The reason was obvious but again largely overlooked: wealth production was increasingly undermined by the growing costs incurred in the production of that wealth. Very few economists noted the effects of technique on the economy that could have made this entirely predictable. It led to the disastrous throwback to monetarism.

As an autonomous discipline, economics also could not grasp the dependence of economies on "human nature." As people changed the economy, this economy simultaneously helped to change people. The *homo economicus* that emerged in the first generation of industrializing societies paved the way for an entirely new kind of "human nature" that began to emerge in the second generation of industrializing societies. These have also been referred to as mass societies characterized by mass production, mass consumption, and mass advertising. The "consumer" was very different from a *homo economicus* by having almost insatiable needs that could be moulded to what was technically possible and thus socially desirable, according to the second generation of secular myths. With the rise of anti-societies, their other-directed personalities began to take on the character of a *homo informaticus*. Once again, this had a fundamental influence on the economic behaviour of the members of these societies.[30] Moreover, the discipline of economics could not explain in economic terms the reversal that gradually took place in the relationship between the economy and society within a lived unity.

In all traditional societies, the culture-based connectedness had ordered the technology-based connectedness by enfolding it into the social fabric. However, the first generation of secular myths implied that humanity had been searching for meaningful and good lives on this planet by going in the wrong direction. Priority had to be given to the technology-based connectedness because the resulting material progress would inevitably produce social progress and happiness. The material world was thus endowed with non-material powers, and a materialistic "human nature" was born. As a result, the economy no longer needed to be governed by cultural values. What was required was so self-evident that everyone could agree on it: A steady improvement of the technology-based connectedness organized by the Market would ensure an almost limitless goodness. This created the illusion that every society had an economic base or core, as proclaimed by both Karl Marx and Adam Smith – in ways that were fundamentally not as different as we commonly presuppose, relative to the human history that went before.[31] The successive new "human natures" that accompanied all this were the result of technology and the economy changing people in such a dramatic way that it would have been inconceivable without the new secular myths. Moreover, without these changes, the highly significant role that needed to be played by mass advertising relative to mass production and mass consumption could not be understood, because it was an integral part of the emerging integration propaganda that was required to supplement the increasingly desymbolized cultures and thus their ability to give meaning, direction, and purpose to the lives of their members.[32] This integration propaganda associated products with symbols, and later with the logos of corporations backed by limitless technique, with the result that they became endowed with some of the characteristics of these symbols and technical powers. Material objects and services could thus bestow social and even spiritual benefits on people's lives, as was clearly shown in the advertising of those days.

As the new secular myths destroyed the myths of all traditional societies around the globe, all economies became oriented in the same direction, and were thus classifiable as so-called underdeveloped, developing, or developed, which makes no sense whatsoever without these new myths.[33] As a result, these new economies were thus integrated into the global economy according to their technical success and the powers this bestowed on them. In addition, this integration of national economies paved the way for free trade agreements. These permitted corporations to relocate their production where they could pay even lower wages and significantly increase the social costs they could externalize because of differences in labour, health, social, and environmental standards. Trade was thus beneficial for corporations but not for the people who had to adjust to this massive economic reorganization of global production enabled by these free trade agreements. The new secular myths made it possible for the global technical order with its constituent economies to coexist with politically sovereign state-nations.[34]

From this brief overview, we may conclude that if we embarked on a comprehensive resymbolization of economic phenomena relative to all other categories of phenomena during the last two hundred years, any plausibility of the autonomy of any of these categories of phenomena would surely disappear, and with it the scientific validity of all the corresponding disciplines, especially that of economics. The resymbolization of economic phenomena relative to all other important categories of phenomena would take us back to economics taking on a form not unlike that of political economy, which examined economic phenomena against a background of all other phenomena with which it interacted.

In the meantime, our pseudo-science of economics and its apparent support for an economic worldview remains the only economic game in town for most of our politicians, especially those who may be referred to as market fundamentalists, monetarists, or populists. The opposition voices a deafening silence because there is no adequate scientific understanding of our economies that could sustain an adequate critique, or any alternative proposals based on it. If there ever was a sign of our economic and political alienation, surely this is one, given the vast influence all this has, especially on the weakest and most vulnerable portion of humanity.

A few additional issues need to be briefly highlighted. Possibly one of the most fundamental human implications of most of our economies is their contribution to the production of inequalities. It is difficult to overstate the devastating and widespread consequences of high levels of social inequality on human life and society and thus for the economy. The economic costs are enormous, as any comparisons between the most equal societies (including Finland and Japan) with some of the most unequal societies (including the United States and Canada) have demonstrated.[35] It may well not be an exaggeration to suggest that the best economic policy we could pursue is to lower the costs we incur in running our societies by making them more equal, thus reversing much of what is currently happening. Such a reversal would reduce all manner of "us versus them" attitudes in favour of more inclusive and cooperative approaches and may thus pull us back from the brink of complete political polarization in North America. At present, there is a complete lack of understanding that what divides us is dwarfed by the crises we face. We are electing politicians who have no idea what to do about them, other than to stay in power by a divide-and-conquer strategy. Strangely enough, such politicians would not be in power were it not for the support of the Christian Right, which has its own agenda: to impose a few ideas of its own by means of political power if necessary. Apparently, the ends legitimate the means; but all this gives a whole new interpretation to loving your neighbour as yourself. It is the result of the fact that Christianity is firmly in the grip of the spirit of our age.

In order to understand this spirit of our age relative to the problem of growing social inequality, we need to briefly return to Adam Smith. He was probably the first person in human history to proclaim that, under the circumstances of his day, the self-interested behaviour of wage earners and producers would create the best possible world for most people thanks to the great invisible hand of the Market – to which he refers only once.[36] All the other Market forces, which were resulting in accumulations of the many market externalities, are not mentioned. Hence, this description of a Market economy is disastrously incomplete, but we have never learned the lessons from this unfortunate oversight. These Market forces are capable of pushing an economy in directions that were unintended and even unknown by their participants, with the result that self-interested behaviour accomplished what all religious traditions, without exception, have always warned against: self-interested behaviour destroys the fabric of any community as well as its ability to live justly with other communities, and endangers all life in the biosphere. There is every reason not to endow a Market economy with the magical powers of providence and an ability to set things right as long as we do not interfere with "natural" markets. Such confidence in these beliefs is a testimony to the fact that members of the first generation of industrial societies were enslaved to their secular myths.

The problem of self-interested behaviour has reached a new summit in our days, thanks to technique. First, technique, by desymbolizing all cultures to unprecedented levels, has decisively weakened all remnants of traditional moral and religious attitudes capable of offering any resistance to whatever is becoming a technique, because it comes with fewer and fewer limits. Second, as I have noted, when discipline-based technique is interpreted as the most significant collective behaviour of humanity, then this conduct is psychopathic and sociopathic in character to a degree that would not be condoned in a person. As further noted, technique seeks to achieve the absolute efficiency of any transformation of inputs into a desired output, without any consideration being given to the general relativity of human life and the world from which these inputs are obtained, and to which the desired output is returned. It transforms the general relativity of human life and the world in a piecemeal fashion, one transformation, one technique (and its associated discipline) at a time, thereby distorting this general relativity as if it were expressible in terms of mathematics.

One of the fundamental structural and organizational characteristics of technique as one of the most significant collective actions of humanity is to disregard the consequences it has on human life, society, and the biosphere. This psychopathic and sociopathic characteristic is intrinsic to the structure and organization of technique, even more than it is the result of the intentions of powerful actors because of their alienation by technique. Everyone involved in technique on a high level, especially the senior executives of large corporations, has no choice but to develop these psychopathic and sociopathic tendencies in their personalities – but that does not absolve these people from their responsibilities.[37] Their potential for doing so is enormously facilitated by their other-directed personalities, which scan the relevant aspects of the reality that surrounds these people in order to discover what is normal, which is then taken to be normative. If all senior executives behave like this, why would aspiring executives not behave this way and attempt to do better in order to be promoted? If they allowed any personal values and convictions to interfere with their psychopathic and sociopathic development, they would never reach the top. Once they get to the top, enormous salaries and bonuses with an array of other benefits will ensure that there will be no moderation of their behaviour, which is ultimately sustained by our secular myths. All this is made liveable and is sustained by these myths. Even if all senior executives woke up tomorrow as saints, the pursuit of absolute efficiency and technical power embedded in their methods and approaches would continue, because they know no alternatives.

For the psychopathic and sociopathic behaviour of these powerful actors to be successful, everyone else in our contemporary anti-societies also needs to be possessed by the spirit of our age. When people can save a few dollars by doing business with the "Wal-boxes" of this world, we do so. We thus become their accomplices in exporting the jobs of our neighbours to countries where corporations do not need to pay living wages, benefits, and pensions and can operate to the lowest labour, health, social, and environmental standards. We become accomplices in the creation of the "new" economy, in which most of the new jobs are part-time and have few benefits, and certainly no pensions. We thereby deprive our neighbours of their livelihood, and when they are too old to work, of a humane retirement. Our politicians are equally possessed by this psychopathic and sociopathic spirit when they negotiate new free trade agreements that export the livelihood and future of still more people. We have essentially become a nation of psychopaths and sociopaths, jointly making the world a worse place for everyone.

Many box stores have withdrawn a great deal of business from the downtowns of small communities, to the point of threatening or destroying their viability. In larger communities, these box stores undermine and destroy the liveability of streets and neighbourhoods because they help to reduce the reasons people have for being out on the sidewalks, which is essential to enable strangers to live together in these cities.[38] We are all familiar with the many business practices that cater to our psychopathic and sociopathic tendencies. We respond to sales that all too often are not sales at all, because the goods on offer have been specially produced to be sold at this lower price and thus at the expense of their serviceability and durability. As a result, such sales help to increase the throughput of matter and energy necessary to deliver the goods and services for maintaining our way of life.

This kind of psychopathic and sociopathic behaviour is supposed to be the only way a person can get ahead. Instead, the collective results ensure a world in which we are all worse off. It explains why more equal societies with more cooperative behaviour are much less costly to run while significantly improving the quality of human life.[39] While the evidence for this is overwhelming, it is not about to sustain significant political changes. As a consequence of "technique changing people," we live life relative to reality as opposed to the culture-based connectedness that joined individual and collective human life in the past. The political implications are decisive and far-reaching. A dialectically enfolded culture-based connectedness is capable of sustaining a sense of "being in it together" and of seeing the positive and negative aspects associated with any complex issue as being indissociably linked. As a result, it was once much easier for politicians to cultivate a climate of tolerance and respectful disagreements with people that had opposing interpretations and convictions. It was once possible to believe in a common and public good. The reality relative to which contemporary life is lived tends to sustain the diametrically opposite tendencies. Because of its simple complexity, aspects of complex issues can be added or subtracted at will and reconfigured in almost any possible form to suit the techniques of public relations and integration propaganda. It has also opened the door to carrying our psychopathic and sociopathic spirit to the extreme in a new kind of politics.

In addition to being sustained by a new kind of "cultural soil" made up of what we refer to as reality, our limitless politics has taken on a secular religious character centred on the secular sacred of the state-nation. Once again, it is the extreme Right that has intuited the situation very well and turned it to its own advantage, with great success. There has been a sharp rise in fascistic tendencies as the extreme Right is turning itself into the religious zealots of the new system. They are convinced that they, and they alone, can protect the state-nation from all those who threaten it simply by disagreeing with them. They wrap themselves in the self-righteousness of absolute efficiency, the limitless powers of the market, a "might is right" approach, and an unprecedented rise in psychopathic and sociopathic attitudes toward the most vulnerable and poorest members of society. It is this latter feature that truly unmasks them for what they are: their total devotion to the technical spirit of our age makes everything else simply collateral damage. What is so shocking is that, in North America, the political extreme Right is overwhelmingly supported by conservative Christianity. This shows the extent to which the latter is in the grip of the spirit of our age. They claim to follow Jesus Christ, who has told them that whatever is done to a child (representing extreme vulnerability) is done to him! In the meantime, the extreme Right can claim to have God and history on its side. This also explains why the political centre could not hold. It was unwilling to go to the secular religious excesses of the extreme Right; and this unwillingness to fight fire with fire proved to be its short-term downfall, but its long-term victory. However, if we think back to the rise of fascism in West European so-called Christian societies, we can safely predict an extremely high casualty rate. All of this could have been perfectly predictable if we had symbolized technique and the state-nation in terms of what they can truly deliver for us, where their limits lie, and when to turn to other approaches if we wish to transcend these limits. In this sense, there is nothing new under the sun. Our secular religious attitudes to technique and the state-nation will be extremely costly to ourselves and our planet. A limitless politics can exist only in relation to the limitless state-nation that it serves; and the use of politics beyond its limits will be disastrous. It will not only condone our psychopathic and sociopathic behaviour but will also reward it in an ultimate struggle for power.

Returning to economics as a microcosm of technique, there is another important reason why economics can never take on the form of an autonomous discipline in a genuinely democratic society. It would be impossible to conduct experiments on such a society to verify economic theories. If economists wish to be responsible and accountable, they must engage in the following behaviour, which includes the symbolic equivalent of negative feedback. First, economics must examine the meanings and values of economic phenomena relative to those of all other phenomena, and thus to their meaning and value for human life and society. Second, economics must assess the direction in which an economy is evolving in order to determine whether this is in accordance with the values and aspirations of the society. Third, economics must support interventions in this course of events by means of economic policies that will redirect this course to bring it closer to these values and aspirations. Fourth, it must compare the results of such interventions, in terms of what this economic understanding would have predicted as the outcome, with what actually happened. If there is a significant gap between the two, economics must seek to explain it either as confirmation that this understanding is deficient in some important way or as showing that the policy intervention was faulty in its translation of this understanding into action, or both. Finally, based on an adjusted understanding, a new intervention must be devised to correct the situation in an ever ongoing cycle of such steps, which make up what is known as a praxis. If economics is such a praxis – and this is the only way it can test itself – then it can never be an autonomous discipline because it must be open to everything that is happening in human life and the world, including a society's values and aspirations that presumably guide its way of life. Politics would have to become a great deal more cooperative for this to work. At present, every government appears to be ready to dismantle a great deal of what its predecessor accomplished under its mandate – a mandate typically based on a minority of the popular vote, and thus as a dictatorship of the minority.

Finally, as noted in the previous chapter, discipline-based science begins by translating everything it studies into an architecture that can be expressed with mathematics. It thus multiplies the mathematical games for the scientific understanding and manipulation of everything. The discipline of economics has possibly been more successful in following this path than any other social science. It may even be argued that it has become something of a gold standard for playing these mathematical games. Economics begins by translating what it studies and the results obtained into mathematics, as if human life and society have the corresponding architecture. Next, these findings are mathematically mapped in the domain of its discipline. Consequently, human life and society are examined exclusively in terms of the single category of economic phenomena. This mapping is then turned into a variety of highly simplistic linear mathematical models that can be optimized to illustrate the absolute efficiency that could be achieved by imitating the model. Doing so involves a return to what was originally studied in order to reorganize it in the image of the model demonstrating the possible absolute efficiency.

Many economic games used to simulate all kinds of strategies are based on input-output analysis, which represents the inputs and outputs exchanged by the industries of a society. Although originally invented for an entirely different purpose, to which it was well suited, it is a highly simplistic linear model that resembles the economies of the industrially advanced nations less and less as a result of the above kinds of effects on the economy, not the least of which is the growing global integration. Worse, it has been extended to many other applications; for example, to life-cycle analysis in engineering. Engineering has a perfectly good alternative to this pseudo-science, referred to as Design for Environment, but as a non-discipline-based approach, it is rarely taught. These kinds of extensions take over the highly unscientific character of the original model and make it even worse. We play all manner of economic mathematical games based on this input-output model and its extensions, instead of attempting to gain a good understanding of how economies work, including their dependence on matter and energy borrowed from the biosphere, and on economic activities that are integral to a society. This once again represents our servitude to absolute efficiency and technique, causing incalculable harm to human life and our planet. I cannot resist once again giving the final word to the Nobel prize–winning economist Wassily Leontieff, who put it so well:

> Page after page of professional economic journals are filled with mathematics formulas leading the reader from sets of more or less plausible but entirely arbitrary assumptions to precisely stated but irrelevant theoretical conclusions … Econometricians fit algebraic functions of all possible shapes to essentially the same sets of data without being able to advance in any perceptible way a systematic understanding of the structure and operations of a real economic system.[40]

Since in our daily lives in general, and in our economic behaviour in particular, we no longer benefit from the equivalent of a shared intellectual base map, we are incapable of understanding the relative meaning and value of our knowing and doing, and this removes the possibility of being able to take responsibility for them. Every member of an anti-society is assisted in creating a personal world with their "friends" and their "reality" as a collage of their preferences of goods and services, music, news items, recreational activities, and everything else that they deem necessary to enrich their lives. Doing so further undermines what little remains of the dialectical tensions between individual differences and of a cultural unity that was essential for the evolution of traditional societies in the past.

Our Daily Lives and the Professions of Technique

We will briefly discuss two other professions that deeply influence the daily lives of all of us. Engineering has been chosen because technology in its entirety has an architecture expressible in terms of the principles of non-contradiction, separability, closed definitions, openness to measurement (and thus to quantification and mathematical representation), and a simple complexity. Discipline-based approaches were thus ideally suited to the "world" of technology; but today's technology is a branch of technique, and thus fully shares its incompatibility with human life, society, and the biosphere. We will need to retrace our steps to briefly explain how this happened. The second profession to be briefly commented on is that of medicine because what it deals with, namely, human health and illness, needs to come to terms with an architecture that is entirely unsuited to discipline-based approaches because of the enfolded and dialectically enfolded architectures of human life.

How did engineering move from an almost exclusive focus on the relative efficiency of any technology in a traditional society to an almost exclusive focus on the absolute efficiency of a technology, coupled to an end-of-pipe approach for dealing with the problematic effects of its relative efficiency? This question is paradigmatic of humanity's shift from a complete reliance on culture to a nearly complete reliance on technique. The technologies of traditional societies were the embodiment of the general relativity of human life and society as interpreted by its culture and grounded in its traditional myths. In contrast, the engineering techniques and the technologies they produce in our contemporary anti-societies are a denial of this general relativity. This shift was manifested by essentially abandoning technologies with the best possible relative efficiency to instead focus on the analysis and optimization of any technology by means of discipline-based approaches in order to ensure its absolute efficiency, to the detriment of its relative efficiency. In other words, traditional technologies were designed to fit into and enhance individual and collective human life, while our contemporary technologies are designed to strengthen the technique-based connectedness of our anti-societies in a piecemeal fashion by improving the absolute efficiency, one technology, one discipline, and one category of phenomena at a time. Traditional technologies were focused on human life, while today's technology (as a branch of technique) is the piecemeal improvement of absolute efficiency in denial of that life. It provides no equivalent of a culture-based connectedness in our lives.

How did we back ourselves into this corner, having the best intentions and what often appear as the best possible decisions in the world? I will briefly revisit this journey to remind ourselves that the most significant steps in it involved decisions that are difficult to quarrel with in terms of their making sense at that time. These decisions seem to have overcome the significant problems and limitations characteristic of that time, thus making it difficult to muster significant opposition. It is what human life grounded in myths is able to accomplish: to make self-evident, clear, and obvious what ought to have been the contrary, as becomes evident from a later vantage point grounded in different myths. Does this mean that the first generation of industrializing societies could not help backing themselves into a situation that potentially could threaten all life and the planet? I believe it was avoidable, but only under one condition – and this is not likely to please many people. The condition is that these societies should have done everything in their powers to avoid making anything absolute, limitless, or self-evident. As embodiments of the most influential religious traditions, the Jewish and Christian communities were potentially in a position to understand this and to act on it. However, they had become impotent by being split into their conservative and liberal branches, each of which was perfectly assimilated by what was happening.

Let us briefly sketch the beginning of what may be considered as an intellectual base map of how humanity became increasingly united in its pursuit of a global civilization. It may be interpreted as a long preparation for dealing with the general relativity of human life and the world by relying primarily on technique and only minimally on culture. As noted, the introduction of a technological order had to be regulated by an economic order to fit into the cultural order of an industrializing society. This involved the introduction of making sense of and living in the world by means of an architecture implicit in the visual perception of the world, rather than by the architecture implicit in the traditional approaches based on symbolization, language, and culture. It soon became evident that the cultural approaches of these industrializing societies were ill suited to adapting and evolving these new orders because their architecture was the diametrical opposite of that of the cultural order.

At first, this appeared to be related to the extent to which context had to be taken into account. In the technological and economic orders, the context could be restricted to the local technology-based connectedness of machines and factories. The context beyond the technological and economic "world" could be restricted to those inputs and outputs that linked this world to the broader technology-based connectedness of society. Because of its simple complexity, portions of this technology-based connectedness could be added or subtracted without affecting what had been thought or done previously.

There seemed to be no point in taking into account the fullest possible context, as was necessary in using symbolic cultural approaches. What was hidden behind this self-evidence was the fact that this largest possible context had an entirely different architecture, one that was dialectically enfolded and thus lacking a simple complexity in which things could be added and subtracted at will. What was also hidden was that the shift from a culture-based approach to a rational approach also took a simple complexity for granted, imposing a goal and rearranging the context accordingly. Once again, this context tended to converge with the local technology-based connectedness, thus further reinforcing what appeared to be self-evident. Moreover, all this was so spectacularly successful in the analysis and optimization of the internal workings of machines, in the factories full of these machines, and in the economies dominated by these factories that there was no real contest between these rational approaches and their cultural predecessors. The new approaches gave far better results for the complete reorganization of the technology-based connectedness essential for industrialization.

Nevertheless, collisions began to occur with everything beyond this narrow context, because technology-based connectedness had always been integral to the cultural-based connectedness of any society. With the introduction of the technical division of labour, however, followed by mechanization and industrialization, this was radically changed. Human work was translated from being a dialectically enfolded activity of human life to a construct adapted to the technological and economic worlds. It was the beginning of the dismissal of the culture-based connectedness from human awareness in technological and economic activities; and this eventually became grounded in the first generation of secular myths. They implied that, with hard work, the pursuit of the technology-based connectedness to produce material progress would result in both social progress and human happiness. In other words: seek the improvement of the technology-based connectedness with all your mind and strength, and everything that humanity has always desired will follow, in a way that no traditional society could dream of, let alone accomplish. The technological and economic dimensions of cultural mediation were thus separated from all the other dimensions of cultural mediation. They would become the first two relatively independent constructs as the building blocks of an entirely new order.

The consequences were vast, and contributed to the collapse of traditional cultures everywhere. Instead of the distribution of goods and services being regulated as an integral part of a society's cultural mediation, this distribution was now taken care of in a piecemeal fashion, one good or service at a time, within the limited context of its own market and subject only to its supply and demand. These were brought into equilibrium by the mechanism of the market. Before anything could be traded in these markets, it had to be detached from the general relativity of human life and the world to become a commodity; and in the case of human life and the biosphere, it had to be reified as well. The simple complexity of the technological and economic constructs was thus assured. There was no longer any need to symbolize technological and economic activities as being dialectically enfolded into human life and society and enfolded into the biosphere.

Depending on their socioeconomic position in the first generation of industrializing societies, people's lives were torn to different degrees between participating in the new technological and economic "world" and continuing to participate in the cultural "world" of their society. Living their lives relative to the former was increasingly difficult because these lives working in the background contained few past experiences (and thus the metaconscious knowledge built up with them) that would permit these people to skilfully cope with the new situations they now encountered. Initially, people reacted in the only way they could: they had to think about the situation by relying on their culture-based reasoning. Doing so was very frustrating and did not produce satisfactory results, since they were unaware of the different architectures of these technological and economic constructs. With a little more experience, it became clear that this culture-based reasoning could be greatly simplified by narrowing the context taken into account, which essentially corresponded to the local technology-based connectedness. Moreover, dealing with these situations accordingly did not appear to be affected by what people had arrived at previously, or by what they might deal with next.

In a kind of empirical manner, the new metaconscious knowledge built up from people's experiences of their tentative approaches to these new situations implied that satisfactory results could be obtained by proceeding as if they were now dealing with a simple complexity open to piecemeal approaches. This engagement in the world contributed to the building of a very different kind of metaconscious knowledge, which led to intuitions that the immediate context had an implicit goal and that, once this goal was made explicit, it could be deliberately imposed on the situation to make more sense of it. The context could now be interpreted relative to this goal. Everything immediately relevant to achieving this goal thus became the foreground with everything marginally relevant becoming the background, thereby making everything else irrelevant. As a result, culture-based reasoning gradually mutated into a rational approach.

The consequences of thus rearranging the context by imposing a goal and acting accordingly were not immediately obvious because they began to build up much later and usually in very different areas. Hence, the rational approaches appeared to be superbly effective in all respects. However, there was one overriding limitation: to make sense of a situation in this way was based on the assumption that what could be observed by the senses was indicative of what was happening. In industry, this limitation was first encountered in the chemical and electrical industries. In the chemical industry, what could be observed was often not indicative of what was happening in a chemical reaction; while in the electrical industry, little or nothing could be observed about circuits unless there was a short or an extreme malfunction. What this limitation indicates is that, at this point in their development, these rational approaches remained embedded in experience and culture. In terms of our human development, what was happening gave rise to new clusters of experiences. These were symbolized by corresponding changes to the organizations of people's brain-minds that were differentiated from the clusters associated with their experiences of life lived beyond the technological and economic constructs in the remainder of society. This development did not constitute the beginning of a separation of knowing and doing from experience and culture.

Such a separation of knowing and doing from experience and culture implies a discontinuous development of the technological and economic constructs. When it happened, all economic trends were so deeply affected that the prediction of Karl Marx, that capitalism would self-destruct when competitive pressures would drive it against its own limitations to its breaking point, never occurred.[41] We have failed to appreciate that, according to all the economic trends that Marx had so carefully documented, the likelihood of a collapse of capitalism would have been a new certainty. Marx did not see this coming, and it made his theory of history collapse like a house of cards – but not without tens of millions of people losing their lives.

A new development began in Germany, where discipline-based scientific approaches were being applied in industry. Their success, attributable to rational approaches embedded in experience and culture, quickly yielded a predominance of German patents, especially in the chemical and electrical industries. In contrast, in England, these discipline-based approaches were not so readily adopted, and they did not exercise a significant influence until well after Marx's death. As a result, Germany rapidly took over as the leading industrial power, as its military performance during the so-called Great War demonstrated, because of the dependence of all its combatants on their industrial infrastructure. In France, the separation of knowing and doing from experience and culture produced much less of a discontinuity because, from the very beginning, its political centre regarded industrialization as an integral part of the building of a rational society. In fact, the vocabulary of the French language has not made any distinction between technology and technique.[42]

From a historical and sociological perspective, the separation of knowing and doing from experience and culture was thus more disruptive and easier to observe in England than it was in Germany and France, where it was much more integral to the transformations of technology and society. Because of our being a symbolic species, as people changed technology, technology simultaneously changed people. Hence, the discontinuity in the development of technology and the society into which it was embedded corresponded to a discontinuity in the way people's lives worked in the background.

As noted, in a traditional society, the organizations of people's brain-minds functioned as a kind of mental map, provided that we remind ourselves that this mental map is unlike any geographical map. This map symbolically represented the dialectically enfolded life of a society as uniquely experienced and lived by each individual. Moreover, it also dialectically enfolded the social self of the individual as its map reader. Consequently, technological doing mediated by the domains of the emerging technological disciplines represented a development within the organizations of people's brain-minds that was discontinuous with the development of the primary and secondary frames of reference, as was previously examined for mathematics and science. The emergence of the tertiary frame of reference was associated with the development of an entirely new kind of metaconscious knowledge, because the associated experiences have a foreground with a simple complexity that is thus discontinuous with the background with a dialectically enfolded complexity. Consequently, these experiences, the metaconscious knowledge built up with them, and their joint constitution of a tertiary frame of reference were an extension of the visual approaches for dealing with human life and the world, which had the diametrically opposite architecture from that of cultural approaches.

This extension of the visual approaches enabled the emergence of the mental equivalent of "scanning" the domain of a discipline. It would later develop into the radar-like behaviour of the other-directed person in a mass society whose highly desymbolized cultures needed to be supplemented by integration propaganda distributed by the mass media in order to provide people with adequate "meaning," "direction," and "purpose" for sustaining their lives. This support, now based on an extension of the visual approaches, replaced those based on symbolization, language, and culture. It was not until the permeation of children's lives by screen-based devices that these developments began to profoundly affect people at a much earlier age. The reality portrayed by the old and new media must be scanned by the tertiary frame of reference, and this capability superimposes itself on making sense of and living in the world based on the primary and secondary frames of reference. It is this latter capability that continues to be enormously desymbolized and curtailed at an ever earlier age. Nevertheless, the primary and secondary frames of reference, by retaining the dialectically enfolded character of human life, can only function with these lives working in the background. Whether this trend can continue indefinitely is as yet impossible to tell. However, it certainly is not inconceivable that eventually language will be treated more and more as built up with image-words. If this development occurs, *homo informaticus* may include personalities that may have difficulties situating human life relative to time.

The implications for "people changing technique" and for "technique changing people" are enormous. We have explained that the development of the technological and economic constructs within the first generation of industrializing societies, and the proliferation of constructs in the second generation of mass societies, may be interpreted (with a great deal of hindsight) as preparing for the information and computer revolution, as a result of more and more of individual and collective human life being treated as having the architecture of reality and of the domains of disciplines. The development of first the technology-based connectedness and then the technique-based connectedness of human life and society occurred at the expense of their culture-based connectedness ever since the introduction of the technical division of labour and everything that followed it. This steady, relentless desymbolization of the culture-based connectedness made the design of any technology in terms of its relative efficiency increasingly difficult. Following the separation of knowing and doing from experience and culture, engineering technology quickly became a branch of technique that concentrated on endlessly improving the absolute efficiency of any technology at the expense of its relative efficiency.

As noted, the study of current undergraduate education clearly showed that apparently, future engineers have no need whatsoever to obtain an intellectual grip on how technology influences human life, society, and the biosphere, and how to use this understanding to ensure that a technology enhances them by balancing its absolute (internal) efficiency with its relative (external) efficiency. All attempts to revive the traditional aspects of design dealing with the relative efficiency of technology have failed. So-called professional ethics has been reduced to essentially an end-of-pipe subject that does not get to the root of the problem: technology has become a branch of technique, and ethics can be learned after graduation to pass an ethics exam. When I was asked to rewrite the textbook for such an exam by one of the professional engineering organizations, I refused because it would first have to commit to restructuring the curriculum before a genuine ethics as a praxis could be contemplated. In the meantime, engineering design remains more or less a rational approach to stating a problem in a manner that facilitates the reorganization of something to open it up to analysis and optimization, in order to achieve its absolute efficiency.

The transition of many technologies embedded in experience and culture to technologies separated from experience and culture to become techniques received an enormous boost during the Second World War. The situations people encounter during the conduct of a war are conducive to being dealt with by extending their visual approaches, especially when confronted with an immediate danger to their lives. This imposes a focus of attention on killing or being killed. A rational approach thus becomes necessary, which can best be sustained by extending the visual approaches rather than the symbolic ones. As a result, every military operation had to be reorganized in order to achieve the greatest possible absolute efficiency, which meant that its relative efficiency became mere collateral damage. Operations research received an enormous boost, and demonstrated its advantages in such activities as the sinking of submarines. At the same time, it was equally necessary to develop human techniques to do everything possible to ensure that teams of recruits who had come from all walks of life could effectively work together despite a lack of shared experience. After the war, many of these techniques were extended not just into industry but also into society, education, and government. The striving for absolute efficiency began to permeate all of human life.[43]

Technology becoming a branch of technique had vast implications for all our lives. Before the dominance of technique relative to culture, people lived alienated lives because their culture had to be grounded in traditional myths. Today, our lives continue to be grounded in our secular successors to these myths, but superimposed on the alienation they produce is the reification of life caused by technique. An alienated life can be liberated from whatever enslaves it. However, a reified life is a non-life that can only be recreated in the likeness of life. Technique thus involves us in an ultimate confrontation with our finitude and the relativity of our lives as a symbolic species. In engineering, this struggle plays itself out through the impossibility of simultaneously paying attention to the absolute efficiency of the "internal workings" of technology and the relative efficiency of its "external functioning" within the general relativity of human life and the world. This situation could have been partly accomplished through preventive approaches. However, this potential has been blocked by our secular myths.

In discipline-based medicine, we can recognize similar issues, provided that we face the fact that our contemporary anti-societies do not have healthcare, but only end-of-pipe disease care. This observation does not in any way diminish our considerable medical accomplishments. It simply is impossible to prevent all disease, and whatever cannot be prevented must be dealt with as best we are able. Except in countries like the United States, which still does not have a comprehensive healthcare system, we can expect the best possible care when we face a serious crisis such as a heart attack.

A comprehensive healthcare system, however, goes well beyond what all the medical specialities can individually and jointly accomplish. Together they represent a medical division of labour that implies an underlying metaconscious knowledge of health and illness built up with the "experiences" of the domains of medical disciplines. Each specialty proceeds as if each tissue, organ, or ensemble of organs is busy turning certain inputs into desired outputs that can be measured by laboratory tests or observed by imaging techniques. When the numbers representing these outputs fall within a certain range, the performance is deemed optimal, but suboptimal when the numbers are too low or too high. The medical arts have thus been turned into medical science and technique.

What is almost entirely missing in this medical division of labour is the ability to assess the dialectically enfolded and enfolded character of our social, mental, and physical lives. As noted, every constituent of our bodies has come into being through stem cells becoming unique manifestations of the whole, expressed through the DNA, by becoming brain cells, heart cells, lung cells, and so on. Similarly, the organizations of our brain-minds are an individually unique expression of the society and culture in which we have grown up. What are the medical implications of our lives being biologically enfolded and socioculturally dialectically enfolded? Some of these are clear, but not widely recognized and accepted. For example, there can be no simple causality or an equivalent of negative feedback in living beings, the way they occur in any entity with a simple complexity such as that of a classical or information machine. We need to recognize that everything is both "internally" and "externally" related to everything else. The full implications of the complex interactions that are thus possible still remain almost entirely beyond our understanding.

Let us assume a situation in which measurements show that the upper and lower values of a patient's blood pressure fall outside of the normal range. A comprehensive healthcare system would then need to investigate what could cause this abnormal performance of the heart and circulatory "system," by fully examining what is happening in the person's life. With the aid of social epidemiology, it may be possible to connect the blood pressure numbers to stresses imposed by dietary intake, occupation, recreational activities, residential urban form, lifestyle, social support (including loneliness), and exposure to screen-based devices and other technologies. Internal causes would also have to be explored, including a possible genetic predisposition that may make a person more vulnerable to external stresses. The stresses would have to be assessed over time to determine whether or not they have likely translated into strain, with its known association to disease. None of this rules out a temporary end-of-pipe treatment involving the prescription of the usual medications, but a preventive approach would have to consider the root causes.

From the perspective of our present anti-societies, the massive influence of technique and its omnipresent collisions with human life, society, and the biosphere must be regarded as a major producer of physical and mental illness. It is almost certain that its production of disease far outpaces our ability to deal with it in an end-of-pipe manner, given the growth of our medical budgets and the enormous pressures on these budgets. If we are to control the costs of disease care, we will need to take a careful look at the ways in which we produce disease. There is a great deal of evidence that much of it could be prevented if our striving for absolute efficiency were not systematically done at the cost of relative efficiency.

For example, the movie *Super-Size Me* was largely misunderstood thanks to the enormous public relations efforts of the fast food industry. The experiment depicted in the movie showed how high-dosage short-term exposure to fast food was exceedingly ruinous to a person's health. The point I am making is not that no one would consume this much fast food in such a short time. Perhaps the real point is that if this fast food had been a drug, it would have been tested in a manner similar to that depicted in the movie, and it would never have been licensed.

We know very well that our workplaces have become major producers of physical and mental illness. At one point, Blue Cross was the largest supplier of health care to what, at that time, was the largest corporation in the world. Apparently, sick workers were its most valuable output.[44] Similarly, our urban habitat in its present forms imposes many stressors on human lives and communities, many of which have been correlated with the production of disease.[45] We have also discussed the likely consequences of excessive exposure to screen-based devices in children. In other words, we must regard technique as a major producer of disease in all areas of our contemporary anti-societies. With the aid of social epidemiology and an exploration of how this production of disease can be managed in a cost-effective manner by reducing stress through preventive approaches, societies could set effective and practical labour, social, health, and environmental standards. Doing so would in turn compel corporations to adopt preventive approaches and thus reduce the social costs they export into societies.[46] Once again, it must be remembered that, prior to the widespread proliferation of free trade agreements, many corporations were experimenting with work redesign. Such efforts could have been extended into many other areas.[47] Consequently, a comprehensive healthcare system would be a radical challenge to technique, in favour of resymbolization and thus of culture.

In the meantime, technique continues to create a broad-spectrum attack on our health. It is wholly or partly responsible for some of the more complex diseases that appear to be almost impossible to diagnose, especially when they are chronic and have a wide diversity of symptoms that are also next to impossible to diagnose, one medical specialty at a time. In many such cases, the medical system goes through an initial period of denial. These kinds of diseases are almost certainly a manifestation of the limits of what can be achieved by means of medical science and technique.

On the opposite end of the spectrum, emergency medicine is often spectacularly successful. In an acute crisis, discipline-based approaches work very well, because everything not immediately relevant to the crisis must be neglected (at least for now) if the patient is to survive. Moreover, when the treatment resembles a kind of re-engineering of a damaged "part" of the body, the results are often nothing short of astonishing. In other words, the successes and failures of our current medical science and technique appear to correlate rather well with the strengths and weaknesses of their discipline-based approaches.

Can alternative medical arts be developed that can grow beyond the limits imposed by a primary reliance on discipline-based approaches? Will such approaches be able to deal with the enfolded and dialectically enfolded architectures of our lives? Will such approaches be able to develop the equivalent of an intellectual base map to which all specialties can add their results, so as to indicate their relative significance to each other and the whole? At the same time, as technique increasingly overshadows more societies around the globe, will the comparative role of social epidemiology become diminished in its effectiveness?

It is quite astonishing how little our future doctors learn about healthcare relative to disease care. When this changes (after extensive reforms in our post-secondary system), they will be able to recognize how architects, engineers, urban planners, urban geographers, landscape architects, and other professionals could be welcomed to the healthcare team. We will never be able to prevent all disease, but a shift toward a more comprehensive healthcare will challenge much of how we currently go about our lives.

Technique and Non-Life

In our anti-societies, where technique has overshadowed culture and thus undermined our being a symbolic species, the question must be raised as to human survival. Are we now acting and living as if there no longer is any human life, or as if what little is left of it does not count for much? The possibility of a reification of human life is not being considered as one of the major difficulties we face. Individual and collective human life, as well as the biosphere, have become the human, social, and natural resources for building the global technical order. We have grown so accustomed to the commoditization and reification of life, which began with the development of a Market economy, that the omnipresence of dealing with life as non-life appears so self-evident that we cannot think or live as if things could possibly be any different. Consequently, we behave as if our lives are a collection of reified aspects that can be turned into resources to produce whatever outputs are required to build the universal technical order. We are no longer shocked when people talk about human, social, or natural "capital." Nor are we shocked that we have human resources departments in all of our organizations. All this and more is the manifestation of "technique changing people."

To brush all this off as an exaggeration, a misplaced pessimism, an ideology, or anything that trivializes what confronts us is entirely understandable. An incredible hope is required, based on a conviction that technique is not the destiny of humanity. Our assertions to the effect that we are "moving forward" are hardly convincing, since, physically, we generally move forward, not backwards. It would appear, therefore, that the assertion of "moving forward" implies an admission of a kind of existential void in which we have lost our way and are taking things one step at a time. There are many indicators to this effect, including an unprecedented incidence of anxiety, depression, and mental illness. We may be facing signs of anomie.

Technique as Response to Relativism, Nihilism, and Anomie

For all of human history, groups and societies have respected the relativity of human life and the world to the greatest extent possible by relying on symbolization, language, and culture to make sense of and live in the world. These approaches are ideally suited to doing so when everything is related to everything else, and thus evolves relative to everything else. What no groups and societies could accept, however, was the finitude of human life. The way they symbolized the general relativity of life and the world by means of a culture had to be the only possible way to avoid plunging the community into relativism, nihilism, and anomie. What was known and lived constituted the only possible absolute vantage point for living in the world, the only possible way of symbolizing an ultimately unknowable universe, and the only viable orientation for journeying in time and for making history. In this matter, each group and society created something absolute in order to make the finitude of life bearable and liveable – as long as the members of the community abided by what they had absolutized. This, however, imposed all manner of obligations on them. Alienation was the unavoidable price that had to be paid to make the general relativity of life and the world bearable.

Absolutizing elements of a general relativity inevitably leads to a complete distortion of it. In turn, such distortions lead to burdening human life with endless contradictions. Each and every daily-life experience is a confrontation between what people take to be absolute in their lives and its inability to deliver, since it too is an integral part of a general relativity. Fortunately for our lives, the overwhelming majority of these contradictions go unnoticed.

Nevertheless, these contradictions accumulate in the general relativity of human life in a world grounded in and distorted by myths, and this accumulation gives rise to emerging difficulties of liveability. What (with hindsight) turns out to be the end of a historical epoch in the journey of a society is a time when these contradictions begin to overwhelm life – to the point that some people begin to seek to make more sense of their experiences by reinterpreting them in a different way. When this is successful and becomes widely shared, a new culture-based connectedness may begin to emerge that will be grounded in a different set of myths that will displace the old myths, and will sustain human life for the next historical epoch. When these efforts are not successful, a community will disintegrate and disperse, because its members no longer share a "body of common sense."

With the benefit of a great deal of hindsight, the last two hundred years may be interpreted as a comprehensive preparation for something new taking over from culture as the way groups and societies had always faced their general relativity and that of the world in which they lived. Much of this preparation paved the way for technique, which was supposed to be able to do a better job; and yet technique is the diametrical opposite of culture. It overcame the limits of culture by organizing everything as if there no longer was any human life and as if the world never had a general relativity. It is as if humanity is somehow in denial of itself and of the world in which it lives. The unlimited use of discipline-based approaches for human knowing, doing, and political organizing is an implicit affirmation that human life never had a relative character. We have finally realized that all of it was merely the leftovers of a superstitious, moral, and religious past.

Humanity has finally learned that everything could be grasped through the principle of non-contradiction and thus by mathematics. Everything can be separated into autonomous categories of phenomena that can be relegated to scientific disciplines in order to become fully known. Everything can be captured in terms of closed definitions, making it thinkable not only that everything has an absolute efficiency but also that achieving it is the only possible course of action. To do all this effectively requires that everything be measured, quantified, and mathematically represented in order to most effectively achieve this absolute efficiency. Finally, all this can be achieved in a piecemeal fashion, one category of phenomena, one autonomous discipline, and one application area at a time. There is no need to even consider the possibility of this course of action getting us into a great deal of trouble because of an underlying general relativity. As far as I am able to understand it, this is where humanity, of which I am an integral part, finds itself in the early twenty-first century.

From this perspective, the fundamental role of creativity and imagination in our growing up and living in the world takes on another dimension. Since human life in a general relativity can never repeat itself in quite the same way, the living of a life involves a constant creative and imaginative adaptation of that life to situations that are similar to and yet different from what went before.[48] This recognition alone possibly constitutes one of the greatest condemnations of our present education system, which has become one of the most uncritical promoters of information technology, screen-based devices, and Google classrooms.

Living with an element of our general relativity as if it were absolute, and thus introducing all manner of contradictions, greatly reinforces our dependence on creativity and imagination as a dimension of the symbolization of all our experiences (regardless of which frame of reference is used) in order to fit it into an alienated, and now reified, life in the world. These creative and imaginative efforts slowly but surely bring to an end the ability of children to play in the world without any care about its general relativity. However, when something does not work out, when they are stuck, when they are hurt, or when they encounter any other negative surprises, all this can be fixed by a loving parent who puts the world right again with a big hug and a kiss. It is this nurturing love that permits them to play as they do, before their creativity and imaginations are also drawn into an enslavement to the myths of their community.

Technique may thus lead us to one of the more drastic confrontations with our finitude and the general relativity of our lives in the world. Behaving as if "all that" no longer matters, or as if "all that" was nothing more than the superstitions of the past, will not work. As a symbolic species, we simply are not built that way. Nor is our world built that way. Consequently, the clash between technique and whatever residues of the general relativity of our life in the world remain could lead to an acute crisis of who we are and a realization that naming ourselves through our works as a *homo informaticus* was a dreadful mistake.

Perhaps humanity cannot reach that point because the desymbolization of our species will further desymbolize what little is left of our cultures, until our deepest metaconscious knowledge can no longer form the secular myths for grounding our lives in this universe. At this point, we will have brought down upon ourselves a secular hell characterized by a complete cultural chaos that will make any relationship and the lives built up from such relationships radically impossible. It would mark the end of our history as a symbolic species. However, my reason for writing this book is to point out that this is very far from our destiny. The limitless character of our discipline-based approaches in our knowing, doing, and political organizing will dissipate like a mist under the sun the moment we stand up and deny these approaches their god-like status as an ultimate illusion of the general relativity of our lives in the world. But, the more we desymbolize our capabilities as a symbolic species as a consequence of our global project of technique, the more difficult it will be to undertake what each and every society and civilization has always accomplished when it reached the end of one of its epochs in its historical journey. In this sense, technique is more of a threat than the kinds of challenges to a society or civilization that Arnold Toynbee and others have studied.[49] If a society or civilization failed to meet such a challenge, it collapsed; but all the other societies and civilizations continued. Increasingly, humanity is wagering its future by staking everything on technique. For my part, phenomena such as the boundless secular faith of cognitive psychology, the expectations of artificial intelligence based on deep learning, the confidence of eternal life on the Net, the hope in the emergence of a super-intelligence on the Net, dreams of a world run by big data while ignoring the "weapons of math destruction," and similar visions of an ultimate *homo informaticus* represent such a denial of who we are and our dependence on being a symbolic species that they appear to be the pursuit of absolute efficiency taken to the highest level of absurdity. There are still enough people who do not buy into this dream that there remains a beachhead within humanity for waking up, symbolizing what is happening to us, and doing what humanity has always done: deal with the general relativity of our situation as best we can in order to determine what our works may be able to deliver and what they can never deliver to our lives.

Epilogue: Possessed by Secular Myths

Endangered by Secular Religious Attitudes

Let me begin where we hopefully have come to agree: that this analysis of technique and culture confronts us all with a deeply disturbing situation. If we wish to struggle against the alienation and reification that we have unleashed on ourselves as participants in our contemporary anti-societies "moving forward" with a limitless knowing, doing, and organizing of human life and the world, we confront the following choice. Either we expect that our current situation is temporary and that soon we will have acquired even more power, which will enable us to make a decisive difference and turn things around; or we recognize that this way of pursuing absolute efficiency can do nothing but further tear apart the fabric of all life, including our own. If we strongly believe that the first possibility is the way out of our dilemma, I would simply ask that we carefully think about the emerging nanotechnology, biotechnology, and the latest information technology. Are they not the latest discipline-based approaches, and is there any reason to believe that their effects on everything will be any different from those of their precursors? If we are willing to entertain this possibility, we can take another step in our journey together.

If we wish to struggle against the alienation and reification that we have brought on ourselves as members of our contemporary anti-societies, which entrust their future to the limitless exploitation of discipline-based approaches to knowing, doing, and organizing – what can we possibly do that may make any difference? Can we hope to think outside the box that we have created together as a consequence of what we have referred to as "technique changing people"? Have we now trapped ourselves in this "existential box" because of our alienation and reification?

Can we continue this discussion by acknowledging the possibility that we may be endangered by science, technique, and the state-nation? If we live as if our discipline-based approaches to scientific knowing have no limits, our lives are wide open to secular religious attitudes to this knowing. If we live as if discipline-based approaches to technical doing have no limits, we are again wide open to secular religious attitudes that will permeate all our doing. If we live as if politics has no limits, we surrender our lives to the state-nation, which will not hesitate one moment to squander or sacrifice them while demanding our complete and total devotion and love. All of human history shows what happens when people become possessed by religious attitudes and, more recently, by secular religious attitudes.

The combat against religious or secular religious attitudes must deal with the objects of such attitudes, which result from individual and collective human life being suspended in traditional or secular myths. Western civilization, which gave birth to our present global one, offers two sources of inspiration. Observers who lived in the first generation of industrializing societies began to describe what was happening to individual and collective human life in terms of the newly created concept of alienation. What some observers were arguing may be expressed as follows: the influence individual and collective human life had on the building of the new industry and its economy increasingly appeared to be dwarfed by the influences this industry and economy were having on human life and society. This latter influence appeared to go so deep that it could best be described in terms of possessing that life in a way comparable to the way masters possessed the lives of their slaves in antiquity. It was becoming obvious to some observers that the new factory work, now organized on the basis of the technical division of labour, had a very different effect on people's lives from human work that was not organized in this manner. For example, in England, the factory workers were as poor as many of the peasants living on the land; and there was no question that poverty deeply affected all their lives. However, factory workers appeared to suffer a great deal more. Their participation in this work was essentially restricted to the movements of their hands, with the result that they had to learn to suppress themselves in their work. This suppression did not magically disappear when they went home. It appeared to affect all their relationships – with their partners, children, family, friends, and neighbours. While the poverty of the peasants was a financial one, the poverty of the factory workers was multi-dimensional: they were deprived of having the kind of marriage, family, and social relations that other people of that time, place, and culture could expect. Their poverty was thus existential as well as financial. Some observers even claimed that marriage and family life had become the privilege of the wealthy who were not involved in industrialization.

In the same vein, the participation of the owners in this factory work was constrained in the opposite way: they were involved mostly through their mental functions and thus had little or no experience of what was happening on their factory floors. The accusation was made that this new organization of human work alienated the rich and poor alike: both the factory owners and their wage earners. Other observers took this diagnosis of alienation a step further. The new technological and economic forms of the first generation of industrializing societies, by alienating the lives of everyone involved, were themselves alienating and thus capable of possessing human life the way masters had possessed the lives of their slaves in antiquity. There was thus something structural about these forms that made them capable of enslaving human life.

Despite this harsh diagnosis, there was one observer who went beyond this analysis, based on the first generation of secular myths, to promise that this enslavement of humanity would not last long. Karl Marx announced that capitalism would be driven against its own limits and that this would soon cause its collapse.[1] All communities would then take over the new technology and economy, to produce what he referred to as a socialism capable of ending all alienation. This assumption was a dreadful mistake, for a number of reasons. First, soon after Marx's death, the limits of capitalism were essentially eliminated by the separation of knowing and doing from experience and culture.[2] Second, the Communist government that took control of Russia following the revolution could not escape the need to renew and accumulate the capital necessary to build the new industry and its massive infrastructure. It therefore adopted a kind of state capitalism where cost prices had to be calculated in the five-year plans in order to avoid the loss of capital. The ideological veil drawn over all this, to the effect that alienation was being avoided because this capital was being accumulated on behalf of the people and for the people, gradually became less and less tenable. For example, eventually more and more observers recognized that workers on the assembly lines in the former Soviet Union were just as alienated as the assembly line workers in the West.

In the former Czechoslovakia, a Communist government attempted to get to the bottom of it all following the Second World War. Marx's thought was re-examined for the twentieth century. What came to light was that science and technology had become the most decisive influences on production, having gradually dwarfed that of capital.[3] The result was that to avoid alienation, the study recommended very different kinds of policies – a heresy against Communism that was soon stamped out by the Soviet tanks.[4] In the West, John Kenneth Galbraith came to similar conclusions when he examined the decisive role played by highly specialized knowledge in American industry and the economy.[5] I have shown that all this and more can be directly attributed to all human knowing and doing being reorganized in the form of autonomous disciplines that are separated from experience and culture.[6] In other words: somewhere along the way, the political Communist Left had completely gone out of the business of eliminating alienation. The thought of Karl Marx became a secular religious justification for totalitarian systems in the business of seizing and maintaining power. The records of Communist regimes and their "people's republics" speak for themselves.[7] At the same time, as academia was taken over by discipline-based approaches, it also mostly went out of the business of understanding and attempting to deal with alienation in order to invent alternative technological and economic forms.

It is important not to surrender the concept of alienation to the political Left, which has mostly dropped it in its struggle for power. The Left has long abandoned the defence of our humanity. Avoiding the trap of this politicization of the concept of alienation becomes easier when we recognize that other observers, as members of the first generation of industrializing societies, began to be concerned about the alienating powers of the new industrial centres. As people changed their urban habitat, this habitat simultaneously influenced people through a variety of "urban stressors" that included sensory overload, social overload, crowding, noise, pollution, and the effects of tall buildings.[8] The fact that the literature on cities at the time was either wildly optimistic or deeply pessimistic implies that everyone recognized the depth of their influence on the lives of people, creating new urban personalities and a new way of life. The possibility of cities "possessing" human lives to the point of alienating them must therefore be given serious consideration. In terms of the political implications, I have not yet been able to find observers claiming that, in Communist nations, cities have no such alienating effects. The issue of alienation is not nearly restricted to work and the economy and is not at all a matter of private or public property.

The modern city is very much a creature of industrialization, which is indissociably linked to urbanization. The urbanized portion of humanity has been growing ever since the great cultural divide, with the result that any discussion regarding human freedom and alienation will have to include the effects our urban-technical habitat is having on all of us. This habitat has become a kind of physical integrator of everything technical as well as the seat of most of our technical endeavours, which are beginning to include our dreams of creating smart cities through big data. Its influence has gone well beyond what can be explained in terms of the other-directed personality. We have created a new technical life-milieu that interposes itself between people and their former life-milieu of a highly transformed nature. Like its predecessors, it correlates well with our dominant myths and the moralities and religions anchored in them. This technical life-milieu will decisively influence the liveability and sustainability of our future as a symbolic species. The alienating powers of this life-milieu should be carefully considered. It is not only the political spectrum from one end to the other that has deserted the business of confronting alienation, it has also been lost in the process of city building.

A very long time ago, the Jewish and Christian traditions were in the business of living out of freedom rather than alienation. I will attempt to speak for the latter community in order to explain the untenable situation in which organized Christianity finds itself. A significant portion has turned itself into the judge and jury of us all, proclaiming our condemnation and threatening us with hell. It represents a form of love for one's neighbour that has gone so awry that we do not need to discuss it here, other than to recognize that this portion of organized Christianity, now flexing its political muscle, has given almost everyone a very bad taste in the mouth. I can assure my reader that this is hardly what genuine Christianity is all about.

During the period of the first generation of industrializing societies, the so-called Ten Commandments were extensively relied on as the golden rule for society. The embarrassing difficulty of this attempt to impose Christianity on society is that these "commandments" are actually promises because, in the original Hebrew text, they are written in the future tense. This is obscured in the English translations. What the first three "promises" tell us has to deal with what I refer to as our being a symbolic species, suspended in languages and cultures grounded in traditional or secular myths. The promises essentially tell us that the day will come when we will no longer need to make false gods for ourselves, when we will no longer need to represent these gods through idols or make any religious use of God's name and thus of everything that name means for humanity. In other words, these three promises appear to imply that one day we will be completely liberated from the threat of cultural chaos; and this redemption will completely transform what we now understand by our being a symbolic species, whose relativity must be grounded in myths that must be served through moralities and religions that enslave us.[9]

My Christian readers will surely remember that the opening chapters of both the Jewish and the Christian Bibles attribute the break between God and humanity to the intervention of the serpent, which was a common deity in the cultures of the people that surrounded Israel at the time these texts were written. This is confirmed by the fact that these texts mount a critique of religion, morality, and magic.[10]

It should also be noted that, much later in the history of the Jewish people, Qohelet (commonly translated as Ecclesiastes) critically examined this history, including its religious and cultural life, with an eye on these texts from Genesis. The book opens with the its conclusion: "Vanity of vanities, all is vanity."[11] For contemporary readers, this can best be understood in terms of human life being suspended in a language and culture grounded in myths. These distort the meaning and value of everything in the general relativity that is being symbolized, and plunge human life into endless contradictions.[12] Some readers may also remember that the so-called Ten Commandments follow a preamble that tells the people who are about to receive them that they have thereby been set free from the double anguish of life and death, symbolized by their enslavement in Egypt. In other words, they ought not to squander this liberation by turning these promises into a morality or religion that would tell them what to do. These promises, and everything else in the Jewish Torah, were summarized in terms of love; and this is further confirmed in the Christian Bible. Surely, people who are in love hardly require a morality and religion to tell them how to behave toward each other. It confirms the conclusion of Qohelet: that plunging human life into vanity can be avoided only by following God's way of liberation, love, and non-power. Relationships of love are the antithesis of the relationships we know based on the exercise of power, hierarchy, domination, exploitation, taking advantage of the weaknesses or frailties of others, and so on.

Ironically, as soon as Christianity became the official religion of the Roman Empire, it ceased to be the follower of Christ's way because the church had to deal with millions of new Christians, which it did by creating a huge organization based on hierarchies and the exercise of power. Prior to these developments, Christians had been widely accused of being dangerous anarchists who subverted the Roman order, which they refused to serve as functionaries or soldiers.[13] They refused all religion, restricting themselves to establishing new kinds of relationships that contested all forms of power and hierarchy. Fellow Greeks and Romans thus regarded Christians as atheists and irreligious people who were against all cults, including a place for their god in the Roman pantheon.[14] They did not regard themselves as having a superior religion competing with pagan ones, but of refusing to regard anything as sacred along with any religious attitudes toward it. Christianity thus threatened the traditional myths that grounded the cultures of that time, and because of this Christians were brutally persecuted. By the time Christianity faced the conditions that followed the great cultural divide, however, it had long ago ceased to be in the business of combating alienation by establishing new kinds of relationships based on freedom, love, and non-power.

Ironically, the influences of Judaism and Christianity on the history of Western civilization had contributed to the very possibility of the concept of alienation becoming thinkable. It should be noted that this concept, despite its ongoing presence and its role in the social sciences, is not scientific in character. It implies a norm: that alienated individual and collective human life is not acceptable. Of course, today most of humanity recognizes that slavery is an unacceptable form of human life.

For Judaism and Christianity, this unacceptability was fundamental. Sin was compared to a condition of being enslaved to powers that dominated human life the way a master dominated a slave in antiquity. These traditions were concerned with the redemption of humanity, which is a concept borrowed from Roman law. It obliged all Romans, and later the Roman state, to purchase Roman prisoners of war who had been turned into slaves by an enemy. Following this purchase, the redeemed slave was brought before a magistrate to make arrangements to have the debt repaid, or else to be forgiven the debt and, in any case, be restored to full citizenship.[15]

Having abandoned the business of combating alienation, organized Christianity increasingly identified with turning an architecture of relationships of love into one that required a Christian morality and religion. Doing so caused terrible anxiety and unhappiness because Christians were made to feel morally and religiously guilty for their sins, which were reinterpreted as the actions of slaves rather than the condition of slavery itself. It was the master stroke of the new morality and religion that dominated the first generation of industrializing societies. It amounted to holding slaves morally and religiously accountable for their condition. This shift in the focus of Christianity, from an enslavement to the powers that possessed humanity to the moral and religious actions of enslaved people, changed everything. During the time that this reinterpretation of Christianity spread and took hold, the awareness of a growing alienation of human life as a consequence of industrialization, urbanization, rationalization, and secularization gradually lost ground with the rise of the second generation of secular myths. The accompanying secular political religion of democracy declared, against all evidence, that everyone was free and proceeded accordingly.

Our present situation may thus be characterized as follows: In the days of our youth we have all encountered a science that has the potential of knowing everything, and a technique that potentially can do everything, along with the assurance that this potential is entirely at our disposal because of our limitless politics. Few of us, including Jews and Christians, were brought up with the warning that, if this appears too good to be true, it almost certainly is. Probably even fewer of us were brought up to understand the deep religious significance of it all: that we were being confronted with the same temptation to submit to a religion, and should not be fooled by its having taken on a new secular form. As human creations, science and technique cannot possibly be limitless, any more than the traditional gods of past cultures were omnipotent. Nor can the state be "all for all." It was the same old story now dressed up in secular garb.

Organized Christianity had also been seduced by this new secular religion.[16] In order to have it all, namely the new religion during the week and the Christian one on Sundays (and perhaps in a personal entirely private life during the week), Christianity transformed itself into both a conservative way and a liberal way of accomplishing just that! Both forms of Christianity were justly and severely criticized by many observers for their contribution to the alienation of human life. However, organized Christianity was too busy having it all to take any notice. A new relationship between this Christianity and some of our contemporary societies began to emerge, especially in North America.[17] It became detrimental to all parties, including the people within Christianity. They were not immune from all the turmoil that was happening to everyone's life during the week.

Is Humanity Truly against Enslavement?

If humanity has indeed reached a consensus that all forms of slavery represent unacceptable forms of human life, then all of us ought to be in the business of combating alienation. However, I have attempted to show that the entire political spectrum, from left to right, has abandoned any such effort. I have argued that universities, being in the grip of discipline-based approaches grounded in our present secular myths, can hardly be regarded as operating on the basis of genuine academic freedom, let alone combating alienation. Surely, this is something we ought to expect of our public universities. I have also argued that Judaism and Christianity have abandoned the business of establishing the kinds of relationships that are contrary to any sociocultural architecture grounded in myths, whether traditional or secular. Hence, even if we all get out of our trenches to have a discussion about this, the most essential perspectives are almost certainly going to be absent.

This prediction is confirmed by the very possibility of our contemporary multicultural societies having a democratic political system. In North America, almost all of us come from religious and political traditions brought by immigrants, often fleeing countries where religious intolerance reigned. The possibility of what, for most of human history, was impossible has come from our making sense of and living in the world by means of three frames of reference. These frames of reference are coordinated by and anchored in the second generation of secular myths. Consequently, we may well experience a great deal of anxiety and a lack of purpose and direction because our society is building a global technical order by means of discipline-based approaches, but because these are separated from culture, they fail in their attempts to make sense of anything. It is an order of non-sense and non-life that excludes us all if we attempt to make any sense of it, other than in our private personal lives; and the results are to be held as our subjective opinions and convictions. All this we readily leave behind when we participate in our anti-societies.

What our second generation of secular myths makes possible is that we can have it all: our servitude of a limitless science, technique, and politics on the one hand, and our personal and private religious or political convictions on the other hand. Even when we attempt to express our personal and private convictions during elections, we discover that only what has become knowable and doable according to our second generation of secular myths is represented as a political choice, while everything else simply does not appear on the ballot. Hence, there is no possible way that we can vote for political parties that seek to establish anything but the kinds of relationships grounded in our secular myths. Even on the margins of our anti-societies, it is very difficult to find anyone who is in the business of combating alienation as a way to weaken what we have made autonomous and limitless by means of our secular religious approaches. For example, our chances of confronting a crisis such as global warming in an effective and timely fashion while clinging to our secular myths is essentially nil. We know this from the evidence of the way humanity behaved after the establishment of a growing consensus that we must find more sustainable ways of life – a warning urged by a global commission on our common future. We sprinkle the words green and sustainable everywhere, environmental programs and centres have been introduced at our universities, corporations have created "green" products, and governments have passed a great deal of environmental legislation – yet the trends that have made our ways of life unsustainable have generally been little affected.

Our ability to have our personal private beliefs and convictions on the one hand and our servitude of our secular myths on the other hand is also reflected in the kinds of discussions mentioned earlier, where professional audiences were asked what science could never know and what technology would never be able to accomplish for us. Apart from some silly personal answers, it is clear that we are in the grip of our secular myths.

Nevertheless, I know from experience that it is entirely feasible to get into the business of combating alienation by questioning everything that appears to be totally self-evident, certain, and impossible to doubt, while at the same time refusing anything autonomous or limitless or that in any other way has been withdrawn from the general relativity of human life and the world through secular religious attitudes. I have been able to document these efforts in many volumes inquiring into the relationship of technique relative to culture, and of highly desymbolized cultures relative to technique. In order to obtain a better grip on what is happening to our lives and our world, there appears to be no other possibility than to resymbolize them in order to complement what we learn from our discipline-based approaches when used within their limits, and by reinterpreting their findings when they transcend these limits. We need to resymbolize everything relative to everything else in order to create a kind of dynamic intellectual map of how our present world and human life within it has gradually emerged following the great cultural divide. The findings can then immediately be applied to developing approaches that can make sense of and deal with situations in which more than one category of phenomena make significant contributions to the point that they cannot be neglected. Once this is successful, it can be taught in different ways and on different levels throughout our educational system. Integral to this would be the identification of what entities we treat as autonomous and limitless as a consequence of secular religious attitudes directed toward them, in order to make ourselves aware that our lives are necessarily grounded in myths to avoid cultural chaos. By introducing a more critical attitude toward such entities, we can begin to create a little "play" in the "system." Such an approach to teaching, although weakening the grounding of our lives, will make new possibilities thinkable and doable and will thus potentially contribute to the creation of a more liveable and sustainable future.

Although I am convinced that our getting into the business of combating alienation is entirely feasible, essential, and urgently necessary, its success is increasingly threatened by the ongoing desymbolization of our languages and cultures by technique. This desymbolization is possibly the greatest threat to humanity's ability to get a better grip on what is happening in order to deal with it in a head-on fashion instead of with end-of-pipe approaches, and thus to have a chance at confronting all our crises, which, almost without exception, are rooted in the limits of our discipline-based approaches. If humanity has indeed accepted that slavery represents an unacceptable form of human life, we are restricting this to ancient practices. We are refusing to see that we are all possessed by our secular myths, which we slavishly serve with all our hearts, minds, and strength because our personal convictions are kept to ourselves. Hopefully, this work is a wake-up call to all of us.

Can our educational systems be enlisted to combat against our alienation by our secular myths, which has resulted in our indiscriminate use of discipline-based approaches beyond their limits? It should come as no surprise not only that our faculties of education, especially our institutes for research in education, are in the grip of discipline-based approaches, but that the applications of these approaches are deeply influenced by an understanding of human life and the world that is grounded in these myths. To lessen this grip, the previously noted distinction between numeracy and literacy may be helpful.[18] Numeracy may be defined as including all disciplines whose domains are built up through mathematics, and thus with the principle of non-contradiction, to yield a simple complexity. Literacy can be defined as referring to all subjects that are built up by means of the primary and secondary frames of reference and thus through language and culture, to yield a dialectically enfolded complexity. Hence, numeracy and literacy have diametrically opposite architectures. The architecture of the numeracy subjects is suited to the "worlds" of technology and technique and thus of non-life. The architecture of the literacy subjects is suited to the "worlds" of natural and cultural life, which are now in urgent need of expansion with subjects having a preventive orientation.[19] Our second generation of secular myths embraces numeracy and all but excludes literacy, thereby building a global technical order that excludes all sense. It is an order in which we cannot participate as human beings. We must attempt to "live" as best we can outside of it through our personal subjective lives. Despite all the rhetoric of our education specialists, the desymbolization of our symbolic species continues, because our educational systems, in their enslavement to the second generation of secular myths, prioritize numeracy over literacy.

Reversing this situation is intellectually and practically feasible and not particularly difficult if educators were a little more critical of their being possessed, like all of us, by an enslavement to our new secular gods. It is also very clear that, in North America, the vast and extensive Jewish and Christian educational systems are not in the business of combating alienation. If these trends continue, they will above all sustain the ongoing desymbolization of our anti-societies, and thus close off the very possibility that we could effectively deal with our crises and survive on this planet. If we were not in the grip of our secular approaches, it would be very easy to determine what we could expect from our discipline-based approaches when applied within their scientific and technical limits, and how we could avoid participating in the destruction of all life by exceeding these limits. At the same time, prevailing political illusions would vanish like a mist, to expose the enslavement of the entire political spectrum to the spirit of our age.

Such efforts will crack open the door to relativism, nihilism, and anomie. In other words, any commitment to converting the intelligence by which we make sense of and live in the world must be sustained by strong personal commitments to keep nihilism at bay. It is the strength of such commitments that may determine how deeply we are able to confront our secular myths without plunging our lives into chaos. For this very reason, we cannot have it all. Whatever commitments we make in our personal lives will have to confront the implicit commitments we make in our collective life through our secular myths. We cannot serve two masters any more than we can completely and unreservedly love several partners, regardless of what the media may have us believe. If we are fortunate at all, we may meet others who are also in the business of attempting to get a better grip on their lives by resymbolizing them. I cannot and will not promise any utopias. The struggle is very difficult and frequently a lonely business. However, in my experience, it is also one of the most rewarding tasks that a human being can undertake. It makes a lot more sense than anything many people live by, it does not gamble with our future, and it may leave a liveable planet for our children. It would prepare our highly desymbolized "cultural soil" to grow more understanding and sustain more activities for life, thereby crowding out the highly disturbing neo-fascist tendencies that are on the rise. I would not want to have lived my life in any other way; and I hope this work may assist my readers in taking a good look at the business of combating alienation.

There is nothing any of us can undertake that will more effectively reduce the war that we are conducting against ourselves and all life, and support the battle we are waging for our symbolic species, than to question our commitment to anything that is limitless, autonomous, or all-powerful. Only our secular religious attitudes can make this self-evident, obvious, and undoubtable. They are as destructive as such attitudes have always been in human history. We need to symbolize technique relative to culture, and resymbolize our cultures relative to technique. It should be the one priority that could address a great many of our crises, thus reducing our enslavement to create a little bit of freedom relative to the global technical order within which neither we nor any other forms of life can live and have a future. Our possession by our myths makes any significant alternative unthinkable and unliveable for the time being. Our political options will thus rule out what we truly need; our discipline-based knowing will exclude what we need to know; and our discipline-based doing will prevent us from doing what is urgently required for a liveable and sustainable future. Our state-nations will do nothing else but contribute to the building of the global technical order – the order of non-life and non-sense. It ought to make all of us, and all life, tremble before it. These are my conclusions to seven volumes seeking to resymbolize human life and our world, with the hope that we put the findings into practice to serve a future and a planet we can all share. A strategy of freedom, love, and non-power is the only one against which our traditional or secular myths cannot prevail. The tunnel into which we have thrust ourselves by our second generation of secular myths, including the myth of history, allows for nothing else but a blind moving forward without sense. This tunnel is long and dark, but its end can be guaranteed by refusing all secular religious attitudes and as much as possible returning everything to the general relativity of human life and the world.

Notes

Introduction

1 Willem H. Vanderburg, *The Growth of Minds and Cultures: A Unified Interpretation of the Structure of Human Experience*, 2nd ed. (Toronto: University of Toronto Press, 2016).
2 Willem H. Vanderburg, *The Labyrinth of Technology* (Toronto: University of Toronto Press, 2000).
3 Ibid.
4 Ibid.
5 Willem H. Vanderburg, *Our War on Ourselves: Rethinking Science, Technology and Economic Growth* (Toronto: University of Toronto Press, 2011), chapter 5.
6 Vanderburg, *The Growth of Minds and Cultures.*
7 Ibid.
8 Ibid.
9 Willem H. Vanderburg, *Living in the Labyrinth of Technology* (Toronto: University of Toronto Press, 2005); *Our Battle for the Human Spirit: Scientific Knowing, Technical Doing, and Daily Living* (Toronto: University of Toronto Press, 2016).
10 Ibid.
11 Vanderburg, *The Growth of Minds and Cultures.*
12 Ibid.
13 Arnold Toynbee, *A Study of History*, abridged version, ed. D.C. Somervell (New York: Dell, 1978).
14 Vanderburg, *The Growth of Minds and Cultures.*
15 Jacques Ellul, *The New Demons*, trans. C. Edward Hopkin (New York: Seabury, 1975).
16 Ibid.

17 Jacques Ellul, *Perspectives on Our Age: Jacques Ellul Speaks on His Life and Work*, 2nd ed., ed. Willem H. Vanderburg (Toronto: Anansi, 2004), chapter 4.
18 Ibid.
19 C. Lévi-Strauss, *The Savage Mind* (Chicago: University of Chicago Press, 1966).
20 Richard Stivers, *Evil in Modern Myth and Ritual* (Athens: University of Georgia Press, 1982).
21 Vanderburg, *The Growth of Minds and Cultures*.
22 Ibid.
23 Ibid.
24 Ibid.
25 Ibid.
26 Ibid.
27 Vanderburg, *Living in the Labyrinth of Technology*, Part 1.
28 Norman Doidge, *The Brain That Changes Itself: Stories of Personal Triumph from the Frontiers of Brain Science* (New York: Penguin Group, 2007).
29 Vanderburg, *Our Battle for the Human Spirit*, see Introduction.
30 Milan Kundera and Michael Henry Heim, *The Unbearable Lightness of Being* (New York: Harper and Row, 1987).
31 Lucien Malson, *Les enfants sauvages* (Paris: Union Générale d'Editions, 1964). A good account in English can be found in Roger Shattuck's *The Forbidden Experiment* (New York: Washington Square, 1981).
32 Ibid.
33 Ibid.
34 Lewis Mumford, *The Myth of the Machine: Technics and Human Development* (New York: Harcourt Brace Jovanovich, 1967).
35 Jacques Ellul, *Histoire des institutions: l'antiquité* (Paris: Presses Universitaires de France, 1961).
36 Ibid.
37 Vanderburg, *Our War on Ourselves*.
38 Vanderburg, *Our Battle for the Human Spirit*.
39 Werner Heisenberg, *Physics and Beyond: Encounters and Conversations* (New York: Harper and Row, 1971), 63.
40 Thomas S. Kuhn, *The Structure of Scientific Revolutions*, 2nd ed. (Chicago: University of Chicago Press, 1970).
41 Ibid.
42 Raymond Aron, *Progress and Disillusion* (New York: Praeger, 1968); William E. Connolly, "Essentially Contested Concepts," in *The Terms of Political Discourse* (Princeton: Princeton University Press, 1984), 10–44.
43 Vanderburg, *Living in the Labyrinth of Technology*, Part 3.

44 Hubert L. Dreyfus, *Being in the World: Commentary on Heidegger's Being and Time, Division I* (Cambridge, MA: MIT Press, 1991).
45 Hubert L. Dreyfus, *What Computers Still Can't Do: A Critique of Artificial Reason* (Cambridge, MA: MIT Press, 1992).
46 For an insightful application of some of the key ideas of Søren Kierkegaard to our uses of the Internet, see Hubert L. Dreyfus, *On the Internet* (New York: Routledge, 2001). Although the second edition brings the book up to date regarding advances made in the design of search engines, the first edition has a more comprehensive discussion of the relevance problem, which is fundamental for understanding this work.
47 Ellul, *Perspectives on Our Age*.
48 Dreyfus, *Being in the World*.
49 Hubert L. Dreyfus, "Why Heideggerian AI Failed, and How Fixing It Would Require Making It More Heideggerian," *Artificial Intelligence* 171 (2007): 1137–60.
50 www.deeplearningbook.org, Part III: Deep Learning Research
51 Jan Hendrik van den Berg, *The Changing Nature of Man* (New York: Delta, 1961).
52 Ibid.
53 Vanderburg, *Our War on Ourselves*.
54 Ibid.
55 Ibid.
56 Robert Karasek and Tores Theorell, *Healthy Work: Stress, Productivity and the Reconstruction of Working Life* (New York: Basic Books, 1990).
57 Vanderburg, *Living in the Labyrinth of Technology*, chapter 1.
58 Ibid., chapter 2.
59 Karl Polanyi, *The Great Transformation* (New York: Farrar and Rinehart, 1944).
60 Vanderburg, *Living in the Labyrinth of Technology*, chapter 1.
61 Ibid.
62 Ibid.
63 Vanderburg, *Our War on Ourselves*, chapter 3.
64 Ibid., chapter 4.
65 Ibid.
66 Vanderburg, *Living in the Labyrinth of Technology*, chapter 3.
67 Ibid.
68 Jacques Ellul, *The Technological System*, trans. Joachim Neugroschel (New York: Continuum, 1980).
69 Vanderburg, *Living in the Labyrinth of Technology*, Part 2.
70 Ibid.

71 Vanderburg, *Our War on Ourselves.*
72 Willem H. Vanderburg, *Secular Nations under New Gods: Beyond Christianity's Subversion by Technology and Politics* (Toronto: University of Toronto Press, 2018).
73 Jacques Ellul, *The Technological Society*, trans. John Wilkinson (New York: Vintage Books, 1964), chapter 4.
74 Adam Smith, *An Inquiry into the Nature and Causes of the Wealth of Nations*, 2 vols. (Chicago: University of Chicago Press, 1976).
75 Vanderburg, *Our Battle for the Human Spirit*, chapter 3.
76 Vanderburg, *Our War on Ourselves.*
77 H.H. Gerth and C. Wright Mills, eds., *From Max Weber: Essays in Sociology* (New York: Oxford University Press, 1963); Rogers Brubaker, *The Limits of Rationality: An Essay on the Social and Moral Thought of Max Weber* (London: Allen and Unwin, 1984).
78 Vanderburg, *Living in the Labyrinth of Technology*, chapter 2.
79 Sigfried Giedion, *Mechanization Takes Command* (New York: Norton, 1969).
80 Raymond Williams, *Keywords: A Vocabulary of Culture and Society* (London: Fontana Paperbacks, 1983).
81 Jacques Ellul, *Métamorphose du bourgeois* (Paris: Calmann-Lévy, 1967).
82 Vanderburg, *Living in the Labyrinth of Technology*, chapter 2.
83 Ibid., chapter 3.
84 Gerth and Mills, eds., *From Max Weber*; Brubaker, *The Limits of Rationality.*
85 John Kenneth Galbraith, *The New Industrial State* (New York: New American Library, 1985). This was more fully developed in Vanderburg, *Living in the Labyrinth of Technology*, chapter 6.
86 Michael Hammer and James Champy, *Reengineering the Corporation: A Manifesto for Business Revolution* (New York: HarperCollins, 1993); Thomas H. Davenport, *Process Innovation: Reengineering Work through Information Technology* (Boston: Harvard Business School Press, 1993).
87 Vanderburg, *Living in the Labyrinth of Technology*, Part 3.
88 Cathy O'Neil, *Weapons of Math Destruction: How Big Data Increases Inequality and Threatens Democracy* (New York: Penguin Random House, 2016).
89 Ellul, *The Technological Society*; *The Technological System.*
90 Vanderburg, *Our War on Ourselves*, chapters 3, 4.
91 Vanderburg, *Our Battle for the Human Spirit*, chapter 4.
92 Ibid.
93 Jacques Ellul, *Propaganda: The Formation of Men's Attitudes*, trans. Konrad Kellen and Jean Lerner (New York: Vintage Books, 1965).
94 Vanderburg, *Living in the Labyrinth of Technology*, Part 2.

95 M. McCloskey, "Intuitive Physics," *Scientific American* 248 (April 1983): 122–30. For evidence of an "intuitive arithmetic," see J. Lave, "The Values of Quantification," in *Power, Action and Belief*, ed. J. Law (London: Routledge and Kegan Paul, 1986), 88–111.
96 Kuhn, *The Structure of Scientific Revolutions*.
97 Vanderburg, *Our War on Ourselves*, chapter 5.
98 Ibid.
99 Ibid.
100 Ellul, *The Technological Society*; *The Technological System*; Vanderburg, *Our Battle for the Human Spirit*.
101 I have slightly modified the definition of technique that Jacques Ellul gave in his note to the reader in *The Technological Society* (xxv).
102 Vanderburg, *Our Battle for the Human Spirit*.
103 Ibid.
104 Jacques Ellul, *The New Demons*, trans. C. Edward Hopkin (New York: Seabury, 1975).
105 Stivers, *Evil in Modern Myth and Ritual*.
106 Nick Turse, *The Complex: How the Military Invades Our Everyday Lives* (New York: Henry Holt and Company, 2008).
107 Jacques Ellul, *The Political Illusion*, trans. Konrad Kellen (New York: Alfred E. Knopf, 1967).
108 Kevin M. Kruse, *One Nation under God: How Corporate America Invented Christian America* (New York: Basic Books, 2015).
109 Ellul, *The New Demons*.
110 Braden R. Allenby and Deanna J. Richards, eds., *The Greening of Industrial Ecosystems* (Washington, DC: National Academy Press, 1994), Introduction.
111 Galbraith, *The New Industrial State*; Radovan Richta, *Civilization at the Crossroads: Social and Human Implications of the Scientific and Technological Revolution* (Prague: International Arts and Sciences Press, 1969).
112 Vanderburg, *The Labyrinth of Technology*, chapter 3.
113 Willem H. Vanderburg, "Placing Engineering and Other Professions under Public Oversight," *Bulletin of Science, Technology & Society* 32, no. 2 (March 2012): 171–80.
114 See, for example, James Downey and Lois Claxton, *Inno'va-tion: Essays by Leading Canadian Researchers* (Toronto: Key Porter Books, 2002).
115 Ellul, *Histoire des institutions: l'antiquité*.
116 Vanderburg, *The Growth of Minds and Cultures*; see 272–8.
117 Ellul, *The Technological Society*, chapter 4.
118 Vanderburg, *Our War on Ourselves*, chapter 5.

119 Jennifer Chandler, "The Autonomy of Technology: Do Courts Control Technology or Do They Just Legitimize Its Social Acceptance?" *Bulletin of Science, Technology & Society* 27, no. 5 (October 2007): 339–48.
120 Joel Bakan, *The Corporation: The Pathological Pursuit of Profit and Power* (Toronto: Penguin Canada, 2004).
121 Willem H. Vanderburg, "Technology and the Law: Who Rules?" *Bulletin of Science, Technology & Society* 27, no. 4 (July 2007): 323–34.

1 Our Physical Embodiment within the Relativity of Life and the World

1 David Bohm, *Wholeness and the Implicate Order* (New York: Routledge, 2002).
2 Georges Devereux, *From Anxiety to Method in the Behavioral Sciences* (New York: Humanities Press, 1967).
3 Willem H. Vanderburg, *Our War on Ourselves: Rethinking Science, Technology and Economic Growth* (Toronto: University of Toronto Press, 2011); *Our Battle for the Human Spirit: Scientific Knowing, Technical Doing and Daily Living* (Toronto: University of Toronto Press, 2016).
4 Willem H. Vanderburg, *Living in the Labyrinth of Technology* (Toronto: University of Toronto Press, 2005), chapter 10.
5 Jacques Ellul, *The Technological Society*, trans. John Wilkinson (New York: Vintage Books, 1964); *The Technological System*, trans. Joachim Neugroschel (New York: Continuum, 1980).
6 Vanderburg, *Living in the Labyrinth of Technology; Our War on Ourselves.*
7 Adam Smith, *An Inquiry into the Nature and Causes of the Wealth of Nations*, 2 vols. (Chicago: University of Chicago Press, 1976); Jacques Ellul, *La pensée marxiste*, comp. and ed. Michel Hourcade, Jean-Pierre Jézéquel, and Gérard Paul (Paris: La Table Ronde, 2003).
8 Dwight D. Eisenhower, Military-Industrial Complex Speech, Public Papers of the Presidents, 1960, 1035–40; http://coursesa.matrix.msu.edu./~hst306/documents/indust/html.
9 John Kenneth Galbraith, *The New Industrial State* (New York: New American Library, 1985).
10 Nick Turse, *The Complex: How the Military Invades Our Everyday Lives* (New York: Metropolitan Books, 2008); Joseph E. Stiglitz, *The Three Trillion Dollar War: The True Cost of the Iraq Conflict* (New York: W.W. Norton, 2008); Wendy Kaminer, *It's All the Rage: Crime and Culture* (Reading, MA: Addison-Wesley, 1995).
11 Jacques Ellul, *The Subversion of Christianity*, trans. Geoffrey W. Bromiley (Grand Rapids, MI: Eerdmans, 1986).

12 Jacques Ellul, *La pensée marxiste*, comp. and ed. Michel Hourcade, Jean-Pierre Jézéquel, and Gérard Paul (Paris: La Table Ronde, 2003).
13 Max Weber, *The Protestant Ethic and the Spirit of Capitalism* (New York: Routledge, 2013).
14 Will Herberg, *Protestant, Catholic, Jew: An Essay in American Religious Sociology* (Garden City, NY: Anchor Books, Doubleday, 1960); Kevin M. Kruse, *One Nation under God: How Corporate America Invented Christian America* (New York: Basic Books, 2015).
15 Radovan Richta, *Civilization at the Crossroads: Social and Human Implications of the Scientific and Technological Revolution* (Prague: International Arts and Sciences Press, 1969).
16 Jacques Ellul, *Les successeurs de Marx*, comp. and ed. Michael Hourcade, Jean-Pierre Jezequel, and Gérard Paul (Paris: La Table Ronde, 2007).
17 Galbraith, *The New Industrial State*.
18 Vanderburg, *Living in the Labyrinth of Technology*, chapter 6.
19 Galbraith, *The New Industrial State*.
20 H.H. Gerth and C. Wright Mills, eds., *From Max Weber: Essays in Sociology* (New York: Oxford University Press, 1963); Rogers Brubaker, *The Limits of Rationality: An Essay on the Social and Moral Thought of Max Weber* (London: Allen and Unwin, 1984).
21 Ellul, *The Technological Society*.
22 Willem H. Vanderburg, *Secular Nations under New Gods: Beyond Christianity's Subversion by Technology and Politics* (Toronto: University of Toronto Press, 2018).
23 Vanderburg, *Our Battle for the Human Spirit*.
24 Jacques Ellul, *Reason for Being: A Meditation on Ecclesiastes*, trans. Joyce Main Hanks (Grand Rapids, MI: Eerdmans, 1990).
25 Vanderburg, *Secular Nations under New Gods*, Epilogue.
26 Robert Karasek and Töres Theorell, *Healthy Work: Stress, Productivity and the Reconstruction of Working Life* (New York: Basic Books, 1990); Willem H. Vanderburg, *The Labyrinth of Technology* (Toronto: University of Toronto Press, 2000), chapter 10.
27 Karasek and Theorell, *Healthy Work*.
28 Willem H. Vanderburg, *The Growth of Minds and Cultures: A Unified Interpretation of the Structure of Human Experience*, 2nd ed. (Toronto: University of Toronto Press, 2016).
29 Ibid.
30 Vanderburg, *Living in the Labyrinth of Technology*; see Part 2; Hubert L. Dreyfus and Stuart E.. Dreyfus, with Tom Athanasiou, *Mind over Machine:*

The Power of Human Intuition and Expertise in the Era of the Computer (New York: Free Press, 1986).

31 For an introductory overview, see Hubert L. Dreyfus's introduction to Samuel Todes, *Body and World* (Cambridge, MA: MIT Press, 2001); Hubert L. Dreyfus, *What Computers Still Can't Do: A Critique of Artificial Reason* (Cambridge, MA: MIT Press, 1992).
32 Hubert Dreyfus, *What Computers Still Can't Do.*
33 Mark Johnson, *The Body and the Mind: The Bodily Basis of Meaning, Imagination and Reason* (Chicago: University of Chicago Press, 1987, 1992).
34 Ibid.
35 Cornelius Castoriadis, *L'Institution imaginaire de la société* (Paris: Seuil, 1975).
36 Johnson, *The Body and the Mind.*
37 Thomas Kuhn, *The Structure of Scientific Revolutions*, 2nd ed. (Chicago: University of Chicago Press, 1970).
38 Johnson, *The Body and the Mind.*
39 Herberg, *Protestant, Catholic, Jew.*
40 Vanderburg, *Secular Nations under New Gods*, chapter 6 and Epilogue.
41 Vanderburg, *Our War on Ourselves.*
42 Vanderburg, *Our Battle for the Human Spirit.*
43 Vanderburg, *The Growth of Minds and Cultures.*
44 Ibid.
45 Ludwig Wittgenstein, *Philosophical Investigations*, trans. G.E.M. Anscombe (Oxford: Blackwell Publishing, 1953).
46 Vanderburg, *Our Battle for the Human Spirit.*
47 Ibid.
48 See, for example, Sherry Turkle, *The Second Self: Computers and the Human Spirit* (New York: Simon and Schuster, 1984); *Life on the Screen: Identity in the Age of the Internet* (New York: Simon and Schuster, 1995).
49 Vanderburg, *Our War on Ourselves.*

2 Our Social and Cultural Embodiment in the Relativity of Human Life in the World

1 Willem H. Vanderburg, *The Growth of Minds and Cultures: A Unified Interpretation of the Structure of Human Experience*, 2nd ed. (Toronto: University of Toronto Press, 2016), chapter 8.
2 I am taking a somewhat different approach to accounting for the emergence of the dimensions of cultural mediation from Gilbert Simondon in *Du mode d'existence des objets techniques* (Paris: Aubier Montaigne, 1969).

3 Vanderburg, *The Growth of Minds and Cultures*.
4 C. Lévi-Strauss, *The Savage Mind* (University of Chicago Press, 1966).
5 Jacques Ellul, *The Subversion of Christianity*, trans. Geoffrey W. Bromiley (Grand Rapids, MI: Eerdmans, 1986).
6 Jacques Ellul, *The New Demons*, trans. C. Edward Hopkin (New York: Seabury, 1975).
7 Jacques Ellul, *Perspectives on Our Age: Jacques Ellul Speaks on His Life and Work*, ed. Willem H. Vanderburg, 2nd ed. (Toronto: Anansi, 2004), chapter 4.
8 Jan Hendrik van den Berg, *The Changing Nature of Man* (New York: Delta, 1961).
9 Vanderburg, *The Growth of Minds and Cultures*, chapter 8.
10 Ibid.
11 Ibid.
12 Willem H. Vanderburg, *Living in the Labyrinth of Technology* (Toronto: University of Toronto Press, 2005).
13 Willem H. Vanderburg, *Our War on Ourselves: Rethinking Science, Technology and Economic Growth* (Toronto: University of Toronto Press, 2011).
14 Willem H. Vanderburg, *Our Battle for the Human Spirit: Scientific Knowing, Technical Doing and Daily Living* (Toronto: University of Toronto Press, 2016).
15 Willem H. Vanderburg, *The Labyrinth of Technology* (Toronto: University of Toronto Press, 2000).
16 Jacques Ellul, *The Political Illusion*, trans. Konrad Kellen (New York: Alfred E. Knopf, 1967).
17 Vanderburg, *Our Battle for the Human Spirit*.
18 Ellul, *The New Demons*.
19 Ibid.
20 Jacques Ellul, *The Technological Society*, trans. John Wilkinson (New York: Vintage Books, 1964), chapter 4.
21 Joel Bakan, *The Corporation: The Pathological Pursuit of Profit and Power* (Toronto: Penguin Canada, 2004).
22 Bernard Lietaer, *The Future of Money: A New Way to Create Wealth, Work, and a Wiser World* (New York: Random House, 2001).
23 Ibid.
24 Philip Coggan, *Paper Promises: Money, Debt and the New World Order* (New York: Penguin Books, 2011).
25 Ibid.
26 Lietaer, *The Future of Money*.
27 For a discussion of the Tobin tax from the perspective presented in these works, see Vanderburg, *Our Battle for the Human Spirit*, chapter 3.

28 Ralph Estes, *Tyranny of the Bottom Line: Why Corporations Make Good People Do Bad Things* (San Francisco: Berrett-Koehler Publishers, 1996); David K. Johnston, *Free Lunch: How the Wealthiest Americans Enrich Themselves at Government Expense (and Stick You with the Bill)* (New York: Penguin Group, 2007).
29 Herman E. Daly and John B. Cobb Jr., *For the Common Good: Redirecting the Economy toward Community, the Environment, and a Sustainable Future* (Boston: Beacon Press, 1989).
30 Herman E. Daly, *Beyond Growth: The Economics of Sustainable Development* (Boston: Beacon Press, 1996).
31 Jacques Ellul, *Propaganda: The Formation of Men's Attitudes*, trans. Konrad Kellen and Jean Lerner (New York: Vintage Books, 1965).
32 Vanderburg, *Our Battle for the Human Spirit*, chapter 4.
33 Ibid.
34 Ibid.
35 Ibid.
36 Ellul, *Propaganda*; Neil Postman, *Amusing Ourselves to Death: Public Discourse in the Age of Show Business* (New York: Penguin Group, 1985).
37 Vanderburg, *Our Battle for the Human Spirit*, chapter 4.
38 Jeffrey Freed and Laurie Parsons, *Right-Brained Children in a Left-Brained World: Unlocking the Potential of Your ADD Child* (New York: Simon and Schuster, 1997).
39 For a detailed study of the phenomenon of the separation of knowing and doing from experience and culture, see Vanderburg, *Living in the Labyrinth of Technology*, Part 2.
40 Ibid.
41 Vanderburg, *The Labyrinth of Technology*, chapter 3.
42 Willem H. Vanderburg, *Secular Nations under New Gods: Beyond Christianity's Subversion by Technology and Politics* (Toronto: University of Toronto Press, 2018).
43 Vanderburg, *Our Battle for the Human Spirit*.
44 R. Shayna Rosenbaum, Stefan Köhler, Daniel L. Schachter, Morris Moscovith, Robyn Westmacott, Sandra B. Black, Fuquiang Gao, and Endel Tulving, "The Case of K.C.: Contribution of a Memory-Impaired Person to Memory Theory," *Neuropsychologia* 43 (2005): 989–1021.
45 Vanderburg, *Our Battle for the Human Spirit*, chapter 2.
46 Sherry Turkle, *The Second Self: Computers and the Human Spirit* (New York: Simon and Schuster, 1984); Sherry Turkle, *Life on the Screen: Identity in the Age of the Internet* (New York: Simon and Schuster, 1995); Sherry Turkle, *Alone Together: Why We Expect More from Technology and Less from Each Other* (New York: Basic Books, 2011).

47 Craig Brod, *Technostress: The Human Cost of the Computer Revolution* (Reading, MA: Addison-Wesley, 1984); Gary Small and Gigi Vorgan, *iBrain: Surviving the Technological Alteration of the Modern Mind* (New York: HarperCollins Publishers, 2008); Larry D. Rosen, *iDisorder: Understanding Our Obsession with Technology and Overcoming Its Hold on Us* (New York: Palgrave Macmillan, 2012).

3 Living with a Dual Relativity beyond Cultural Embodiment

1 Willem H. Vanderburg, *Our War on Ourselves: Rethinking Science, Technology and Economic Growth* (Toronto: University of Toronto Press, 2011), chapters 3 and 4.
2 Willem H. Vanderburg, *Living in the Labyrinth of Technology* (Toronto: University of Toronto Press, 2005), chapter 1.
3 Herman E. Daly and John B. Cobb Jr., *For the Common Good: Redirecting the Economy toward Community, the Environment, and a Sustainable Future* (Boston: Beacon Press, 1989); Vanderburg, *Our War on Ourselves*, chapter 4; Willem H. Vanderburg, *Our Battle for the Human Spirit: Scientific Knowing, Technical Doing and Daily Living* (Toronto: University of Toronto Press, 2016), chapter 3.
4 Karl Polanyi, *The Great Transformation* (New York: Farrar and Rinehart, 1944).
5 Vanderburg, *Our War on Ourselves*, chapters 4 and 5.
6 Vanderburg, *Living in the Labyrinth of Technology*, chapter 4.
7 Ibid.; Michael Hammer and James Champy, *Reengineering the Corporation: A Manifesto for Business Revolution* (New York: HarperCollins, 1993); Thomas H. Davenport, *Process Innovation: Reengineering Work through Information Technology* (Boston: Harvard Business School Press, 1993).
8 E.O. Wilson, "Is Humanity Suicidal?" *New York Times Magazine*, 30 May 1993, 24–9.
9 Arnold Toynbee, *A Study of History*, abridged version, ed. D.C. Somervell (New York: Dell, 1978).
10 Willem H. Vanderburg, *The Growth of Minds and Cultures: A Unified Interpretation of the Structure of Human Experience*, 2nd ed. (Toronto: University of Toronto Press, 2016).
11 The legal dimension of cultural mediation serves as a key example. Jacques Ellul never published his doctoral course on law, which explored the universality of legal institutions from a cultural-anthropological and historical perspective. He gave me permission to publish an outline of it based in part on what he had published and on my recordings of that course. I included it in *The Growth of Minds and Cultures*, chapter 8. These recordings are now published in Jacques Ellul, *Philosophie du droit* (Paris: La Table Ronde, 2022).

12 Gilbert Simondon, *Du mode d'existence des objets techniques* (Paris: Aubier-Montaigne, 1969), Part 3.
13 Ibid.
14 Richard Stivers, *Technology as Magic: The Triumph of the Irrational* (New York: Continuum, 1999).
15 Jacques Ellul, *The New Demons*, trans. C. Edward Hopkin (New York: Seabury, 1975).
16 Simondon, *Du mode d'existence des objets techniques*; see Part 3.
17 Ibid.
18 Ibid.
19 Ibid.
20 Vanderburg, *The Growth of Minds and Cultures*, chapter 8.
21 Jacques Ellul, *The Technological System*, trans. Joachim Neugroschel (New York: Continuum, 1980), Introduction.
22 Jack P. Manno, *Privileged Goods: Commoditization and Its Impact on Environment and Society* (Boca Raton, FL: CRC Press, 2000).
23 Jacques Ellul, *The Empire of Non-Sense: Art in the Technological Society*, trans. Michael Johnson and David Lovekin (Winterbourne, UK: Papadakis, 2014).
24 Siegfried Giedion, *Mechanization Takes Command* (New York: Oxford University Press, 1948).
25 Jacques Ellul, "Remarks on Technology and Art," *Bulletin of Science, Technology and Society* 21 (February 2001): 26–37.
26 Ibid.
27 Jan Hendrik van den Berg, *The Changing Nature of Man* (New York: Delta, 1961).
28 Ellul, "Remarks on Technology and Art."
29 Vanderburg, *Our Battle for the Human Spirit*.
30 Jacques Ellul, *Propaganda: The Formation of Men's Attitudes*, trans. Konrad Kellen and Jean Lerner (New York: Vintage Books, 1965).
31 Ellul, "Remarks on Technology and Art."
32 Ibid.
33 Jane Jacobs, *The Death and Life of Great American Cities* (London: Pelican, 1961).
34 Willem H. Vanderburg, *The Labyrinth of Technology* (Toronto: University of Toronto Press, 2000), chapter 11.
35 Ellul, "Remarks on Technology and Art."
36 Ellul, *The Technological System*.
37 Ellul, *The Empire of Non-Sense*.
38 Ibid.
39 Ellul, "Remarks on Technology and Art."
40 Ibid.

41 Ellul, *The Technological System*.
42 Ellul, *The New Demons*; Richard Stivers, *Evil in Modern Myth and Ritual* (Athens: University of Georgia Press, 1982).
43 Ellul, *The Empire of Non-Sense*.
44 Ibid.
45 Jacques Ellul, *La pensée marxiste*, comp. and ed. Michel Hourcade, Jean-Pierre Jézéquel, and Gérard Paul (Paris: La Table Ronde, 2003).
46 Radovan Richta, *Civilization at the Crossroads: Social and Human Implications of the Scientific and Technological Revolution* (Prague: International Arts and Sciences Press, 1969).
47 John Kenneth Galbraith, *The New Industrial State* (New York: New American Library, 1985). For a critical analysis, see Vanderburg, *Living in the Labyrinth of Technology*, chapter 6.
48 Vanderburg, *Living in the Labyrinth of Technology*, chapter 3.
49 Vanderburg, *Our Battle for the Human Spirit*, chapter 5.
50 Ibid.
51 Eli Parisier, *In the Filter Bubble: What the Internet Is Hiding from You* (New York: Penguin Press, 2011).
52 Cathy O'Neil, *Weapons of Math Destruction: How Big Data Increases Inequality and Threatens Democracy* (New York: Penguin Random House, 2016).
53 Ibid.
54 Vanderburg, *Living in the Labyrinth of Technology*.
55 Ibid.
56 Vanderburg, *Our Battle for the Human Spirit*, chapter 4.
57 Hubert L. Dreyfus, *On the Internet* (New York: Routledge, 2001).
58 Galbraith, *The New Industrial State*.
59 Ellul, *Propaganda*; Vanderburg, *Our Battle for the Human Spirit*, chapter 4.
60 Ellul, *The New Demons*; Stivers, *Evil in Modern Myth and Ritual*.

4 Mathematics as the Non-Language of Science and Technique

1 Jan Hendrik van den Berg, *The Changing Nature of Man* (New York: Delta, 1961).
2 Ibid.
3 Ibid.
4 Philip J. Davis and Reuben Hersh, *The Mathematical Experience* (Boston: Houghton Mifflin, 1981).
5 Ibid.
6 Jeremy Campbell, *The Improbable Machine: What the Upheavals in Artificial Intelligence Research Reveal about How the Mind Really Works* (New York: Simon and Schuster, 1989).

7 John R. Milton, "The Origin and Development of the Concept of the 'Laws of Nature,'" *Archives Européennes de sociologie* 22 (1981): 173–95.
8 M. Kline, "The Mathematization of Science," in *Mathematics: The Loss of Certainty* (New York: Oxford University Press, 1960).
9 Ibid.
10 Kline, "The Mathematization of Science."
11 Milton, "The Origin and Development of the Concept of the 'Laws of Nature.'"
12 Jacques Ellul, *The Technological Society*, trans. John Wilkinson (New York: Vintage Books, 1964).
13 Willem H. Vanderburg, *Living in the Labyrinth of Technology* (Toronto: University of Toronto Press, 2005), chapter 3.
14 Willem H. Vanderburg, *Our War on Ourselves: Rethinking Science, Technology and Economic Growth* (Toronto: University of Toronto Press, 2011), chapters 3 and 4.
15 Willem H. Vanderburg, *Our Battle for the Human Spirit: Scientific Knowing, Technical Doing and Daily Living* (Toronto: University of Toronto Press, 2016).
16 Ibid.
17 Vanderburg, *Living in the Labyrinth of Technology*, Part 2.
18 Ibid., chapter 2.
19 Vanderburg, *Living in the Labyrinth of Technology*.
20 Vanderburg, *Our Battle for the Human Spirit*.
21 Willem H. Vanderburg, *The Labyrinth of Technology* (Toronto: University of Toronto Press, 2000); Vanderburg, *Our War on Ourselves*, chapter 5; Vanderburg, *Our Battle for the Human Spirit*.
22 Vanderburg, *Living in the Labyrinth of Technology*.
23 Wolfgang Schivelbusch, *The Railway Journey: The Industrialization and Perception of Time and Space in the 19th Century* (New York: Berg, 1986).
24 Georg Simmel, "The Metropolis and Mental Life," in *The Sociology of Georg Simmel*, ed. Kurt Wolff (Glencoe, IL: Free Press, 1950).
25 Jacques Ellul, *Métamorphose du bourgeois* (Paris: Calmann-Lévy, 1967).
26 Vanderburg, *Living in the Labyrinth of Technology*, chapter 2.
27 Ibid., Part 2.
28 Harry Redner, *The Ends of Science: An Essay in Scientific Authority* (Boulder, CO: Westview Press, 1987).
29 John Horgan, *The End of Science: Facing the Limits of Knowledge in the Twilight of the Scientific Age* (Reading, MA: Helix Books, 1996).
30 Thomas S. Kuhn, *The Structure of Scientific Revolutions*, 2nd ed. (Chicago: University of Chicago Press, 1970).

31 Ludwik Fleck, *Genesis and Development of a Scientific Fact* (Chicago: University of Chicago Press, 1979).
32 Kuhn, *The Structure of Scientific Revolutions.*
33 Ibid.
34 Jacques Ellul, *The Humiliation of the Word*, trans. Joyce Main Hanks (Grand Rapids, MI: Eerdmans, 1985).
35 Kuhn, *The Structure of Scientific Revolutions.*
36 Fleck, *Genesis and Development of a Scientific Fact.*
37 Vanderburg, *Our War on Ourselves*; Vanderburg, *Our Battle for the Human Spirit.*
38 Ibid.
39 Georges Devereux, *From Anxiety to Method in the Behavioral Sciences* (New York: Humanities Press, 1967).
40 See especially David Bohm, *Wholeness and the Implicate Order* (London: Routledge, 1980).
41 Bernard D'Espagnat, *In Search of Reality* (New York: Springer-Verlag, 1983).
42 Horgan, *The End of Science.*
43 Davis and Hersh, *The Mathematical Experience.*
44 Ibid.
45 Ibid.
46 Michael Polanyi, *Personal Knowledge* (Chicago: University of Chicago Press, 1962).
47 Hubert L. Dreyfus and Stuart E. Dreyfus, with Tom Athanasiou, *Mind over Machine: The Power of Human Intuition and Expertise in the Era of the Computer* (New York: Free Press, 1986). I do not believe that Stages 6 and 7 that were later added by Hubert Dreyfus are applicable to our global civilization in general, and to skills based on discipline-based approaches, in particular. See Hubert L. Dreyfus, *On the Internet* (New York: Routledge, 2001).
48 Davis and Hersh, *The Mathematical Experience.*
49 Hubert L. Dreyfus, *What Computers Still Can't Do: A Critique of Artificial Reason* (Cambridge, MA: MIT Press, 1992).
50 Didier Nordon, *Les mathématiques pures n'existent pas!* (Le Paradou: Editions Actes Sud, 1981).
51 Ibid., 19.
52 Jacques Ellul, *La pensée marxiste*, comp. and ed. Michel Hourcade, Jean-Pierre Jézéquel, and Gérard Paul (Paris: La Table Ronde, 2003).
53 Radovan Richta, *Civilization at the Crossroads: Social and Human Implications of the Scientific and Technological Revolution* (Prague: International Arts and Sciences Press, 1969).

54 Kuhn, *The Structure of Scientific Revolutions*.
55 Nordon, *Les mathématiques pures n'existent pas!*.
56 Ellul, *The Technological Society*.
57 Ellul, *The Technological System*, trans. Joachim Neugroschel (New York: Continuum, 1980).
58 Nordon, *Les mathématiques pures n'existent pas!*.
59 Ibid.
60 Ellul, *The Humiliation of the Word*.
61 Nordon, *Les mathématiques pures n'existent pas!*.
62 Ibid.
63 Ibid., 152 (my translation from the original).
64 Cathy O'Neil, *Weapons of Math Destruction: How Big Data Increases Inequality and Threatens Democracy* (New York: Penguin Random House, 2016).
65 Richard Wilkinson and Kate Pickett, *The Spirit Level: Why More Equal Societies Almost Always Do Better* (London: Penguin Books, 2009).
66 Vanderburg, *Living in the Labyrinth of Technology*.
67 Ibid.
68 Willem H. Vanderburg, *Secular Nations under New Gods*.
69 Vanderburg, *Our War on Ourselves*.
70 Ellul, *The Technological Society*, chapter 5.
71 Vanderburg, *Our Battle for the Human Spirit*, chapters 4 and 5.
72 Nicholas Carr, *The Shallows: What the Internet Is Doing to Our Brains* (New York: W.W. Norton and Company, 2011).
73 Fleck, *Genesis and Development of a Scientific Fact*; Kuhn, *The Structure of Scientific Revolutions*.
74 Vanderburg, *Our Battle for the Human Spirit*, chapters 4 and 5.

5 Human Knowing and Discipline-Based Science

1 Jacques Ellul, *La pensée marxiste*, comp. and ed. Michel Hourcade, Jean-Pierre Jézéquel, and Gérard Paul (Paris: La Table Ronde, 2003).
2 H.H. Gerth and C. Wright Mills, eds., *From Max Weber: Essays in Sociology* (New York: Oxford University Press, 1963); Rogers Brubaker, *The Limits of Rationality: An Essay on the Social and Moral Thought of Max Weber* (London: Allen and Unwin, 1984).
3 Willem H. Vanderburg, *Our War on Ourselves: Rethinking Science, Technology and Economic Growth* (Toronto: University of Toronto Press, 2011).
4 Thomas S. Kuhn, *The Structure of Scientific Revolutions*, 2nd ed. (Chicago: University of Chicago Press, 1970).

5 Jacques Ellul, *The New Demons*, trans. C. Edward Hopkin (New York: Seabury, 1975).
6 Kuhn, *The Structure of Scientific Revolutions*.
7 Ludwik Fleck, *Genesis and Development of a Scientific Fact* (Chicago: University of Chicago Press, 1979).
8 Willem H. Vanderburg, *Our Battle for the Human Spirit: Scientific Knowing, Technical Doing and Daily Living* (Toronto: University of Toronto Press, 2016), chapter 4.
9 Willem H. Vanderburg, *Secular Nations under New Gods: Beyond Christianity's Subversion by Technology and Politics* (Toronto: University of Toronto Press, 2018), chapter 1.
10 Willem H. Vanderburg, *The Growth of Minds and Cultures: A Unified Interpretation of the Structure of Human Experience*, 2nd ed. (Toronto: University of Toronto Press, 2016), chapter 8.
11 Fleck, *Genesis and Development of a Scientific Fact*.
12 Stuart E. Dreyfus, "System 0: The Overlooked Explanation of Expert Intuition," chapter 2 in *Handbook of Research Methods on Intuition*, ed. M. Sinclair (Cheltenham: Edward Elgar Publishers, 2014).
13 Fleck, *Genesis and Development of a Scientific Fact*.
14 Ibid.
15 Ibid.
16 Ibid.
17 Ibid
18 Ibid
19 Michael Polanyi, *Personal Knowledge* (Chicago: University of Chicago Press, 1962).
20 Fleck, *Genesis and Development of a Scientific Fact*.
21 Kuhn, *The Structure of Scientific Revolutions*.
22 Hubert L. Dreyfus and Stuart E. Dreyfus, with Tom Athanasiou, *Mind over Machine: The Power of Human Intuition and Expertise in the Era of the Computer* (New York: Free Press, 1986).
23 Vanderburg, *The Growth of Minds and Cultures*.
24 Sherry Turkle, *The Second Self: Computers and the Human Spirit* (New York: Simon and Schuster, 1984); Sherry Turkle, *Life on the Screen: Identity in the Age of the Internet* (New York: Simon and Schuster, 1995); Sherry Turkle, *Alone Together: Why We Expect More from Technology and Less from Each Other* (New York: Basic Books, 2011).
25 Vanderburg, *Our War on Ourselves*, chapter 2.
26 Ibid., chapter 1.
27 Vanderburg, *Secular Nations under New Gods*, chapter 1.

28 Thomas S. Kuhn, *The Essential Tension: Selected Studies in Scientific Tradition and Change* (Chicago: University of Chicago Press, 1977).
29 Kuhn, *The Structure of Scientific Revolutions.*
30 John Kenneth Galbraith, *The New Industrial State* (New York: New American Library, 1985).
31 Jacques Ellul, *The Technological Society*, trans. John Wilkinson (New York: Vintage Books, 1964).
32 James Downey and Lois Claxton, *Inno'va-tion: Essays by Leading Canadian Researchers* (Toronto: Key Porter Books, 2002).
33 David Bohm, *Wholeness and the Implicate Order* (New York: Routledge, 2002).
34 Vanderburg, *The Growth of Minds and Cultures.*
35 Werner Heisenberg, *Physics and Beyond: Encounters and Conversations* (New York: Harper and Row, 1971), 63.
36 Jacques Ellul, *Perspectives on Our Age: Jacques Ellul Speaks on His Life and Work*, ed. Willem H. Vanderburg, 2nd ed. (Toronto: Anansi, 2004).
37 Jacques Ellul, *The Technological System*, trans. Joachim Neugroschel (New York: Continuum, 1980).
38 Willem H. Vanderburg, *Living in the Labyrinth of Technology* (Toronto: University of Toronto Press, 2005), chapter 10.
39 Gerth and Mills, eds., *From Max Weber*; Brubaker, *The Limits of Rationality.*
40 Ellul, *The Technological System.*
41 Vanderburg, *The Growth of Minds and Cultures.*
42 Vanderburg, *Our War on Ourselves*, chapter 5.
43 Sheila Tobias, *They're Not Dumb, They're Different: Stalking the Second Tier* (Tucson: Research Corporation Foundation for the Advancement of Science, 1990); Benson Snyder, "Literacy and Numeracy: Two Ways of Knowing," *Daedalus* 119 (1990): 233–56; Hubert L. Dreyfus and Stuart E. Dreyfus, "What Is Morality? A Phenomenological Account of the Development of Ethical Expertise," in *Universalism versus Communitarianism*, ed. David M. Rasmussen (Cambridge, MA: MIT Press, 1990), 237–64. For a synthesis, see Willem H. Vanderburg, *The Labyrinth of Technology* (Toronto: University of Toronto Press, 2000), chapter 3.
44 Vanderburg, *Our War on Ourselves*, chapter 5.
45 Vanderburg, *Our Battle for the Human Spirit*, chapters 2, 3.
46 Tobias, *They're Not Dumb, They're Different*; Snyder, "Literacy and Numeracy"; Dreyfus and Dreyfus, "What Is Morality?"; for a synthesis, see Vanderburg, *The Labyrinth of Technology.*
47 Ibid.
48 Roger S. Jones, *Physics as Metaphor* (Annapolis: University of Minnesota Press, 1982).

49 Owen Barfield, *Saving the Appearances: A Study in Idolatry* (Middletown, CT, Wesleyan University Press, 1988).
50 Jacques Ellul, *On Freedom, Love and Power,* expanded edition; comp., ed., and trans. Willem H. Vanderburg (Toronto: University of Toronto Press, 2015), Part 1.
51 Jones, *Physics as Metaphor.*
52 Ibid.
53 Ibid.
54 Kuhn, *The Structure of Scientific Revolutions.*
55 Will Herberg, *Protestant, Catholic, Jew: An Essay in American Religious Sociology* (Garden City, NY: Anchor Books, Doubleday, 1960); Kevin M. Kruse, *One Nation under God: How Corporate America Invented Christian America* (New York: Basic Books, 2015).
56 Jacques Ellul, *Reason for Being: A Meditation on Ecclesiastes,* trans. Joyce Main Hanks (Grand Rapids, MI: Eerdmans, 1990).

6 Human Doing, Technique, and the Living of Our Lives

1 David Riesman, Nathan Glazer, and Reuel Denney, *The Lonely Crowd: A Study of the Changing American Character* (Garden City, NY: Doubleday Anchor, 1950).
2 Jacques Ellul, *Propaganda: The Formation of Men's Attitudes,* trans. Konrad Kellen and Jean Lerner (New York: Vintage Books, 1965).
3 Jacques Ellul, *The New Demons,* trans. C. Edward Hopkin (New York: Seabury, 1975).
4 Willem H. Vanderburg, *Our War on Ourselves: Rethinking Science, Technology and Economic Growth* (Toronto: University of Toronto Press, 2011); *Our Battle for the Human Spirit: Scientific Knowing, Technical Doing and Daily Living* (Toronto: University of Toronto Press, 2016).
5 Willem H. Vanderburg, *The Labyrinth of Technology* (Toronto: University of Toronto Press, 2000); *Living in the Labyrinth of Technology* (Toronto: University of Toronto Press, 2005); *Our War on Ourselves; Our Battle for the Human Spirit; Secular Nations under New Gods: Beyond Christianity's Subversion by Technology and Politics* (Toronto: University of Toronto Press, 2018).
6 Vanderburg, *Our War on Ourselves,* chapter 5.
7 Ibid.
8 Ibid.
9 Ralph Estes, *Tyranny of the Bottom Line: Why Corporations Make Good People Do Bad Things* (San Francisco: Berrett-Koehler Publishers, 1996).
10 Vanderburg, *The Labyrinth of Technology,* chapter 7.

11 Herman E. Daly and John B. Cobb Jr., *For the Common Good: Redirecting the Economy toward Community, the Environment, and a Sustainable Future* (Boston: Beacon Press, 1989).
12 Donald A. Schon, *The Reflective Practitioner: How Professionals Think in America* (New York: Basic Books/Harper Colophon, 1983).
13 Eli Parisier, *In the Filter Bubble: What the Internet Is Hiding from You* (New York: Penguin Press, 2011).
14 Vanderburg, *The Labyrinth of Technology*, chapter 3.
15 Jacques Ellul, *The Technological Society*, trans. John Wilkinson (New York: Vintage Books, 1964).
16 Benson Snyder, "Literacy and Numeracy: Two Ways of Knowing," *Daedalus* 119 (1990): 233–56.
17 Vanderburg, *The Labyrinth of Technology*, chapters 8, 9, 10, 11.
18 Vanderburg, *Our Battle for the Human Spirit*.
19 Vanderburg, *Our War on Ourselves*.
20 Vanderburg, *Secular Nations under New Gods*.
21 Vanderburg, *Our War on Ourselves*.
22 Vanderburg, *The Labyrinth of Technology*, chapters 8, 9, 10, 11; *Our Battle for the Human Spirit*, chapter 3.
23 Jacques Ellul, *La pensée marxiste*, comp. and ed. Michel Hourcade, Jean-Pierre Jézéquel, and Gérard Paul (Paris: La Table Ronde, 2003).
24 C. Cobb, T. Halstead, and J. Rowe, "If the GDP Is Up, Why Is America Down?" *Atlantic Monthly* (October 1995): 59–78.
25 Ellul, *The Technological Society*; *Propaganda*.
26 Tim Jackson, *Material Concerns: Pollution, Profit and Quality of Life* (New York: Routledge, 1996).
27 Max Weber, *Economy and Society: An Outline of Interpretive Sociology* (City: University of California Press, 1978); John Kenneth Galbraith, *The New Industrial State* (New York: New American Library, 1985); Radovan Richta, *Civilization at the Crossroads: Social and Human Implications of the Scientific and Technological Revolution* (Prague: International Arts and Sciences Press, 1969). For an interpretation through the lens of technique, see Vanderburg, *Living in the Labyrinth of Technology*, chapter 6.
28 Vanderburg, *Our War on Ourselves*, chapters 3 and 4.
29 Bernard Lietaer, *The Future of Money: A New Way to Create Wealth, Work, and a Wiser World* (New York: Random House, 2001).
30 Vanderburg, *Our Battle for the Human Spirit*, chapters 3, 4, and 5.
31 Karl Polanyi, *The Great Transformation* (New York: Farrar and Rinehart, 1944).
32 Ellul, *Propaganda*.
33 Vanderburg, *Living in the Labyrinth of Technology*, chapter 3.

34 Ellul, *The New Demons.*
35 Richard Wilkinson and Kate Pickett, *The Spirit Level: Why More Equal Societies Almost Always Do Better* (London: Penguin Books, 2009); Anu Partanen, *The Nordic Theory of Everything: In Search of a Better Life* (New York: HarperCollins, 2016).
36 Adam Smith, *An Inquiry into the Nature and Causes of the Wealth of Nations*, 2 vols. (Chicago: University of Chicago Press, 1976).
37 Joel Bakan, *The Corporation: The Pathological Pursuit of Profit and Power* (Toronto: Penguin Canada, 2004).
38 Vanderburg, *The Labyrinth of Technology*, chapter 11.
39 David Cay Johnston, *Free Lunch: How the Wealthiest Americans Enrich Themselves at Government Expense (and Stick You with the Bill)* (New York: Portfolio Penguin Group, 2007).
40 Wassily Leontieff, letter to *Science* 217, 9 July 1982, 104–5.
41 Ellul, *La pensée marxiste.*
42 Vanderburg, *Living in the Labyrinth of Technology*, chapter 7.
43 Ellul, *The Technological Society.*
44 Robert Karasek and Tores Theorell, *Healthy Work: Stress, Productivity and the Reconstruction of Working Life* (New York: Basic Books, 1990).
45 Vanderburg, *The Labyrinth of Technology*, chapter 11.
46 Vanderburg, *Our Battle for the Human Spirit*, chapter 3.
47 Vanderburg, *The Labyrinth of Technology.*
48 For an alternative interpretation, see Cornelius Castoriadis, *L'Institution imaginaire de la société* (Paris: Seuil, 1975).
49 Arnold Toynbee, *A Study of History*, abridged version, ed. D.C. Somervell (New York: Dell, 1978).

Epilogue

1 Jacques Ellul, *La pensée marxiste*, comp. and ed. Michel Hourcade, Jean-Pierre Jézéquel, and Gérard Paul (Paris: La Table Ronde, 2003).
2 Willem H. Vanderburg, *Living in the Labyrinth of Technology* (Toronto: University of Toronto Press, 2005).
3 Radovan Richta, *Civilization at the Crossroads: Social and Human Implications of the Scientific and Technological Revolution* (Prague: International Arts and Sciences Press, 1969).
4 Jacques Ellul, *Les successeurs de Marx*, comp. and ed. Michael Hourcade, Jean-Pierre Jezequel, and Gérard Paul (Paris: La Table Ronde, 2007).
5 John Kenneth Galbraith, *The New Industrial State* (New York: New American Library, 1985).

6 Vanderburg, *Living in the Labyrinth of Technology*.
7 Jacques Ellul, *Changer de révolution: l'inéluctable prolétariat* (Paris: Éditions du Seuil, 1982); *Autopsy of Revolution*, trans. Patricia Wolf (New York: Alfred A. Knopf, 1971).
8 Willem H. Vanderburg, *The Labyrinth of Technology* (Toronto: University of Toronto Press, 2000), chapter 11.
9 Willem H. Vanderburg, *Secular Nations under New Gods: Beyond Christianity's Subversion by Technology and Politics* (Toronto: University of Toronto Press, 2018).
10 Jacques Ellul, *On Freedom, Love and Power*, expanded edition; comp., ed., and trans. Willem H. Vanderburg (Toronto: University of Toronto Press, 2015), Part 1.
11 Jacques Ellul, *Reason for Being: A Meditation on Ecclesiastes*, trans. Joyce Main Hanks (Grand Rapids, MI: Eerdmans, 1990).
12 Vanderburg, *Secular Nations under New Gods*.
13 Jacques Ellul, *The Subversion of Christianity*, trans. Geoffrey W. Bromiley (Grand Rapids, MI: Eerdmans, 1986).
14 Ibid.
15 Jacques Ellul, *The Ethics of Freedom*, trans. Geoffrey W. Bromiley (Grand Rapids, MI: Eerdmans, 1976).
16 Jacques Ellul, *The New Demons*, trans. C. Edward Hopkin (New York: Seabury, 1975).
17 Will Herberg, *Protestant, Catholic, Jew: An Essay in American Religious Sociology* (Garden City, NY: Anchor Books, Doubleday, 1960); *One Nation under God: How Corporate America Invented Christian America* (New York: Basic Books, 2015).
18 Benson Snyder, "Literacy and Numeracy: Two Ways of Knowing," *Daedalus* 119 (1990): 233–56; Vanderburg, *The Labyrinth of Technology*, chapter 3.
19 Vanderburg, *The Labyrinth of Technology*.

Index

Milton Keynes UK
Ingram Content Group UK Ltd.
UKHW012229190424
441406UK00003B/277

9 781487 552473